DISCARD

LIBERATION AND THE AIMS OF SCIENCE
An Essay on Obstacles to the Building of a Beautiful World

LIBERATION AND THE AIMS OF SCIENCE

An Essay on Obstacles to the Building of a Beautiful World

BRIAN EASLEA

ROWMAN AND LITTLEFIELD
TOTOWA N.J.

First published in the United States 1973
by Rowman and Littlefield, Totowa, N.J.

INDIANA UNIVERSITY LIBRARY

NORTHWEST

© Brian Easlea 1973

ISBN 0-87471-477-X

Printed in Great Britain

To
A, A, F,
and
K

I say it is useless to waste your life on one path, especially if that path has no heart. . . .

And, the next thing, the path without a heart will turn against men and destroy them. It does not take much to die, and to seek death is to seek nothing. . . .

For me there is only the travelling on paths that have heart, on any path that may have heart. There I travel, and the only worthwhile challenge is to traverse its full length. And there I travel looking, looking breathlessly.

The Teachings of Don Juan:
A Yaqui Way of Knowledge.   CARLOS CASTANEDA

CONTENTS

Acknowledgements — page x
Preface — xi
Introduction — xiii

Chapter 1 Objectivity and Commitment in the Physical Sciences — 1
1. The Popperian View of Science
2. Kuhn and 'The Structure of Scientific Revolutions'
3. Lakatos' Reaction to Kuhn's Views
4. Some Concluding Comments

Chapter 2 The Copernican Revolution: a Study of Paradigms and Commitments — 27
1. Greek Images of Nature
2. The Rise of a Rival Paradigm
3. The Articulation of the Copernican Paradigm
4. Some Concluding Comments

Chapter 3 The Breakdown of the Newtonian Paradigm: a Study of Crisis — 59
1. The Newtonian Paradigm
2. The Einsteinian Paradigm
3. Competition between Paradigms
4. The Ritzian Paradigm
5. The Popper-Kuhn-Lakatos Debate

Chapter 4 Political Commitments in the Ages of Scientific and Industrial Revolutions — 87
1. Utopian Visions in the Age of the Scientific Revolution
2. The Emergence of Capitalist Society
3. The Commitment of Adam Smith: Promotion of the New Social Order
4. The Industrial Revolution (1780–1850): Capitalism in Crisis
5. The Commitment of John Stuart Mill (1806–73): Progress through Reform
6. The Commitment of Marx and Engels: Progress through Revolution

7. The Evolution of English Capitalism (1850–1900)
8. The Commitment of Alfred Marshall (1848–1924): Progress through Perpetuation of the Status Quo.

Chapter 5 **Capitalism in Crisis and the Keynesian Revolution**   115
1. The Great Depression
2. The Theoretical and Political Commitments of John Maynard Keynes
3. The 'Classical' Paradigm
4. The Keynesian Paradigm
5. The Classical versus the Keynesian Paradigm
6. The Inclusion of Government Expenditure and Taxation
7. The New Deal Policy and the Second World War

Chapter 6 **Objectivity and Commitment in the Social Sciences**   149
1. On the Physical and Social Sciences
2. Three Social Scientists and Their Research-Action Programmes
3. On Marxism
4. On 'Value-Free' Social Science

Chapter 7 **Problems of Post-War Capitalism in the United States**   179
1. The Nature of Advanced Capitalist Society
2. The Militarization of the United States Economy

Chapter 8 **Problems of Underdeveloped Countries**   204
1. On Imperialism
2. The Importance of Raw Materials
3. Markets and Investments in the Underdeveloped Countries
4. On Social Reform in the Underdeveloped Countries
5. Some Concluding Comments

Chapter 9 **Socialism in Crisis: the Russian Revolution and the Origins of the Cold War**   224
1. The Russian Revolution
2. The Soviet Union and the West 1941–5
3. The Cold War Years
4. Some Concluding Comments

Chapter 10 **On the Ethical Neutrality of Science—The Indictment of the Scientific Mentality**   248
1. The Rejection of Science
2. The 'New Science' and its Image of Nature
3. The Mechanistic Imperative

4. An Indictment of Science
5. Remarks Towards an Evaluation
6. The Place of Science in a Liberated Society

Chapter 11 **The Scientific Community in Crisis—Scientists as Agents of Change?** 286
1. Comments on Scientists and Society
2. Nuclear Weapons and the Political Role of Scientists
3. Scientists and Engineers for Social and Political Action?

Chapter 12 **Towards a Beautiful World: Some Concluding Thoughts** 317
1. Résumé
2. Towards a Beautiful World

Appendix **Einstein's Political Struggle** 342

Notes 349

Index 367

ACKNOWLEDGEMENTS

The author and publishers wish to thank the following for permission to reproduce copyright material:

The Trustees and Estate of Albert Einstein for extracts from *Einstein on Peace* and *Why Socialism*

Thomas S. Kuhn and The University of Chicago Press for extracts from *The Structure of Scientific Revolutions*, 1970

William Morris Agency, Inc., New York, and Harper and Row, Inc. for extracts from *The Limits of American Capitalism* by Robert L. Heilbroner

Sir Karl Popper and Hutchinson Publishing Group Ltd., London, for extracts from *The Logic of Scientific Discovery*

Sir Karl Popper and Routledge and Kegan Paul Ltd., London, for extracts from *Conjectures and Refutations: The Growth of Scientific Discovery*

Royal Economic Society, St. Martin's Press, Inc., New York and Macmillan, London, for extracts from *Collected Writings: The General Theory* and *Essays in Persuasion* by John Maynard Keynes

University of California Press for extracts from *A Dialogue Concerning The Two Chief Worlds Systems—Ptolemaic and Copernican* by Galileo Galilei translated by Stillman Drake, 1967.

PREFACE

THIS series of sketches has arisen out of a course of lectures I developed for physics students on the subject of 'Science and Society'. These lectures, explicitly value-oriented from the outset, posed the problem as to why it is that scientists have not been able to help build that society in which they can be sure that the science they produce will be used always and only in the service of all mankind. This book poses the same problem.

The various sketches (which retain the informal style of lectures) have been written mainly but not exclusively with first year science students in mind. However, no prior knowledge of any topic is assumed. In particular the two chapters on 'paradigm change' in the physical sciences are written so that they will be accessible not only to the non-physics science student but to the general reader as well. It is, however, important that the chapters of the book be read sequentially as the argument develops chapter by chapter and much would be lost, therefore, if chapters are read in an arbitrary order.

In the text much use has been made of quotations. I have done this for two reasons. Firstly, I believe that more justice is done to authors by letting them speak as much as possible for themselves rather than by précising and paraphrasing their arguments. Secondly, one of the principal themes running through the sketches concerns the necessity of personal commitment to research-action programmes whether in the natural or the social sciences, and this theme is brought out more clearly, I think, by directly quoting protagonists rather than by simply stating their points of view. Of course, argument by quotation is open to serious abuse in so far as one can foist virtually any opinion on to anyone by simply quoting out of context or by referring to a few of those ill-considered or atypical remarks that everyone sooner or later makes. I hope I have avoided these abuses. In any case, I trust that the reader who finds any argument I have quoted to be particularly interesting will refer to the original author's writings and so evaluate for himself the ideas and commitments he finds there.

Simplification of issues has, of course, been unavoidable. However, it is also desirable that students be presented with a simplified framework of ideas in such an introductory work as this is. For once the reader has grasped the main outlines of what I am attempting to say—and he will the more readily be able to do this the more the framework has been trimmed to its barest essentials—then it is up to him to choose either to extend and elaborate the framework of ideas, modifying and enriching the arguments as he proceeds, although perhaps *ultimately* rejecting their overall direction, or to reject this direction at the outset and initiate his own research-action programme. The first three chapter-sketches should make the logic of this approach clear. In particular, each chapter is to be taken as only a jumping-off point for more detailed study. Indeed I trust that students who read this book will go far beyond its principal arguments during their years at university. My principal hope is simply that this book will serve to provide them with a useful *initial orientation* towards a programme of further research and action.

What I have tried to do, then, in this series of sketches is to set the ball rolling for science students. Admittedly in only one direction. However, the important task as I see it is to *initiate* the process of articulation and debate. Tutors who object to the direction in which I am pushing can and no doubt will push in other directions. In what follows they will find plenty of footholds from which to exert counter-pressure. I hope, however, that no matter what their own commitments they will find the topics I have raised worth discussing.

I should also state here that virtually no idea to be found in this book originated with me. If the *synthesis* of these ideas is original in any way, the component ideas are not (as in general I have tried to make plain throughout the book) and I am heavily indebted both for ideas and information to the many authors whose works I have chosen to rely on. These authors include— and let me repeat how indebted I am to them—F. J. Cook, L. S. Feuer, D. F. Fleming, J. K. Galbraith, R. Gilpin, R. L. Heilbroner, G. Holton, D. Horowitz, T. S. Kuhn, I. Lakatos, S. A. Lakoff, H. Magdoff, H. Marcuse, E. M. Rogers, D. Shapere, I. Scheffler, F. A. Yates and also many others. Clearly, since the world of men and nature is so vast, if the attempt is to be made even to *begin* to evaluate it, then it is necessary not only to draw heavily on the work of 'experts' but also to try to think about what one has not been trained to think about. This book represents such an attempt. Nevertheless, despite the book's necessarily derivative character (or perhaps even *because* of it), I am immodest enough to hope that the book will be of interest, even of use, to those many people, but especially to those students of science, who are similarly attempting to evaluate a world so manifestly in crisis.

In conclusion, I should especially like to thank Norman Clark and Chris Freeman for reading an early version of the manuscript, for helpful criticisms, and for their encouragement to publish. To John Adams, David Albury, Roy Edgley and John Mepham, many thanks indeed for detailed criticisms of the manuscript and for much needed encouragement. To other friends and colleagues who have commented on parts of the manuscript, also many thanks. Obviously, I alone am responsible for whatever errors and misjudgements I have managed to make in this book.

Finally, my thanks to Douglas Meyer for drawing the diagrams, to Chris Stevens and Freda Williams for typing several parts of the manuscript, and to an heroic friend for typing the remainder.

INTRODUCTION

ONE of the most difficult aspects of living is the problem of how to make up one's mind, what to believe in, how to act and to know that one is acting correctly. In a world racked with controversy and conflict it seems very desirable to establish a secure foundation on which to base one's actions. But where can such be found? Is a secure foundation to be built on our 'Western civilization'? Opinions differ.

Karl Popper, an eminent Western philosopher, has declared that 'Our present free world, our Atlantic Community . . . ruled by the interplay of our own individual consciences . . . is the best society that has ever existed'.[1] On the other hand, Herbert Marcuse, another eminent Western philosopher, has expressed a very different evaluation: 'The facile historical parallel with the barbarians threatening the empire of civilization prejudges the issue; the second period of barbarism may well be the continued empire of civilization itself.'[2] Popper and Marcuse do not disagree only on an evaluation of Western civilization. Popper is well known for his attack on Hegel and Marx in *The Open Society and its Enemies*, Marcuse for their defence in *Reason and Revolution*.

How is it possible to evaluate these two diametrically opposed positions? Is it not possible to be 'objective', to see the world 'the way it is', and hence to decide unambiguously between the points of view expressed by Popper and Marcuse?

Our discomfort will be increased if we note that another distinguished Western philosopher radically changed the pattern of his commitment in response to the evolving post-war situation. For during several years after the end of the Second World War, Bertrand Russell advocated strong Western pressure against Soviet totalitarianism; in the 1960s, however, while remaining strongly critical of repressive policies of the Soviet government, he condemned what he saw as Western imperialism as the main threat to world peace and was organizer of the tribunal that was to find Johnson, Rusk and McNamara guilty of genocide in Vietnam.

Where does the truth lie? How can truth be distinguished from falsity? Could not an alternative evaluation of the available evidence, or evaluation of further evidence, result in a totally different appraisal of, for example, the actions of the United States in Vietnam?

An evaluation of the world situation is so clearly necessary. How has it come about that some three hundred years after the birth of Newton the very existence of human life is threatened by scientific weapons of mass destructive power? Did not Francis Bacon declare emphatically that the main objective of science was 'the relief of man's estate'? How can it be that after more than three hundred years of scientific progress, while over one half of the human race still lacks basic necessities of life, the advanced industrial nations continue to destroy the environment while consuming so wastefully the earth's finite natural resources? How can it be that the world's second most powerful country has recently invaded its all but defenceless neighbour, that the world's richest, most scientifically advanced country has made war (and at the time of writing is still making war) against one of the world's poorest, most underdeveloped countries? The nuclear arms race continues to escalate. One ecologist has declared that he believes it doubtful that the earth can support on a sustained basis its present population.[3] And yet this population will double in some thirty years and then double again in another thirty if present trends continue. It seems apparent that unless there is radical change soon the world is heading for a series of disasters of catastrophic proportions. Why is it that of all people, scientists, those people supposedly committed to *rational* thought and action, who pride themselves on being members of an international community of scholars, either stand by powerless or even participate in actions that are—to say the least, the very least—in no way conducive to the building of a world in which people can cooperate together to ensure that each individual is able to live a full and creative human life? Are in fact some of the dissenting young justified in their belief that it will never be possible to build a beautiful world unless the scientific enterprise is rapidly brought to an end?

These are some of the more important questions that this series of sketches attempts to answer. In order to do so, however, a long but I hope interesting and informative journey will be necessary. We must look in detail at the nature of science, the nature of revolutionary changes in science, the development and structure of Western societies, the nature of the social sciences and their relation to the physical sciences, problems of 'underdeveloped countries', the development and structure of Soviet society, reasons for the development of the Cold War, the nature of the current attack on the ethos of science, and the possible role of scientists as agents of social change. During this journey answers will gradually emerge to the questions asked above—questions that both underlie and link each chapter of the book. Of course it will be up to the reader to decide on the acceptability of the historical and conceptual framework that this book attempts to establish and, above all, to decide on the appropriateness of the research-action programme underlying this book.

Chapter 1

OBJECTIVITY AND COMMITMENT IN THE PHYSICAL SCIENCES

> If what we are discussing were a point of law or of the humanities, in which neither true nor false exists, one might trust in subtlety of mind and readiness of tongue and in the greater experience of the writers, and expect him who excelled in those things to make his reasoning most plausible, and one might judge it to be the best. But in the natural sciences, whose conclusions are true and necessary and have nothing to do with human will, one must take care not to place oneself in the defence of error; for here a thousand Demostheneses and a thousand Aristotles would be left in the lurch by every mediocre wit who happened to hit upon the truth for himself.
>
> GALILEO GALILEI
> *A Dialogue Concerning the Two Chief World Systems*

SCIENTIFIC knowledge works. The quantum mechanical laws of nature are valid and (it is argued) must be recognized to be valid by all people who wish to understand the structure of matter. Even if, of these people, there are some who would like the laws to be other than they are, that is their personal subjective bias with which science is not concerned. And if such people wish to be scientists then they must ruthlessly eliminate their subjective desires from their evaluation of scientific theories. Such is part of the conventional view of science.

Consider, for example, the following remarks by Israel Scheffler, a philosopher of science. 'A fundamental feature of science is its ideal of objectivity, an ideal that subjects all scientific statements to the test of impartial criteria, recognizing no authority of persons in the realm of cognition.'[1] As a scientist, therefore, one cannot believe whatever one wants to believe: the criteria are 'independent and impartial'. And since (it can be argued) it is a defining characteristic of a *rational* man that he finds *objective* ways to distinguish valid from invalid beliefs, it follows, and Scheffler makes the point very forcefully, that science enjoys 'the distinctive features of the rational quest', indeed that science is not merely one of several rational enterprises but that it 'has given us a new appreciation of reason itself . . . of responsibility of belief, embodied not only in a firm commitment to impartial principles by which one's own assertions are to be measured, but in a further commitment to make those principles ever more comprehensive and rigorous'.[2] Of great human significance, therefore, is the 'moral import of science: its dynamic articulation of the impulse to responsible belief, and its suggestion of the hope of an increased rationality and responsibility in all realms of conduct and thought'. As a contrast with the desirable scientific approach, we are told that 'to propound as true a belief protected from the hard test of experience flowing out of action is from the pragmatist's standpoint, wilful self-assertion or self-deception, and only secondarily irresponsible in dissipating the power to remedy avoidable evils and to render more lovely the scene of human life'.[3]

To render more lovely the scene of human life! It can be, and is, argued that in a world apparently heading for self-destruction whatever is rational needs to be preserved and cherished. The scientific method is an essential feature of the rational quest, it is commonly believed. Certainly if the aim of human beings is to acquire valid rather than invalid theories about the world they live in, then it would appear that it is the scientific method they must understand and practise. However if the aim of human beings is to build a more lovely world out of the existing ugly one—to build a beautiful world—then perhaps it is legitimate to ask to what extent the scientific method would be a useful ally. For that which is beautiful appeals to the emotions rather than to reason. The absence of 'independent and impartial criteria' to enable us to decide objectively between what is beautiful and what is not means, perhaps, that appeals to construct a beautiful world will meet with no favourable response in the world of science. Let us look at some of Popper's remarks in this context.

Like Scheffler, Popper believes that scientific practice is essentially rational in nature. 'The rationalist attitude', he tells us, '. . . is very similar to the scientific attitude, to the belief that in the search for truth we need cooperation, and that, with the help of argument, we can in time attain something like objectivity'. Like Scheffler, Popper is anxious to defend the rational attitude against the advocacy of irrationalism, a conflict which, Popper tells us, 'has become the most important intellectual, and perhaps even moral, issue of our time'.[4]

But what does rationalism mean? Popper explains that it is 'an attitude that seeks to solve as many problems as possible by an appeal to reason, i.e. to clear thought and experience, rather than by an appeal to emotion and passions'. It is 'an attitude of readiness to listen to critical arguments and to learn from experience. It is fundamentally an attitude of admitting that "I may be wrong and you may be right, and by an effort, we may get nearer to the truth".'[5] In contrast, however, to this rational approach to life, states Popper, the irrationalist 'insists that emotions and passions rather than reason are the mainspring of human action'. While Popper agrees that this may be so, he believes it is much to be regretted and insists that 'we should do what we can to remedy it and should try to make reason play as large a part as it possibly can'. In defence of this objective we are told that the 'irrational emphasis upon emotion and passion leads ultimately to what I can only describe as crime'. For 'he who teaches that not reason but love should rule opens the way for those who rule by hate'.[6]

Popper leaves his readers in no doubt as to why he is so firmly committed to reason rather than to emotion. Indeed, according to Popper,

Aestheticism and radicalism must lead us to jettison reason and to replace it by a desperate hope for political miracles. This irrational attitude which springs from an intoxication with dreams of a beautiful world is what I call Romanticism. It may seek its heavenly city in the past or in the future; it may preach 'back to nature' or 'forward to a world of love and beauty'; but its appeal is always to our emotions rather than to reason. Even with the best intentions of making heaven on earth it only succeeds in making it a hell—that hell which man alone prepares for his fellow-men.[7]

It is, I think, fair to say that Popper is somewhat less than enthusiastic over the prospect of explicit commitment to the goal of building a beautiful world.

Arguments and counter-arguments with respect to Scheffler's and Popper's assertions immediately spring to mind. However, their assertions have been stated here (in drastically truncated form) not in order to provoke criticism and counter-criticism at this stage but in order to introduce a rather typical attitude with respect to science and to bring to the fore the concept of rationality and its (problematic) place in science. For both Scheffler and Popper see science as an activity that is of great significance in its welcome promotion and strengthening of rationality. The essential and welcome feature of science is that no matter what statements are made, and no matter who makes them, they are accepted as valid if and only if they satisfy the test of 'independent and impartial criteria'; subjective factors, aesthetic or ethical, play no role and must be allowed no role in the evaluation of scientific theories. Where subjective factors do play an evaluatory role in any sphere of human activity, that sphere forfeits any right even to the claim to scientific status.

In what follows we shall first elaborate an interpretation of the nature of scientific methodology based primarily on Popper's attitude towards the rationality of science, after which we shall subject it to critical scrutiny. In order to do so we concentrate our attention on physics, rightly or wrongly widely regarded as the model science, by virtue perhaps of its seniority, its undoubted successes, and the ability of physicists to take revolutionary changes in their stride (other reasons will be suggested in chapter 10). Hopefully, by looking at the aims of physicists and the criteria they use to choose between competing theories, we shall be able to judge to what extent subjective factors have in practice been eliminated from the activity of physics and therefore to what extent physics can be called an objective enterprise. The question is of great importance, as later chapters in this book will attempt to make clear.

1. The Popperian View of Science

The scientists' professional identity, as a sociologist recently concluded an article, is that of 'a body of men dedicated to the disinterested pursuit of truth'.[8] Such a claim has a reassuring and ennobling ring to it. But what does it mean? And how in practice can we distinguish between the man who is truly a scientist and the man who is only pretending to be—the pseudo-scientist. A brief glance at the history of physics will give us a preliminary perspective.

The success of Newtonian physics convinced many physicists that a part of 'the truth' had been found; foundation stones had been laid upon which the rest of the edifice of physics could be built. Popper describes their feeling of triumph: 'A unique event had happened in the history of thought, one which could never be repeated: the first discovery of the absolute truth about the universe. An age-old dream had come true. Mankind had obtained knowledge, real, certain, indubitable and demonstrable knowledge—and not merely . . . human opinion.'[9] In 1905, in a justly famous work on astronomy, Dreyer wrote that 'from Thales to Kepler philosophers had searched for the true planetary system; Kepler had completed the search; Isaac Newton was to prove that the system found by him not only agreed with observation, but that no other system was possible'.[10] In short, Aristotle and Ptolemy had got it wrong, Newton had got it right—not the whole truth to be sure, but a part of the whole truth, never again to be challenged. Yet but a few years after Dreyer's apotheosis of Newtonian physics, the entire edifice of classical physics lay in ruins; a foundation stone on which the structure of classical physics had been built was removed from under it. Time was declared no longer absolute—the Einsteinian revolution had occurred (see chapter 3) and the intellectual edifice had to be rebuilt. No sooner had physicists successfully completed the rebuilding than yet another foundation stone was removed —an atom, it was declared, cannot be said to have both a definite position and velocity at the same instant of time. With respect to this new revolution, Einstein himself wrote that 'it was as if the ground had been pulled out from under one, with no firm foundation to be seen anywhere, upon which one could have built'.[11] Once again, however, new foundation stones have been laid and the edifice of physics rebuilt; but this time, while constant improvements are being carried out in the main structure, an uneasy watch is kept on the foundations. There is no longer confidence in their permanent reliability.

The 'disinterested pursuit of truth' has had a stormy history. Let us try to see why this might well have been expected.

Clearly scientists are not searching for *any* kind of truth; they would scarcely be grateful for the gift of a large collection of empirically true statements about events observed at random. They are rather searching for *interesting* truth, that is to say, for theories that enable them to explain and predict the behaviour of phenomena in which they are interested. This statement, of course, immediately gives rise to two important questions. Why does a particular group of men find certain phenomena interesting and other phenomena uninteresting? And what is the nature of a scientific theory? We tackle the second question first noting, however, that throughout the entire book we shall be returning time and again to the first question.

Now as soon as we enquire into the nature of a scientific theory we can understand why it is that scientific revolutions are possible. We note first that since some theories typically contain (or imply) universal statements of the form, for example, 'all bodies attract each other with a force (for any two bodies) proportional to the product of the two masses and inversely proportional to the square of their distance apart', we see that in general theories cannot be induced from observed facts: that since no one has ever observed the cited universal statement to be the case, it can only have been imaginatively hypothesized to be part of a theory. Furthermore, we note that the carrying out of a finite number of successful tests can never prove such a theory to be true. There can be no guarantee that the next test, performed at a different time, possibly at a different place on the earth's surface, with possibly greater experimental accuracy than achieved in former tests, will not produce disagreement between predictions made from the theory and the result of the experiment. If and when such experimental results are observed, scientific revolutions follow. Thus, for example, (so the story goes) despite the hitherto overwhelmingly large number of observations and experiments in favour of fundamental principles proposed by Newton and his successors, the result of the Michelson-Morley experiment showed quite conclusively that one or more of these principles were false. The Einsteinian revolution followed.

Considerations such as these enable us, believes Popper, to set up demarcation criteria to distinguish between scientific and non-scientific theories as well as criteria to distinguish between scientists and pseudo-scientists. In particular, they have led Popper to declare that it is a mistake to suppose that the essential feature distinguishing a scientific from a non-scientific theory is that the former is empirically *verifiable*. On the contrary, scientific theories, as opposed to non-scientific, are in principle capable not of empirical verification (as we might naïvely have supposed to be the case) but rather of empirical *falsification*. It is falsifiability, not verifiability, says Popper, 'that is to be taken as a criterion of demarcation' between scientific and non-scientific (or pseudo-scientific) theories.[12] 'A theory which is not refutable by any conceivable event is non-scientific; ... every genuine *test* of a theory is an attempt to falsify it.'[13] And how is it possible on this view to distinguish scientists from those who only pretend to be scientists? Declares Popper: 'Those among us who are unwilling to expose their ideas to the hazard of refutation do not take part in the scientific game.' The scientific method of research, Popper tells us, 'is not to defend [our present conjectures] in order to prove how right we were. On the contrary, we try to overthrow them. Using all the weapons of our logical, mathematical and technical armoury, we try to prove that our anticipations were false—in order to put forward in their stead new unjustified and unjustifiable anticipations, new "rash and premature prejudices" '.[14]

In order to analyse this claim that the true scientist—as opposed to the pseudo-scientist—is a man who strives to prove existing theories false (so that he can make progress by replacing them by theories not in disagreement with experiment) and to examine the difficulties that this Popperian view of correct scientific practice has encountered, I wish first to summarize what might be called the standard view of science, succinctly described by Scheffler as follows. Broadly speaking the standard view claims that science progresses on two different levels, the theoretical and the observational (or experimental). Whereas observational laws 'make reference to perceived things and processes',[15] theories make use of terms and concepts that are not directly observable and thus cannot be *directly* tested; their function is to explain the observational laws that are themselves generalizations of directly observable phenomena. An important feature of this two-tier picture of science, Nagel tells us, is that 'the experimental law, even when explained by a theory, retains a meaning that can be formulated independently of the theory; and it is based on observational evidence that may enable the law to survive the eventual demise of the theory'.[16] Thus, when disagreement with experiment results in the falsification of a theory and its replacement by another theory which is not in conflict with the experimental data in question, 'the genuine facts' accounted for by the superseded theory are not discarded. For although science is not cumulative at the theoretical level (as we have seen) it is so as the observational level; and through the ever-changing flux of scientific *theories* there is therefore 'a solid growth of knowledge which represents progress in empirical understanding'. It is this point of view, affirms Scheffler, that 'understands science to be a systematic public enterprise, controlled by logic and by empirical fact, whose purpose is to formulate truth about the natural world'.[17]

This hypothetico-deductive model of science with the emphasis on falsifiability is surely an attractive one. How a scientist arrives at a particular theory is a matter only of psychological interest; there is no set of rules and procedures for making discoveries and ensuring new insights. As Popper sees it: 'The initial stage, the act of conceiving or inventing a theory seems to me neither to call for logical analysis, nor to be susceptible to it. The question of how it happens . . . may be of great interest to empirical psychology; but it is irrelevant to the logical analysis of scientific knowledge. The latter is concerned only . . . with questions of justification or validity.'[18] The point is that having once made public his theory, no matter how arrived at, the scientist has exposed his hypotheses to *objective* evaluation. Subjective factors undoubtedly play an important role in the invention of theories but they play none whatsoever in their evaluation. For should predictions made from the theory disagree either immediately or eventually with experiment and observation, then the theory has been falsified and scientists must try again.

According to Popper therefore science progresses, or rather should progress, in the following way. T_1, the current theory, is acceptable only for as long as it agrees with relevant observational statements. As soon as there is disagreement, the theory has been refuted and must be ruthlessly rejected. A new theory T_2 must be sought which re-establishes agreement between theory and all relevant observational statements. Clearly, since T_2 is required to explain both the new observations that led to T_1's downfall *and* those observations formerly explained by T_1, it follows that in some sense T_2 must contain the former theory T_1 as a special case. As Popper has written, 'A theory which has been well corroborated can only be superseded by one of a higher level of universality; that is, by a theory which is better testable and which, in addition, contains the old, well-corroborated theory'.[19] Since T_2 is a scientific theory (as opposed to a pseudo-scientific theory) it will be falsifiable and indeed the aim of all scientists will be to subject the initially successful theory to the most searching experimental tests possible in the hope of falsifying it. For 'once a theory is refuted, its empirical character is secure and shines without blemish'.[20] However, we should note that, perhaps a little uneasy at the priority given to attempts at falsification, Popper also tells us that 'we require of a good theory that it should be successful in some of its *new* predictions; secondly, we require that it is not refuted too soon—that is, before it has been strikingly successful'.[21]

Popper's attack on instrumentalism

In his 'Three Views Concerning Human Knowledge', Popper attacks the so-called instrumentalist interpretation of scientific theories, namely, that theories are merely useful tools for ordering certain domains of experience. According to the instrumentalists (and Popper, in particular, refers to Berkeley, Mach, Duhem and Poincaré) it is meaningless to ask whether scientific theories are true or false; theories can only be more useful or less useful, they cannot be either true or false. Theoretical terms are free creations of the human mind which are not related to physical entities and not reducible to observational terms, although they are functionally related to the latter. In opposition therefore to the instrumentalist Popper insists that 'the scientist aims at a true explanation of observable facts'. To be sure Popper concedes that 'though this remains the aim of the scientist, he can never know for certain whether his findings are true, although [and this is the point] he may sometimes establish with reasonable certainty that a theory is false'.[22] Yet instrumentalism, declares Popper, can never get beyond the assertion that *different theories have different ranges of application*. Necessarily, he argues, instrumentalism neglects falsification since 'a mere instrument for prediction cannot be falsified' and in doing so instrumentalism proves itself an obscurantist philosophy 'for it is only in searching for refutations that science can hope to learn and to advance'.[23] Indeed 'in contrast to the highly critical attitude requisite in the pure scientist, the attitude of instrumentalism is one of complacency at the success of applications'.[24] Finally, Popper claims that instrumentalism cannot account for the fact that a scientific theory often predicts events of a new kind, which are then discovered by experimentalists (e.g. radio waves). Does this not suggest, he asks, that theoretical terms are at least partially related to real physical entities?

The instrumentalist would presumably reply that Popper is being unjust, that he is not 'complacent' at the success of applications but is always seeking to order new domains of experience and to extend the *usefulness* of his theory, i.e. to extend its domain of validity. A failure of the theory to be useful in any domain simply means that a new and better theory is needed for which the instrumentalist then searches; the instrumentalist would not declare, however, that his theory has been falsified, only that limitations to its usefulness have been exposed.

One cannot help feeling that to some extent the disagreement between Popper and instrumentalists is merely verbal, despite Popper's polemic to the contrary. The instrumentalist certainly accepts the domain of observable phenomena as real and the corresponding observational terms thus refer to real physical phenomena; however, he does not accept that theoretical terms correspond to real physical entities. But this dichotomy between observational and theoretical terms itself begs a very important question: is there, in fact, a clear distinction between these two classes of terms? Is it possible to distinguish precisely between a theoretical and observational law or hypothesis? The question is more important than the context in which it has been raised, as will become evident later. It is but the first ripple of a tidal wave that threatens to overwhelm the claim that science is a rational enterprise and, with it, the claim that science advances to ever-greater truth through steps objectively controlled by logic and the neutral court of observational appeal.

On the autonomy of observational terms

At first sight, it would seem, the matter cannot be very serious. To be sure, an experimental law may contain terms that are 'theory-laden'; for instance, the experimental law that Millikan helped to establish (namely, that all electric charges are integral multiples of a certain elementary charge) clearly does not refer to entities that are directly observable. Neither, of course, does Kepler's first law. In fact, both so-called laws are in the nature of hypotheses that themselves contain much theorizing; furthermore, Kepler's first law is only approximately true as the Newtonian theory, which explains it, itself predicted. We have, in fact, been too cavalier in postulating a hard and fast separation between theoretical statements and observational laws; the dividing line cannot easily be drawn and in any case may be continually shifting. Indeed Popper himself has emphasized that 'the customary distinction between "*observational terms*" . . . and *theoretical terms* is mistaken, since all terms are theoretical to some degree . . .'[25] Nevertheless, it is in general always possible to distinguish between those hypotheses that are more closely related to observation than others; the former we shall continue to call observational hypotheses, the latter theoretical.

Popperian rationality

Now, however, we see a very serious difficulty. Suppose there is an inconsistency between two sets of hypotheses, one observational, one theoretical—which set is it rational to reject? It surely must not be an arbitrary choice. We need a set of objective rules, Popper states, for limiting the apparent arbitrariness; for a failure to state objective rules 'throws empiricism overboard'.[26] Every system becomes defensible if one is allowed to reject whatever hypothesis one chooses.

Clearly Popper's demarcation criterion between scientific and non-scientific theories is now in considerable difficulties, since if observational statements are to any degree theory-dependent then theories are not compared against neutral experience and therefore cannot be so readily falsified by disagreement with experiment as Popper desires. Perhaps it is the theory of the experiment that needs revision when theory and 'fact' disagree. Perhaps the inconsistencies can be eliminated, as Popper is aware, 'by suggesting *ad hoc* the adoption of certain auxiliary hypotheses or perhaps of certain corrections to our measuring instruments'.[27] Perhaps rational men can disagree on what needs to be done. Popper is appalled at the prospect. There seems only one way out. It is to try to turn this difficulty for his view of rational scientific practice into a strength.

This is exactly what Popper attempts to do. What we have to recognize, he argues, is the danger that precisely because experiments are open to interpretation 'it is always deceptively easy to find *verifications* of a theory' rather than falsifications. Therefore, it follows, argues Popper, turning defeat into victory (?), that we must try all the more sincerely 'to adopt a highly critical attitude towards our theories if we do not wish to argue in circles: the attitude of trying to *falsify* them'.[28] Exactly because 'no conclusive disproof of a theory can ever be produced' it is folly to insist on it and if you do, Popper warns, 'you will never benefit from your experience and never learn from it how wrong you are'.[29] Thus whenever there is clash between a low-level, observational hypothesis and a higher level theory, then, despite the risk, it is the higher level theory which must be rejected. There is no other way out. A scientist must be a man who seeks and welcomes the opportunity to confess the error of his ways. Dogma and commitment must play no part in the scientific method.

The arbitrariness of Popper's methodological rule is, however, blatantly obvious. The claim that science is a rational enterprise appears gravely suspect. It is therefore not surprising that doubt lingers on in Popper and in a later article, significantly entitled 'Truth, Rationality and the Growth of Knowledge', Popper discusses the problem yet again. Informed that it is 'our way of choosing between theories, in a certain problem situation, which makes science rational', we are then told that of two theories T_1 and T_2 the theory T_2 is to be preferred to the theory T_1 which has failed the severe tests the theory T_2 has passed.[30] But again the difficulty is obvious. Since T_1 and T_2 are compared against experimental results that themselves need interpretation, it is only if we consider this experimental 'background knowledge' to be unproblematic that we can unambiguously state that T_1 has failed severe tests that T_2 has passed. Popper is of course aware of this: 'As long as there are no revolutionary changes in our background knowledge, the relative appraisal of our two theories T_1 and T_2 will remain stable.'[31] Yet Popper knows that the 'background knowledge' *is* subject to change and that it is arbitrary, if not irrational, to call it unproblematic. But again Popper insists that a rule is necessary. Rationality in science must be preserved. It is the case, he declares, that 'almost all the vast amount of background knowledge ... will, for practical reasons, necessarily remain unquestioned; and the misguided attempt to question it all—that is to say, *to start from scratch*—can easily lead to the breakdown of critical debate'. Popper hammers home his belief: it is necessary to understand that 'though every one of our assumptions may be challenged, it is quite impossible to challenge all of them at the same time'. The implications of this position are obvious. Popper spells them out. 'Thus all criticism must be piecemeal ... we should stick to our problem ... and try to solve no more than one problem at a time.'[32] (This is also the position which Popper adopts in his social and political philosophy as we shall have occasion to note in chapter 6.)

We are now far from the 'naïve' hypothetico-deductive model with imaginatively constructed theories being systematically and firmly rejected should any of their predictions conflict with the solid rock of direct observational experience. It now appears that having once embarked on a certain course, then we are committed to it, come what may. To be sure, Popper declares that problems must be tackled but they must be tackled one at a time within the general framework *already established*. Revolutions may well happen but the greater amount of background knowledge will always remain common both to the 'revolutionary' theory and its predecessors. Any attempt to return to the journey's starting point and take an altogether different direction is doomed to failure and can result only in disaster. Rationality, it would seem, despite Popper's good intentions, has been reduced to commitment to an arbitrary methodological principle—one problem at a time but never all together!

It was in the context of this perplexing and unsatisfactory situation that in 1962 Kuhn launched his controversial and radical attack not only upon the 'conventional wisdom' that science progresses cumulatively towards an ever-greater understanding of physical reality, step by step, guided by logic and the appeal to a theory-independent empirical basis, but also upon Popper's advice to scientists to subject their theories to ever-more stringent tests and ruthlessly to reject them as soon as disagreement occurs between their theories and the accepted background knowledge.

2. Kuhn and 'The Structure of Scientific Revolutions'

The measure of Kuhn's disagreement with Popper can be gauged from his calm assertion that 'no process yet disclosed by the historical study of scientific development at all resembles the methodological stereotype of falsification by direct comparison with nature'.[33] 'On the contrary,' Kuhn writes, 'it is just the incompleteness and imperfection of the existing data-theory fit that, at any time, define many of the puzzles that characterize normal science. If any and every failure to fit were ground for theory rejection all theories ought to be rejected at all times.'[34]

It is scarcely possible to imagine a more direct and devastating attack upon Popper's already battered criterion that the scientist, as opposed to the non-scientist, is a man who tries to falsify his theories and rejects them ruthlessly when he finds disagreement with experiment. If Kuhn is right, adoption by scientists of Popperian methodology would mean the instant death of science.

Let us therefore examine Kuhn's provocative remarks within the context of his main thesis.

Normal science

'Normal science', Kuhn claims, is to be sharply distinguished from those periods of scientific practice which constitute 'scientific revolutions'. For whereas, Kuhn argues, normal science consists of articulation of the paradigm to which the scientific community is committed, 'scientific revolutions are . . . non-cumulative episodes in which an older paradigm is replaced in whole or in part by an incompatible new one'.[35] Clearly the key concept here is that of a 'paradigm'. However difficult the concept is to define, Kuhn emphasizes that paradigms are readily revealed in scientists' 'textbooks, lectures, and laboratory exercises' and that scientists can and do 'agree in their *identification* of a paradigm without agreeing on, or even attempting to produce, a full *interpretation* or *rationalization* of it'.[36] Kuhn gives several partial definitions of the concept. He tells us, for example, that the paradigm must be seen as 'prior to the various concepts, laws, theories, and points of view that may be abstracted from it',[37] that it consists of a 'strong network of commitments— conceptual, theoretical, instrumental and methodological', thus permitting 'selection, evaluation, and criticism', and that it is 'the source of the methods, problem-field and standards of solution accepted by any mature scientific community at any given time'.[38] *

Normal science, Kuhn goes on to explain, 'is predicated on the assumption that the scientific community knows what the world is like', and is a 'strenuous and devoted attempt to force nature into the conceptual boxes supplied

---

* Readers of Kuhn's essay will notice, as Kuhn now acknowledges, that the term 'paradigm' was used in his essay in at least two quite distinct ways, the first as referring to a concrete example of successful scientific practice, the second in the way we shall use in the text ('exemplar' and 'disciplinary matrix', respectively, in Kuhn's new terminology).

by the paradigm', so much so that 'normal science ... often suppresses fundamental novelties because they are necessarily subversive of its basic commitments'.[39] Indeed the paradigm commitment can 'even insulate the community from those socially important problems ... because they cannot be stated in terms of the conceptual and instrumental tools the paradigm supplies'.[40] For Kuhn, the whole point of normal science is that the fit between the paradigm and reality is never exact, and it is this mismatch which supplies scientists with 'puzzles' for solution. As Kuhn remarks, 'the object of normal science is to solve a puzzle for whose very existence the validity of the paradigm must be assumed' and it is commitment to the paradigm which gives the scientist 'the conviction that, if only he is skilful enough, he will succeed in solving a puzzle that no one before has solved or solved so well'.[41] Furthermore, 'failure to achieve a solution discredits only the scientist and not the theory'.[42] Normal science, Kuhn tells us, 'is a highly cumulative enterprise, eminently successful in its aim, the steady extension of the scope and precision of scientific knowledge'.[43]

Certainly the history of physics shows that at least on some occasions physicists have achieved success by behaving not as dogmatic falsificationists but as 'Kuhnian' scientists. A specific example is the development of Newtonian theory. The paradigm here entails general commitment to Newton's concepts of space, time and matter, Newton's three laws of motion, and the law of universal gravitation. The relevant question posed was whether this paradigm could be further developed and articulated so as to resolve all the numerous puzzles arising from the mismatch between the paradigm itself and reality. As Kuhn has pointed out, the mismatch did not mean rejection of the paradigm. On the contrary, it was the incentive to develop the paradigm further—the disagreement with observation was its very nourishment! And the Newtonians were successful. As Laplace wrote in 1796, they 'turned each new difficulty into a new victory of their programme'.[44] Newton, himself, solved equations for the motions of the planets assuming no interplanetary forces and showed that Kepler's laws could be deduced as a special case from his more general laws. The solution of the relevant equations with interplanetary forces remained a problem for Newton, however, who was led to believe that the planetary system could not be stable. Lagrange was later to show that this was not so. Furthermore, although Newton assumed that the earth's mass acted on a body as if all the mass were situated at the centre of the earth, it was some time before he could justify this mathematically. Initially 'the apple and the moon' calculations did not correspond to Newton's expectations but he later found that the earth's radius had not been accurately determined. Strange behaviour of Jupiter's moons did not mean the falsification of Newton's theory but rather information on the velocity of light! For sixty years after Newton's original computation, as Kuhn points out, the predicted motion of the moon's perigee remained only half that observed; in 1750 Clairaut was able to show that the mathematics of the application had

OBJECTIVITY AND COMMITMENT

been wrong and Newton's theory remained intact. In the 19th century the movements of the newly discovered planet, Uranus, were not in accord with predictions from Newton's theory. Falsification? Not at all, the correctness of Newtonian theory was taken for granted and theoretical astronomers *used the theory* in order to predict and locate the position of a hitherto unknown planet responsible for the anomalous behaviour of Uranus. Thus Neptune was discovered. Similar misbehaviour by the planet nearest the sun, Mercury, gave rise to predictions of another planet, Vulcan, between Mercury and the sun. It was not observed. Falsification? Not at all, merely a puzzle awaiting solution by an ingenious astronomer. The two centuries of development of Newtonian theory are, in fact, an excellent example of paradigm articulation. The paradigm has to be made to fit reality. As Feynman has put it: 'That is the same with all our other laws—they are not exact. There is always an edge of mystery, always a place where we have some fiddling around to do yet.'[45]

We are now at the opposite extreme to Popper's falsificationism. Although Popper has acknowledged that 'the dogmatic attitude of sticking to a theory as long as possible is of considerable significance', for 'without it we could never find out what is in a theory—we should give the theory up before we had a real opportunity of finding out its strength; and in consequence no theory would ever be able to play its role of bringing order into the world, of preparing us for future events, of drawing our attention to events we should otherwise never observe', nevertheless (as Lakatos has pointed out) Popper's emphasis lies very much elsewhere.[46] For when scientists committed to a paradigm meet apparently falsifying experiments with the claim 'that the experimental results are not reliable, or that the discrepancies which are asserted to exist between the experimental results and the theory are only apparent and that they will disappear with the advance of understanding', Popper condemns them for 'adopting the very reverse of that critical attitude which . . . is the proper one for the scientist'.[47] Yet for Kuhn this apparently dogmatic commitment is a necessary feature of normal science without which progress is impossible. 'Without commitment to a paradigm,' he writes, 'there could be no normal science.'[48] The gulf, then, between Popper and Kuhn is very wide; but it is to grow wider still.

Scientific revolutions

As we noted earlier, physics underwent two revolutions in the first half of the 20th century; the conceptual foundations of the Newtonian paradigm were shattered beyond repair and were replaced by altogether different foundations. Why did this happen and what is the nature of a scientific revolution?

According to Kuhn, a scientific revolution will not happen simply because of a mismatch between theory and experiment. For, as we have seen, defenders of a paradigm 'will devise numerous articulations and *ad hoc* modifications of their theory in order to eliminate any apparent conflict'.[49] However, articulation of any paradigm may well lead to a *crisis* if there is a 'persistent failure of the puzzles of normal science to come out as they should'.[50] In this crisis state, Kuhn tells us, a few scientists, usually young ones (and therefore less emotionally committed to the existing paradigm than more senior scientists), will lose faith in that paradigm and consider alternative ones. There is a 'consequent loosening of the rules for normal research . . . the expression of explicit discontent, the recourse to philosophy and to debate over fundamentals; all these are symptoms of a transition from normal to extraordinary research'.[51] Thus 'all crises begin with the blurring of a paradigm . . . and [if there is to be a revolution] close with the emergence of a new candidate for paradigm and with subsequent battle over its acceptance'.[52] This is necessarily so, argues Kuhn, for 'once it has achieved the status of a paradigm, a scientific theory is declared invalid only if an alternative candidate is available to take its place. . . . To reject one paradigm without simultaneously substituting another is to reject science itself.'[53] However once the transition is complete, 'the profession will have changed its view of the field, its method and its goals'.[54] 'Indeed', Kuhn writes, 'the normal-scientific tradition that emerges from a scientific revolution is not only incompatible but often actually incommensurable with that which has gone before.'[55]

'Incommensurable' is, of course, a provocative word to use to describe the relationship—or lack of relationship—between two competing paradigms. Kuhn therefore attempts to clarify what the term means for him and to justify its use.

Firstly, he argues that 'though an out-of-date theory can always be viewed as a special case of its up-to-date successor, it must be transformed for the purpose'.[56] For example, although the Lorentz transformation equations reduce in mathematical form to the Galilean in the limit $v/c \to 0$, nevertheless the symbols still refer to Einsteinian space and time, not Newtonian. To complete the transformation, the symbols must be given their Newtonian meaning. And that this is necessary, argues Kuhn, reflects the fact that the two theories are incommensurable. Secondly, Kuhn tells us that across the revolutionary divide 'proponents of competing paradigms will often disagree about the list of problems that any candidate for a paradigm must resolve. Their standards or their definition of science are not the same.'[57] Thus Leibniz refused to accept the Newtonian theory of gravitation on the grounds that to suppose action at a distance is not a solution to the problem of planetary motion. Thus, when Einstein insisted that an acceptable scientific theory should allow prediction of the time of decay of an individual radioactive atom, he was told by Bohr that according to the new paradigm—the paradigm Einstein rejected—the phenomenon in question was not *in principle* predictable and was therefore in no way problematic. Thirdly, Kuhn writes that the 'most fundamental aspect of incommensurability between competing paradigms' is that 'the proponents of competing paradigms practice their trades in different worlds'; they 'see different things when they look from the same point in the same direction'.[58] Thus, whereas when Aristotle looked at a swinging stone he saw constrained fall, Galileo saw a pendulum. Perception itself, Kuhn argues, is a function of the paradigm.

Competition between paradigms

Since the transition from one paradigm to another is 'a transition between incommensurables' it cannot be made 'a step at a time, forced by logic and neutral experience. Like the gestalt switch, it must occur all at once (though not necessarily in an instant) or not at all.'[58] Once, however, a scientist has undergone this 'change of world view' to a new paradigm, what must he do 'to convert the entire profession or the relevant professional subgroup' to his 'way of seeing science and the world'?[59]

Kuhn's use of the word 'convert' is highly significant. Conversion is a religious, political, essentially mystical experience, it is not a scientific one. Or so we are assured by those who believe the scientific quest to be a rational endeavour. In science, as distinct from religion or politics, faith is unnecessary; in science one can *know* when one is wrong. Kuhn thinks differently. According to him, there are no objective criteria by which the revolutionary scientist can show his colleagues committed to a different paradigm the error of their ways. Since each paradigm will be defended using 'the criteria that it dictates for itself' it follows that two proponents of different paradigms 'will inevitably talk through each other when debating the relative merits of their respective paradigms'.[60] Each 'may hope to convert the other to his way of seeing his science and its problems, [but] neither may hope to prove his case' since 'neither proof nor error is at issue. The transfer of experience from paradigm to paradigm is a conversion experience that cannot be forced.'[61] And this we remember is exactly Popper's nightmare: 'The misguided attempt to question it all . . . can easily lead to the breakdown of a critical debate.'[32]

But be this as it may be (and we will have much to say about it later) how, according to Kuhn, does the 'revolutionary' scientist attempt to *convert* his 'conservative' colleagues? Kuhn answers that 'probably the single most prevalent claim advanced by the proponents of a new paradigm is that they can solve the problems that have led the old one to crisis'. However, we are warned that 'the claim to have solved the crisis-provoking problems is . . . rarely sufficient by itself. Nor can it always legitimately be made.' In any case, Kuhn continues, 'paradigm debates are not really about relative problem-solving ability, though for good reasons they are usually couched in those terms. Instead, the issue is which paradigm should in the future guide research on problems many of which neither competitor can yet claim to resolve completely.' Furthermore, Kuhn tells us, 'the man who embraces a new paradigm at an early stage must often do so in defiance of the evidence provided by problem-solving. He must, that is, have faith that the new paradigm will succeed with the many large problems that confront it, knowing only that the older paradigm has failed with a few. A decision of that kind can only be made on faith.' There must, of course, be a basis for faith in the particular candidate chosen, 'though it need be neither rational nor ultimately correct'; sometimes it is due 'only to personal and inarticulate aesthetic considerations'.[62]

Finally, Kuhn writes, if one paradigm is destined to win its fight, 'the number and strength of the persuasive arguments in its favour will increase', and there is 'an increasing shift in the distribution of professional allegiances' in favour of the winning paradigm. Ultimately 'only a few elderly hold-outs remain'. Nevertheless, there is no point 'at which resistance becomes illogical or unscientific'. At most it is only possible to say that 'the man who continues to resist after his whole profession has been converted has *ipso facto* ceased to be a scientist'.[63]

The writing and rewriting of history

That dissenters will certainly be regarded by the scientific community as non-scientific (or, more unkindly, cranks) is brought about, says Kuhn, in the following way:

> Revolutions close with a total victory for one of the two opposing camps. Will that group ever say that the result of its victory has been something less than progress? That would be rather like admitting that they had been wrong and their opponents right. To them, at least, the outcome of revolution must be progress, and they are in an excellent position to make certain that future members of their community will see past history in the same way. . . .[64] [For] when it repudiates a past paradigm, a scientific community simultaneously renounces, as a fit subject for professional scrutiny, most of the books and articles in which that paradigm had been embodied.

The result is a 'drastic distortion in the scientist's perception of his discipline's past' which he thus sees as a straight line leading to his discipline's present achievements—in short, it is progress![65] And this distortion is necessary if the scientist is to have sufficient confidence in the existing paradigm to ensure its further articulation. Thus, says Kuhn, 'the depreciation of historical fact is deeply, and probably functionally, ingrained in the ideology of the scientific profession'.[66] The typical scientific education is 'narrow and rigid ... more so than any other except perhaps orthodox theology.'[64] Should all this suggest, says Kuhn, that 'the member of a mature scientific community is, like the typical character of Orwell's *1984*, the victim of a history rewritten by the powers that be', the analogy is 'not altogether inappropriate'![65] However, to its credit, the scientific community does strive for general agreement and does not accept a new paradigm unless it seems to solve new problems, not otherwise soluble, and preserves 'a relatively large part of the concrete problem-solving ability that has accrued to science through its predecessors'.[67]

However, after all that has gone before, this qualification appears to be but a concession to the 'conventional wisdom'; the main theme is that a scientific revolution results in change rather than progress but which is experienced as progress by the younger members of the scientific community following the systematic rewriting of the discipline's history.

Scientific and political revolutions

Finally, let us note that Kuhn's use of a political vocabulary is not accidental for he himself draws an analogy between scientific and political revolutions.

Political revolutions, says Kuhn, are inaugurated by the growing sense, at first shared by only a minority of the community, 'that existing institutions have ceased adequately to meet the problems posed by an environment that they have in part created'. Similarly a crisis state develops when anomalies, which result from the articulation of the scientific community's paradigm, do not within a reasonable time prove susceptible to solution within the framework of the paradigm that has brought them into being. Furthermore 'political revolutions aim to change political institutions in ways those institutions themselves prohibit'. Similarly a scientific revolutionary attacks sacrosanct features of the paradigm to which his 'conservative' colleagues are committed. As the political crisis deepens, the society becomes divided into competing camps committed to radically different goals. Political recourse now fails since there is no supra-institutional framework for the adjudication of revolutionary differences, the contest being decided by mass persuasion and the use of force. Similarly with scientific revolutions; each competing group uses its own paradigm to argue in that paradigm's defence—there is no standard higher, Kuhn claims, than the assent of the relevant community. Force, however, is excluded in scientific revolutions (we must be thankful for small mercies!), experimental results, logic and the techniques of persuasive argumentation being the deciding factors, the latter being apparently necessary since, according to Kuhn (let us repeat once more), the 'issue of paradigm choice can never be unequivocally settled by logic and experiment alone'.[68]

We have come far from Popper's earlier version of the hypothetico-deductive model of scientific progress in which stringent experimental testing serves to constrain the scientific imagination. In place of Popper's community of open-minded, reasonable men, rationally evaluating their various theories against all available experimental results, we now have consensus opinion produced by revolution, psychological conversion, rhetorical persuasion, the dying-off of dissidents and the re-training of the young by the winning faction—a world more in keeping, it would seem, with revolutionary political movements than with scientific research and practice. To be sure, Popper and Kuhn repeatedly qualify the main themes of their arguments, as we have seen; but if anything the qualifications simply contradict their basic interpretations of scientific procedure and thus serve only to demonstrate the necessity of a thorough critique of their respective analyses. The question remains—in what sense, if any, can science be said to be a search for truth, a search in which all statements are subjected to rational evaluation, in which objective criteria automatically eliminate the unwarranted influence of subjective factors?

Although at this point the reader is perhaps anxious to look in some detail at one or two revolutions in the physical sciences (the content of the next two chapters), nevertheless before doing so it will be useful to look at a spirited defence of Popper's philosophy of science undertaken by Lakatos and then at some points made by Kuhn in defence of his own views. (Although important attacks on Kuhn's views, such as those by Shapere and Scheffler, are not considered here, nevertheless I hope the ensuing discussion will give at least a hint of the very considerable impact Kuhn's radical views have made in the world of the philosophy of science.)

3. Lakatos' Reaction to Kuhn's Views

In varying degrees, let us note, Kuhn has several allies, of whom perhaps Polanyi is one of the most outspoken. For example, in his *Personal Knowledge* (published in fact four years before Kuhn's essay) Polanyi writes that 'it is the normal practice of scientists to ignore evidence which appears incompatible with the accepted system of scientific knowledge, in the hope that it will eventually prove false or irrelevant' (i.e. anomaly does not immediately provoke crisis) and that 'proponents of a new system can convince their audience only by first winning their intellectual sympathy for a doctrine they have not yet grasped'.[69] Furthermore a decision is reached 'in the knowledge that we have overruled by it conceivable alternatives, for reasons that are not fully specifiable' and it is this 'framework of commitment' which allows us 'to accredit in advance (if anything is ever to be affirmed) affirmations against which objections can be raised that cannot be refuted'.[70] Indeed, for Polanyi, knowledge in science is personal, committing us 'passionately and far beyond our comprehension, to a vision of reality. Of this responsibility we cannot divest ourselves by setting up objective criteria of verifiability or falsifiability, or testability, or what you will. For we live in it as in the garment of our own skin. Like love, to which it is akin, this commitment is a "shirt of flame", blazing with passion and, also like love, consumed by devotion to a universal standard. Such is the true sense of objectivity in science.'[71]

This, then, is the kind of reasoning and rhetoric that has brought down the wrath of several philosophers of science on the heads of Kuhn and Polanyi. In particular, it has provoked Lakatos to make the outburst that 'Kuhn (like Polanyi) suggests that strength of commitment matters more than (possibly even constitutes) truth in science: and thereby lends—no doubt unintentionally—respectability to the political *credo* of contemporary religious maniacs ("student revolutionaries").'[72] Lakatos is firmly convinced in the belief that objective criteria do exist to enable rational choice to be made between competing theories, and therefore strongly endorses Popper's plea for that intellectual honesty which 'consists not in trying to entrench, or establish, one's position but in specifying precisely the conditions under which one is willing to give one's position up'. To be sure, believes Lakatos, Marxists and Freudians refuse to specify such conditions but this, he claims, is precisely 'the hallmark of their intellectual dishonesty'.[73] Science, on the other hand, Lakatos insists, is or ought to be a rational enterprise undertaken by intellectually honest men. But since Kuhn's thesis implies that the transition from one paradigm to another is an irrational process, 'a matter for mob psychology' (writes Lakatos somewhat unkindly but pointedly), the clash between Popper and Kuhn is not merely over a technical point in epistemology, Lakatos warns, it is over our central intellectual values. Indeed, 'the methodological implications of the competing positions reach beyond theoretical physics to the underdeveloped social sciences and even further into moral and political philosophy'.[72] Lakatos therefore undertakes to defend an improved Popperian view of science against Kuhn's onslaught.

In his article 'Criticism and the Methodology of Scientific Research Programmes' Lakatos immediately rejects 'naïve' falsificationism. Its essential (and untenable) feature, Lakatos explains, is the mono-theoretical model of criticism in which it is supposed that 'one single theory is confronted by potential falsifiers supplied by some authoritative experimental scientist'.[74] Thus if falsification is to be possible, science must be divided up into the problematic and the unproblematic, a division which Lakatos calls dogmatic and irrational. As we have already discussed, experimental techniques and results themselves depend upon theory and thus cannot be assumed to be unproblematic. Indeed, 'background knowledge' cannot be taken as unproblematic for it is often not well-corroborated and even if it were well-corroborated it would be dogmatic and irrational to say that it must remain for ever unchallengeable. Lakatos therefore strongly rejects the apparent choice between naïve falsificationism and Kuhnian 'irrationality'. The former is, in any case, he claims, irrational. However, more to the point, there is a third possibility.

The point is, says Lakatos, that we work within a multiplicity of theories. It is not the case that 'we propose a theory and Nature may shout NO. Rather, we propose a maze of theories, and Nature may shout INCONSISTENT.'[75] But this does not take us very far. The basic question remains: where do we go from here? Agassi has claimed that we may 'stick to the hypothesis in the face of known facts in the hope that the facts will adjust themselves to theory rather than the other way round'.[76] But, asks Lakatos, *how* can facts adjust themselves? At its simplest the problem can be stated in the following form: given a potential falsifier B and two theories T_1 and T_2, then B refutes T_2 if T_1 is assumed true, while it refutes T_1 if T_2 is assumed true.* How, therefore, should one rationally proceed?

* For example (Lakatos writes), if B = 'this piece of chlorine is pure and has the atomic weight 35·5', T_1 = 'all atoms are compounds of hydrogen atoms and thus atomic weights of all chemical elements must be expressible as whole numbers', T_2 = 'if the following chemical purifying procedures ... are applied to a gas, what remains will be pure chlorine', then B refutes T_1 if T_2 is assumed true or T_2 if T_1 is assumed true.

The problem is further complicated by the fact that all theories are 'born refuted'. Lakatos completely accepts Kuhn's insistence on this point. As he states, all theories emerge in an ocean of counter-examples but these cannot be allowed to eliminate a theory; otherwise all science would be eliminated instantly! Theories are certainly born *'refuted'* but they are not born *rejected*. 'Refutation' must not automatically lead to rejection. Lakatos therefore dismisses Popper's 'objective' criterion for the rejection of scientific theories.

Nevertheless, Lakatos believes that science does progress and that its progress arises through rational evaluation not of competing theories but of competing series of theories which he calls 'research programmes'. In the above (simplified) dilemma the solution is, says Lakatos, to reject each theory in turn and to see which rejection gives rise to a new theory (i) with greater theoretical content than the former and (ii) backed up by new facts not entailed by the former. Such an improvement Lakatos calls a *progressive shift*. However, since all theories are 'refuted' all the time, they are problematic all the time and thus Lakatos prefers to call the improvement a *progressive problem-shift* (to emphasize that problems are ever-present). On the other hand, if the inconsistency is resolved in a 'linguistic reinterpretation' of the theory which is content-decreasing, then the stratagem is referred to as a degenerating (pseudo-scientific) problem-shift. The important point is that 'not an isolated *theory*, but only a series of theories—or a research programme—can be said to be scientific or unscientific'. (Lakatos writes that Popper has confused his readers on this point, e.g. he has written that 'Marxism is irrefutable' and, at the same time, 'Marxism has been refuted'; but what he means is that, as a theory, 'Marxism has been refuted', but, as a research programme, 'Marxism is irrefutable'.)[77]

Since Lakatos' concept of 'research programme' will be compared with Kuhn's concept of 'paradigm', let us look further at this series of theories implied by the research programme.

Continuity across the series evolves from a genuine research programme articulated at its inception; every research programme has methodological rules which can be divided into two groups: those telling us what possible research procedures are to be avoided (the 'negative heuristic') and those recommending certain procedures to be pursued ('positive heuristic'). It is, of course, mainly the negative heuristic which supplies continuity across the series of theories.

Each research programme can be characterized by a 'hard core' of hypotheses that the *negative heuristic* declares to be sacrosanct; the whole object of the research programme will be to build around the hard core a protective belt of auxiliary hypotheses which will be constantly modified or even occasionally entirely replaced as the programme proceeds. If it is successful, the process will result in a progressive problem-shift; if unsuccessful, a degenerating problem shift results.

The counter-examples are not, however, tackled in a random manner but decided on in advance by the *positive heuristic* of the research programme which specifies in a semi-articulated way the various stages through which the programme will be taken. It is, therefore, the verifications achieved which keep the programme going and maintain the confidence of the programme's protagonists.

We have already described the outstanding example of Newton's research programme. The hard core of the programme consisted of Newton's concepts of absolute space and time, the three laws of mechanics plus the law of gravitation; these were declared 'irrefutable' by the methodological decision of the programme's protagonists. The positive heuristic comprised (according to Lakatos) the following stages: calculations assuming a point planet and fixed point-like sun; then (because this model was ruled out by Newton's third law) a model in which the planets and the sun revolved around the common centre of mass; then a model in which the planets and the sun were mass-balls rather than point-masses (since point-masses were ruled out by observation, although not by the hard core hypotheses); then a model in which interplanetary forces were allowed for, and so on. We have already seen how counter-examples led to consistently greater empirical content in the auxiliary hypotheses protecting the hard core, i.e. the process was one of a continual progressive problem-shift.

The similarity of the 'research programme' with 'paradigm articulation' is striking. The reader is referred again to our earlier discussion of the articulation of the Newtonian paradigm. We merely repeat here Kuhn's description that 'the success of a paradigm . . . is at the start largely a promise of success discoverable in selected and still incomplete examples' and that 'normal science consists in the actualization of the promise . . . achieved by extending the knowledge of those facts that the paradigm displays as particularly revealing, by increasing the extent of the match between those facts and the paradigm's predictions, and by further articulation of the paradigm itself'.[78] Where, therefore, is the crucial difference between Lakatos and Kuhn that enables Lakatos to believe that he can justifiably make the claim that scientific growth is rational? Clearly it lies in the nature of the competition between rival research programmes.

Competition between research programmes

For simplicity let us consider competition between two research programmes R_1 and R_2. At the outset they will probably be concerned with different aspects of the same domain. But, as the rival research-programmes expand, Lakatos tells us that they will 'gradually encroach on each other's territory',[79] an encroachment that is clearly impossible for Kuhn since in a truly revolutionary situation the competing paradigms envisaged by Kuhn will be 'incommensurable'. (Whether, historically, proponents of competing paradigms have managed to find common ground will be explored in chapters 2 and 3.) Lakatos continues that this overlapping of R_1 and R_2 eventually results in the first *battle* between the two programmes in which the nth version of R_1 will be 'blatantly, dramatically inconsistent' with the mth version of R_2. The battle is, say, won by programme R_1. But the *war* is not yet over since 'any research programme is allowed a few such defeats'.[80] All that is necessary for R_2 to make a comeback is to produce a $(m + 1)$th version in which greater theoretical content is matched by some empirical verification. Should, however, the proponents of R_2 fail to produce the required improvement 'after sustained effort', the war is lost and R_1 is declared the winner by the scientific community. Proponents of R_2 who remain loyal to the defeated programme must after this point be deemed guilty of irrational behaviour.* Furthermore at this point it becomes rational to regard the experimental result which led to the clash between R_1 and R_2 as 'crucial'.

Lakatos concludes his paper by emphasizing once again that 'purely negative, destructive criticism, like "refutation" or demonstration of an inconsistency does not eliminate a programme'; 'one must treat budding programmes leniently'. Research programmes can be undermined but it is usually only *constructive criticism*, together with the help of rival programmes, which can achieve major successes, although visible as such 'only with hindsight and rational reconstruction'.[81]

Lakatos' blind spot

Clearly a major difficulty with Lakatos' analysis, and one that Kuhn and other critics have not failed to notice, is that Lakatos fails to specify the point at which it becomes irrational for proponents of the losing research programme to attempt to continue its articulation. Should, in the course of the programme's initial implementation, each shift be *consistently* progressive in empirical content and at least *intermittently* progressive in corroboration, then, says Lakatos, the achievement of intermittent empirical confirmation 'gives sufficient scope for dogmatic adherence' to the programme 'within the bounds of rationality'.[82] Yet even without intermittent (?) empirical successes an apparently defeated programme may offer resistance for a long time, holding out, as Lakatos puts it, with 'ingenious content-increasing innovations unrewarded with empirical success'.[83] As he admits, it is very difficult to defeat a research-programme supported by talented, imaginative scientists and, most certainly, a programme meeting with preliminary defeats in battles

---

* Should such a die-hard be one of the world's greatest scientists such as Einstein (in his refusal to accept 'uncertainty' as inherent in nature) then perhaps a polite epitaph will be written, such as ter Haar's 'Einstein . . . was a genius, a virtuoso, who in later life tended to become conservative . . . he was extremely naïve . . .' Or perhaps bewilderment and anguish will be expressed such as Max Born's '[Einstein] was a pioneer in the struggle for conquering the wilderness of quantum phenomena. Yet later, when out of his own work a synthesis . . . emerged which seemed to be acceptable to almost all physicists, he kept himself aloof and sceptical. Many of us regard this as a tragedy—for him, as he gropes his way in loneliness, and for us who miss our leader and standard-bearer. I shall not try to suggest a resolution of this discord. We have to accept the fact that even in physics fundamental convictions are prior to reasoning, as in all other human activities.'

with rival programmes should not be regarded as '*objectively* eliminated' just because most scientists shift allegiance to the winning research programme offering 'cheaper successes'.[84] But, then, at exactly what point does it become unreasonable for a scientist to remain committed to the hard core of his research programme? Lakatos argues that the war between two rival research programmes must be objectively declared over and the winner announced when one programme, defeated in battle, fails to make a progressive comeback 'after sustained effort.' But what constitutes 'sustained effort'? The Aristarchean theory of the world system made a progressive comeback after some 2 000 years of seemingly permanent defeat! It is difficult to see how a triumphant return of an apparently defeated research programme, possessing the same negative heuristic as before but a better articulated or even different positive heuristic, can ever be objectively ruled out. But if it is to make a triumphant return, then it will have to account for at least most of the successes enjoyed by other research programmes while it was in a state of hibernation and in addition produce some new ones of its own. However it cannot make a return unless there are *some* scientists seeking to develop its positive heuristic. And Lakatos conspicuously fails to tell us at what point it becomes irrational to continue to make such efforts. Feyerabend drives the point home: 'If you are permitted to wait, why not wait a little longer?' he asks.[85]

Furthermore, Lakatos does not explain why scientists commit themselves in the first place to their respective research programmes. Is it rational for a scientist to commit himself to any research programme whatsoever until, 'after sustained effort' to no avail, he must admit defeat? Lakatos claims that it is.

Some comments by Kuhn on his critics

Kuhn is amused neither by his critics, nor by some of his supporters. To Popper's accusation that normal science is 'a danger to science and, indeed, to our civilization',[86] Kuhn replies calmly that he is only describing the nature of scientific activity, expressing neither approval nor disapproval of it. To Scheffler's and Lakatos' frequent taunts that he is defending irrationality, Kuhn replies that the label of irrationality is a mere shibboleth. In any case, as he points out, Lakatos' views are very similar to his own, although he appreciates that Lakatos has not yet realized the fact. But it is for Feyerabend's argument in defence of irrationality in science that Kuhn reserves his strongest condemnation. He finds this point of view 'not only absurd but vaguely obscene'.[87] The 1970 Kuhn wishes to reassure the over-hasty readers of his 1962 essay that 'no process essential to scientific development can be labelled "irrational" without vast violence to the term'.[88] On the contrary, the next step in the research programme must be to show that 'existing theories of rationality are not quite right and that we must readjust or change them to explain why science works as it does'.[87] No matter, then, what procedure scientists find it necessary to adopt in order to choose between competing paradigms, the process is to be defined as 'rational'! Of course there are good reasons, Kuhn writes, for choosing one paradigm rather than another. Kuhn insists that he never meant to imply otherwise. Judgements are made and must be made as to the 'accuracy, scope, simplicity, fruitfulness and the like' of each paradigm.[89] Nevertheless, the point is, Kuhn insists, that different scientists will in general reach different personal judgements with respect to each paradigm. Can a scientist, however, remain committed to his favourite paradigm, come what may? He agrees with one of his critics that 'there must be a critical level at which a tolerable turns into an intolerable amount of anomaly'.[90] Whereas, however, Kuhn's critics want that level to be made non-subjective, Kuhn insists that, on the contrary, it 'ought not to be the same for everyone, nor need any individual specify his own tolerance level in advance. He need only be certain that he has one, and aware of some sorts of discrepancies, which would drive him towards it.'[90] Certainly, agrees the 1970 Kuhn, as the paradigm debate proceeds, 'as argument piles on argument and as challenge after challenge is successfully met, only blind stubbornness can at the end account for continued resistance' to the triumphant paradigm.[91] But this is not the point at issue, he attempts to explain: 'The scientific community cannot wait for history, though some individual members do. The needed results are instead achieved by distributing the risk that must be taken among the group's members.'[92] Finally Kuhn insists that successive scientific paradigms are not ever-better approximations to 'the truth'. In examining successive paradigms in the physical sciences he is able to see no convergence to a definite ontology. And, since there are no pure observational terms, Kuhn maintains that he was justified in calling different paradigms incommensurable. But this is not to say that communication across the revolutionary divide is impossible, he writes. Only that it is at best partial. More than that, Kuhn now claims, he never meant to imply. At the time of writing the debate is still in full swing. The reader is recommended to follow it.

4. Some Concluding Comments

We are now at the stage when a direct look at one or two scientific revolutions would help clarify matters considerably. Let me therefore bring this chapter to a close with the following remarks.

As far as physics is concerned, Kuhn's schematization of scientific activity into normal and revolutionary science is, I believe, a very useful one. Clearly, however, both these modes of scientific activity need very careful consideration. Normal science can, perhaps, best be regarded (in retrospect) as that paradigm articulation in which all competing sub-paradigms have certain hard core hypotheses in common; the overall paradigm is thus defined by an inner hard core of hypotheses consisting of the overlapping region of the hard cores of the various competing sub-paradigms. Many of these inner hard core hypotheses will, of course, remain imprecisely articulated. A crisis can then be said to occur when 'after sustained effort' none of the competing subparadigms is able to achieve a progressive problem shift and the articulation of a rival paradigm (research programme) is undertaken whose hard core hypotheses imply the negation of at least one of the inner hard core hypotheses characterizing the period of normal science. This negation, which creates perhaps for the first time an explicit awareness of the hypothesis in question, causes a philosophical and psychological upheaval in the scientific community referred to by Kuhn as a state of crisis. Should, however, the new paradigm achieve progressive problem-shifts with respect to those counter-examples that had baffled the protagonists of the normal science paradigm, and at the same time achieve new corroborated successes, then allegiance of the scientific community will begin to swing towards it. To be sure, as Kuhn has quite rightly stressed, the original protagonists of the new paradigm must have faith in their research programme, perhaps in the teeth of much empirical evidence, knowing only that the counter-examples abounding have resisted the sustained efforts of ingenious scientists committed to the inner hard core hypotheses characterizing the normal science period. Only ensuing action and positive results will prove that faith justified.

But faith *is* necessary. None of Kuhn's critics has been able to show that subjective factors do not play an important part in the *evaluation* of competing paradigms. To be sure, when the crisis period is long past and one paradigm has been successfully articulated far beyond those problems its one-time competitors were unable to resolve, then commitment to this paradigm will seem the only rational course of action. But to say this is not to say much. Philosophers of science, as Keynes once said about economists, 'set themselves too easy a task if in tempestuous seasons they can only tell us that when the storm is long past the ocean is flat again'.[93] It is the crisis state that is of absorbing interest and it is to such states we turn in the next two chapters.

We state below the major paradigms we shall be discussing, giving at the same time a rough indication of those hard core hypotheses negated by the rival revolutionary paradigms:

Ptolemaic paradigm	*Copernican paradigm*
The earth is immobile and the centre of the universe	The earth is NOT immobile and is NOT the centre of the universe
Newtonian paradigm	*Einsteinian paradigm*
Time is absolute: velocities add in a Galilean way	Time is NOT absolute: velocities do NOT add in a Galilean way
Newtonian paradigm	*Quantum mechanical paradigm*
An atom has at any instant both a definite position *and* velocity	An atom has NOT at any instant both a definite position *and* velocity

In what follows we shall not only be looking at certain aspects of these paradigms and revolutions in some detail but we shall also be concerned with the reasons for different commitments on the part of the principal protagonists.

The sociologist A. Inkeles has written that 'in a civilized world a man should be free to choose the position he finds congenial'.[94] But our world is not civilized and few positions seem 'congenial'; in this world commitment would appear necessary, indeed unavoidable, and yet it appears that we can never be sure that we are right or that we are acting most effectively. In 1964 C. H. Townes, the then President of the American Physical Society, stated the dilemma in the following way: 'How can a person plan his life and act vigorously, perhaps at great cost and with great effort, without saying "I know the truth and this is the only right thing to do"? This,' Townes continued, 'is a difficult and yet a very important problem—to make the best judgement of ourselves and the world around us and be ready to bet our best efforts and even our lives on this judgement, and yet to recognize that it can only be tentative.'[95] The whole point, of course, of looking in detail at the physical sciences is to try to throw some light on this most difficult problem by examining the nature of action and evaluation in what would appear to be one of man's most successful ventures so far in his attempt to orientate himself in his environment. Yet though our tentative conclusions have been helpful, they have not been as conclusive as we might, perhaps, have initially expected. Science progresses through competition between research programmes characterized by (at times) radically different sets of hard core hypotheses; but often there is, it seems, for appreciable periods of time, no rational way of knowing to which research programme a scientist should commit himself. Yet in a more civilized world (one at the very least not threatened with nuclear extinction and experiencing mass malnutrition, disease and illiteracy), this would scarcely matter—on the contrary, the more research programmes, the merrier. Research in the physical sciences would become a kind of play affording intellectual challenge, aesthetic pleasure and a means of meeting one's fellow human beings in the common quest to understand nature. But this more civilized world has yet to be achieved. Herein lies the dilemma. Just how is it to be achieved?

Certainly we shall have to compare the physical against the social sciences and, in particular, compare the determinants and results of commitment to 'research programmes' in these two different spheres of activity. Before doing so, however, we must get a clearer idea of the structure of revolutions in the physical sciences and of the motivations behind commitment to different research programmes. This is the aim of the next two chapters.

Chapter 2

THE COPERNICAN REVOLUTION: A STUDY OF PARADIGMS AND COMMITMENTS

SAGREDO O Nicholas Copernicus, what a pleasure it would have been for you to see this part of your system confirmed by so clear an experiment!

SALVIATI Yes, but how much less would his sublime intellect be celebrated among the learned! For as I said before, we may see that with reason as his guide he resolutely continued to affirm what sensible experience seemed to contradict. I cannot get over my amazement that he was constantly willing to persist in saying that Venus might go around the sun and be more than six times as far from us as at another, and still look always equal, when it should have appeared forty times larger.

GALILEO GALILEI
A Dialogue Concerning the Two Chief World Systems

In this chapter I wish to sketch some reasons why in the first place men should ever have conjectured that the earth moves and, in the second place, why certain of these men committed themselves to *proving* that the earth moves in the way they conjectured. For all appearances were to the contrary. That the earth is at rest there could be no cause for doubt. Nevertheless, men at first doubted, then became certain in their own minds and finally tried to convince others that indeed the earth moves. The relevant question to ask, therefore, from this point of view is not why such an unlikely conceptual revolution did not happen in other civilizations such as the Indian and Chinese, but why it happened at all! The Chinese, for example, were among the most careful observers of the heavens. But since science is not founded solely on observation, this alone did not suffice. Yet the Chinese also asked questions. The Taoist, Chuang Tzu, observed first the obvious: 'How (ceaselessly) heaven revolves! How (constantly) earth abides at rest!', and then asked: 'Do the sun and the moon contend about their respective places? Is there someone presiding over and directing these things? Who binds and connects them together? Who causes and maintains them, without trouble and exertion? Or is there perhaps some secret mechanism, in consequence of which they cannot but be as they are? Is it that they move and turn without being able to stop of themselves?' More questions followed and then the defiant and challenging assertion, 'I venture to ask about the causes'.[1] But all this, too, was not enough. Something extra was needed to direct men's attention towards the opposite of what appeared obvious: the constant immobility of the earth and the ceaseless motion of the heavens. Around 400–300 B.C. the Mohists in China had stated that motion is eternal unless an opposing force causes its cessation, a fact, they declared, 'as true as that an ox is not a horse'.[2] Yet despite this insight the earth, as was obvious, remained at rest. Nevertheless, at about the same time as the Mohists were groping their way towards a qualitative expression of Newton's first law, Aristarchus of Samos was arguing that the earth rotates once a day and orbits the sun once a year, ideas rejected by his contemporaries but taken up some two thousand years later by Copernicus in what was to herald the dawn of the 17th-century scientific revolution. Why did this happen? And, above all, why did men commit themselves to the truth of such ideas, and to action to make manifest this truth to their contemporaries, when all appearances apparently indicated just the opposite?

1. Greek Images of Nature

The Pythagorean paradigm

In the 5th century B.C. the Pythagoreans convinced themselves not only that the earth moves but also that an unobservable 'counter-earth' accompanies the earth in its daily journey around a central fire. That these extraordinary ideas were conceived at all was due to prior commitment to a view of the world no less extraordinary and contradictory to common experience than the results the Pythagoreans deduced from it. Their great metaphysical principle was no less than the incredible assertion that the key to the riddle of the universe lay in mathematics, that the universe was a pattern of numbers and that geometrical and numerical harmony prevailed in the heavens as on earth. As an example of the latter harmony they pointed to the fact that the lengths of strings making musical harmonies are in simple numerical ratios to each other. But above all it was the heavens, the Home of the Gods, which of necessity had to manifest harmony. Thus it was evident to the Pythagoreans that there must exist an additional body in the universe in order to bring the observable total of nine (the 'sphere of the fixed stars', Saturn, Jupiter, Mars, the sun, Venus, Mercury, the moon and the earth itself) to the sacred number of the decad. But the supposed existence of this 'counter-earth' did not establish complete harmony in the heavens. Far from it. To achieve greater harmony the Pythagoreans were led to postulate that the earth, together with the counter-earth, is not at rest but moves once a day around a central fire. Let us follow their reasoning in some detail for it contains, in part, the embryo of the revolution that was to follow two thousand years later.

From observations the Greeks hypothesized that there is an outer sphere of stars which never change their respective positions. However, there are also five wandering stars (Saturn, Jupiter, Mars, Venus and Mercury) which, together with the sun and moon, move round the heavens from west to east in their own periodic times, the moon in a month, the sun in a year, Saturn for example, in thirty years. However, once a day the sphere of the fixed stars, the moon, the sun and the five wandering stars rotate around the earth in the opposite direction from east to west. The existence of such motions in competition with each other was considered most unaesthetic by the Pythagoreans. For why should Mars move round the earth once a day from east to west only to be at the same time making a journey across the heavens in the opposite direction from west to east once every two years? And the same with respect to the moon, the sun and the other wandering stars? This was certainly not harmonious and geometrically pleasing behaviour. But, on the other hand, harmony could be restored by denying the evidence of the senses. By supposing the earth was not at rest but moved daily from west to east across the *motionless* heavens, just as the moon and sun and the wandering stars similarly moved from west to east in their respective times, all bodies were seen to share the same direction of motion from west to east, the apparent daily motion from east to west being nothing but an illusion caused by falsely supposing the earth to be at rest!

Thus reasoned the Pythagoreans, an 'unscientific' mode of reasoning which brought a sharp rebuke from Aristotle: 'But in this they are not seeking explanations and causes for what one can observe so much as trying to force the phenomena into the framework of their own views and make them fit that way'[3]—ironically enough, an accusation frequently to be made against Aristotle and his followers by the 17th-century scientific revolutionaries.

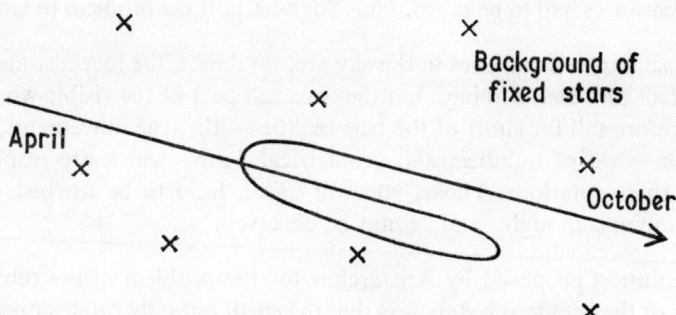

FIG. 2.1 A typical retrograde motion of Mars.
(Adapted from T. S. Kuhn's *The Copernican Revolution*.)

The problems of the Pythagoreans were, however, not at an end. For the wandering stars were divine and therefore, reasoned the Pythagoreans, ought to move in perfect circles at uniform speeds about the central fire. Unfortunately, however, far from always moving in such an acceptable way, the wandering stars periodically behaved in a most irregular and unseemly manner. Approximately every two years, for example, Mars actually stops in its journey across the heavens from west to east, reverses direction, stops again and then resumes its original west to east journey. Jupiter likewise performs retrograde motion, although eleven times every orbit and Saturn twenty-nine times every orbit.

To the Pythagoreans this was preposterous behaviour, both aesthetically and socially unacceptable. As Geminus explained in 70 B.C.:

> It was the Pythagoreans, the first to approach these questions, who laid down the hypothesis of a circular and uniform motion for the sun, moon, and planets. Their view was that, in regard to divine and eternal things, a supposition of such disorder as that these bodies should move now more quickly and now more slowly, or should even stop as in what are called the stations of the planets is inadmissible. Even in the human sphere such irregularity is incompatible with the orderly procedure of a gentleman. And even if the crude necessities of life often impose upon men occasions of haste or loitering, it is not to be supposed that such occasions inhere in the incorruptible nature of the stars. For this reason they defined their problem as the explanation of the phenomena on the hypothesis of circular and uniform motion.[4]

Social and economic influences on the Pythagoreans are here very apparent; the irregular and disorderly behaviour of the planets conformed neither to their paradigm of geometrical harmony and perfect circular motion in the heavens, nor to their belief in the divine status of the heavenly bodies. On the contrary such motion resembled more the movements of servants and slaves. The appearances had to be saved. Thus Socrates puts the problem to Glaucon:

> Those intricate traceries in the sky are, no doubt, the loveliest and most perfect of *material* things, but they are still part of the visible world and therefore fall far short of the true realities—the true movements, in the *ideal* world of numbers and geometrical figures which are responsible for these rotations. Those, you will agree, have to be worked out by reason and thought, and cannot be observed.[5]

One solution proposed by Aristarchus for the problem of the retrograde motions of the wandering stars was that the earth not only rotates once a day from west to east (a development of the early Pythagorean idea of daily west to east motion about a central fire) but also orbits the sun once a year, the daily rotation taking place about an axis not quite perpendicular to the plane of the earth's orbit around the sun. This, then, was the 'Copernican' system of the universe proposed nearly two millenia before Copernicus and no doubt for essentially the same reasons that motivated Copernicus. Although Aristarchus' manuscript has not survived, the similarity of his proposals to those of Copernicus is made clear by the comments of contemporary and later thinkers. However, his ideas met the same fate to be accorded to all such 'earth-moving' theories. They were carefully considered and rejected for sound physical reasons. As Ptolemy explained with reference to the rotating earth theory of the later Pythagorean thinkers: 'These persons forget, however, that while so far as appearances in the stellar world are concerned, there might, perhaps, be no objection to this simpler theory . . . yet to judge by [terrestrial] conditions affecting ourselves and those in the air about us, such a hypothesis must be seen to be quite ridiculous.'[6] The aesthetic advantages of the 'Copernican' system we shall examine in detail later. But first we should consider the solid, one might say, scientific reasons the Greeks had for rejecting Aristarchus' bold speculation.

In the first place the problem of retrograde motion could be solved in a far less radical way, one that neither violated common sense (as the Aristarchean theory did) nor the requirements of circular, uniform motion for the wandering stars. Since Mars was shining most brightly at the point of maximum retrograde motion, the Greeks supposed that at this point Mars was at its closest distance to the earth and that therefore its orbit was similar to the looped curve shown in Fig. 2.2. The anomalous retrograde motion remained but it

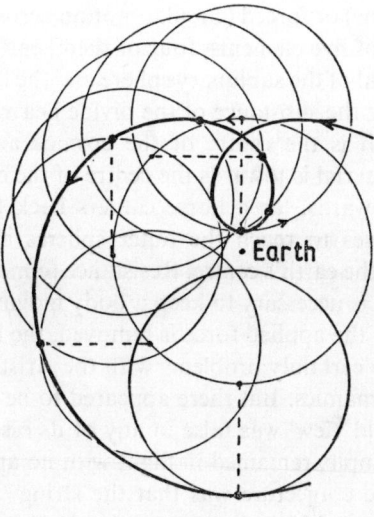

FIG. 2.2

was at least possible to explain it as the combination of *two* circular motions: each wandering star was therefore considered to move in a circle (epicycle) whose centre moved in a yet larger circle (deferent) about the earth. Judicious choice of periods and radii for the two circular movements was all that was necessary to reproduce the retrograde motions characteristic of each wandering star. The two circular motions required, for example, to approximate the orbit of Mars are as shown in Fig. 2.2. Thus, the Pythagorean-Platonic requirement that the appearances be saved in terms of circular, uniform motion is met. In the second place, as we have already indicated, Plato's pupil, Aristotle, did not subscribe to the Pythagorean-Platonic emphasis on mathematical harmony; consequently his influential geocentric world view, organismic rather than mathematical, underlay the Greeks' rejection of Aristarchus' heliocentric theory of the cosmos. Sound common-sense prevailed and the speculative idea of the earth's motion through the heavens was firmly rejected.

The Aristotelian paradigm

Stripped to some of its barest essentials the Aristotelian theory of the universe is the following. The earth is the centre of the cosmos, about which revolve spheres in which the moon, sun, wandering stars and fixed stars are embedded. On earth, change and decay occur; in the heavens, change and decay are impossible. Thus there is a qualitative difference between terrestrial and celestial phenomena, a dichotomy of the cosmos delineated by the sphere in which the moon is embedded. Whereas in the divine, eternal heavens all motion is circular, in the sub-lunary sphere motion can be either natural (towards or away from the centre of the earth—and therefore not eternal as in the divine heavens) or forced (circular motion across the earth's surface). The cosmos is made of five elements, four of them, earth, water, air and fire, making up the material of the sublunary sphere and the fifth, a pure unchangeable element, forming the substance of the divine heavens. The natural place for the element earth is the centre of the cosmos and hence the natural motion of earthly material is towards the centre of the earth, to which it thus returns if thrown upwards, 'as a horse canters back to its stable'. On the other hand, fire strives to reach the outer spheres and hence its natural motion is away from the earth's centre. Resistance to motion is always present and therefore a force is necessary to keep a body moving horizontally on the earth's surface; when the applied force is removed, the body's motion ceases. Of course, there were certainly problems with the Aristotelian paradigm and especially with his dynamics. But there appeared to be none so serious as to suggest that the 'world view' was false in any of its essential features. To be sure, arrows, for example, remained in flight with no apparent force propelling them on, but one conjecture was that the string of the bow sets air in motion behind the arrow which thus propels it onwards until it falls naturally to the ground. Other explanations were proposed. There seemed to be no problem which demanded *rejection* of the Aristotelian paradigm and, in any case, there was no equally comprehensive rival to take its place.

The rejection of the heliocentric theory

Now the Greek philosophers realized that the radius of the sphere of the fixed stars was very much greater than the radius of the earth, since otherwise the relative positions of the fixed stars would appear to change as an observer travelled over the earth's surface. Indeed it was this absence of measurable parallax motion that had led to the rejection of the Pythagorean daily motion of the earth around a central fire in favour of daily rotation of the earth at the centre of the cosmos. Furthermore, it was later to prove an important piece of evidence against Aristarchus' theory. For although the geocentric theory of the cosmos required the radius of the sphere of the fixed stars to be large compared with the radius of the stationary earth, the absence of detectable parallax motion meant that if Aristarchus was correct the distance of the fixed stars from the sun would have to be very much greater than the radius of the earth's orbit around the sun (a distance which Aristarchus himself had estimated). This was too much for most Greeks to accept and furthermore there was, it seemed, no good reason for accepting it. For Aristarchus was hypothesizing the motion of the earth only on aesthetic grounds. He appeared to be quite ignoring the evidence of his own eyes, the violation done to common-sense, and the problem of explaining how it was that, if the earth was rushing through space with an enormous velocity, birds, clouds and all loose objects on the earth's surface somehow or other managed to keep up with the earth. Where, the Greeks asked, was the force keeping them moving with the earth and why did not they feel it acting on them? And what was the nature of the force keeping the earth itself moving? And if the earth was rotating once a day from west to east, why did it not disintegrate? And why was there not a perpetual wind from east to west? Since there were no answers to these damning questions, it is scarcely surprising that Ptolemy would later judge all 'earth-moving' theories to be 'quite ridiculous'.

We have not, however, returned to our original question: how it was that men could possibly have conceived the earth to be in motion? This question has been answered. Men conceived of the earth's motion, despite all sense-experience to the contrary, as a way of showing that the heavens were geometrically harmonious. But whatever gain there was in geometrical simplicity and harmony did not outweigh the abundance of evidence against the idea of a moving earth. Of the two competing paradigms, the geocentric stationary-earth theory won the allegiance of most Greeks. Where the two paradigms clashed over terrestrial phenomena, the geocentric had clear superiority over its rival; and with respect to celestial phenomena the edge if anything was with the geocentric theory. For the absence of parallax motion amounted, it seemed, to clear disproof of the heliocentric system of the cosmos. It but remained to articulate the geocentric theory so as to enable more accurate predictions to be made concerning the motions of the sun, moon and wandering stars, a task to which Ptolemy dedicated himself in the *Almagest*.

The Ptolemaic paradigm

The task, however, was far from easy and Ptolemy found himself forced to make use of a large number of epicycles, some versions of the Ptolemaic system containing up to 40 epicycles. Furthermore, in order to account for the observed speed of the moon, one set of epicycles was used but in order to account for the change of diameter another incompatible set was used. To be sure, the appearances were saved but as Geminus wrote in 80 B.C.:

> By assuming that the circular orbits describe epicycles, the seeming irregularities in their motions will be accounted for. But this is not enough! We must also look into the question in how many different such ways the observed phenomena could be brought about; so that we may bring our mathematical theory of the planets into line with an explanation of the underlying physical causes which is theoretically admissible.[7]

Not only, however, was the accuracy of Ptolemy's predictions of planetary motion not as good as perhaps had been hoped, but an explanation of the underlying physical causes seemed as far away as ever. The appearances were saved but the physical causes of the motions lay shrouded in mystery. Furthermore, the geometrical complexity of the system was great. The ideal of constant velocity in a circle had had to be abandoned and replaced by the concept of constant angular velocity about a point Q in space no longer the centre C of the deferent. For example, the orbit of Mars was described by Ptolemy as shown in Fig. 2.3. Not only does the centre C of the deferent not co-

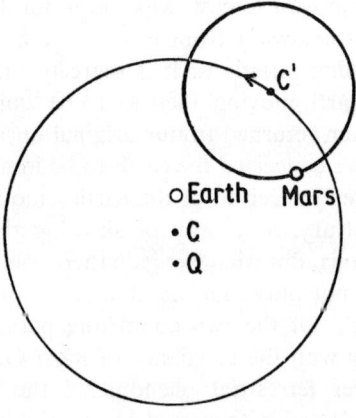

Fig. 2.3

incide with the earth but the centre C' of the epicycle moves around the deferent with a variable velocity such that its angular velocity is constant about the equant position Q. Furthermore, for each wandering star Ptolemy had to use a different centre C and a different equant point Q. No wonder that in the 13th century Alfonso the Wise of Castille declared that if the Lord God had consulted him before constructing the universe he would have advised something simpler! Finally, we remark that there was no obvious way of ascertaining the order of the planets in the celestial hierarchy—there had been much debate, for example, as to the order of Mercury, Venus and the sun with respect to the earth; there seemed to be no way of explaining why it was that Mercury and Venus never strayed far from the sun (a fact accounted for by tying the centre of each planet's epicycle to the line joining the earth and the sun) whereas Mars, Jupiter and Saturn could wander far from the sun—indeed, to the opposite side of the heavens; and why it was that whenever Mars, Jupiter or Saturn was at its position of maximum retrograde motion the sun was always directly opposite the planet in the heavens; and why it was that the retrograde motions of these three planets decreased in amplitude but increased in frequency from Mars to Jupiter to Saturn. The paradigm still had its puzzles to solve! Clearly the sun was in some way a controlling influence over the planets but the Greeks did not see how. The earth, declared Smyrna, a contemporary of Ptolemy, is the centre of the universe as the navel is of man but so also the sun is its centre just as the heart is of man! The paradigm was, however, not further articulated; the puzzles resisted solution. With the rise of Christianity and the collapse of the Roman Empire the Ptolemaic paradigm was itself abandoned.

2. The Rise of a Rival Paradigm

Not until about the 9th century did European philosophers once again generally believe the earth to be spherical, the centre of the cosmos, and surrounded by rotating concentric spheres containing the other celestial bodies. By the 12th century Arab translations of Aristotle were introduced into France via Spain and, despite initial resistance from the Church, Aristotelian cosmology became firmly incorporated into Christian teaching by Aquinas. Thus, according to Aquinas,

> The material of the heavens is, by its intrinsic nature, not susceptible to generation and corruption. . . . Motion is the only sort of change they [the heavens] experience. . . . Futhermore, among the sorts of motion which they might experience, theirs is circular, and circular motion is the one which produces the very minimum of alteration because the sphere as a whole does not change place.'[8]

Whereas this Aristotelian cosmology was regarded as philosophically true, i.e. absolutely true, the Ptolemaic system of epicycles was regarded as only mathematically true and therefore subject to possible revision. Although it was the case, stated Aquinas, that Ptolemy's constructions saved the appearances, this did not mean that they could not be saved by quite different hypotheses. What, however, could not be disputed was that the earth was the degenerate, corrupt centre of the cosmos surrounded by concentric rotating spheres bearing the incorruptible, unchanging celestial bodies.

Thus the 13th century saw the restoration of the Aristotelian-Ptolemaic paradigm; Europe had recovered its intellectual heritage. But with an important difference. The immobility of the earth was now sanctified by the Christian religion; for it was God Himself who had 'made the round world so fast that it cannot be moved'! Moving the earth would now mean not only a major scientific upheaval but a theological crisis as well.

And yet men were to persist in believing the earth to move. And like the Pythagoreans before them they were to do so on the grounds that the Aristotelian-Ptolemaic system of the world was so geometrically unappealing that it could not possibly correspond to the true physical reality. So strong was their belief in a mathematically harmonious and well-ordered cosmos that they were prepared to deny the evidence of their own senses and to commit themselves to the articulation of a new paradigm that would make the motion of the earth seem no longer absurd but, on the contrary, self-evident. It is not my intention to trace here the rise of this recommitment to mathematical harmony. Suffice it to say that in the Renaissance the quest for geometrical harmony showed itself in the writings and works of leading artists. According to Wittkower, the Renaissance architecture 'with its strict geometry, the equipoise of its harmonic order, its formal serenity . . . revealed the perfection, omnipotence and goodness of God'.[9] Alberti, in his *On Painting* (1453), desired first of all the painter to know geometry while Leonardo da Vinci (1452–1519) commanded in his *Treatise on Painting*: 'Let no one who is not a mathematician read my work.'[10] In his *Ten Books on Architecture* Alberti emphasized that a building has beauty only if there is 'a Harmony of all the Parts . . . fitted together with such Proportion and Connection, that nothing could be added, diminished or altered, but for the Worse'. An architect seeking beauty would therefore build 'in such a Manner as to join and unite a certain Number of Parts into one Body or Whole, by an orderly and sure Coherence and Agreement of all those Parts'. Alberti was of the opinion that 'When we lift up our eyes to Heaven, and view the wonderful Works of God, we admire him . . . for the beauties which we see.'[11]

But for Copernicus, in these most important of all works, the work of God himself, there was no harmony to be found. Therefore either God had made a monstrously complicated universe or the system of the world was not that described by Aristotle and the Ptolemaic astronomers. For Copernicus there was no choice as to where the blame should fall. 'With them,' he wrote, referring to the astronomers who had tried to improve the Ptolemaic system, 'it is as though an artist were to gather the hands, feet, head and other members for his own images from diverse models, each part excellently drawn, but not related to a single body, and since they in no way match each other, the result would be a monster rather than a man.'[12] His assistant Rheticus thought the same way:

> We fully grant these distinguished men their due honour, as we should. Nevertheless, we should have wished them, in establishing the harmony of the motions, to imitate the musicians who, when one string has either tightened or loosened, with great care and skill regulate and adjust the tones of all the other strings, until all together produce the desired harmony, and no dissonance is heard in any.[13]

THE COPERNICAN REVOLUTION

The aim of Copernicus, therefore, was to look at the universe in a new way and so reveal the mathematical harmony which he knew the universe must have. Since if the earth moved, the universe possessed mathematical harmony, it was therefore obvious to Copernicus that the earth was not the centre of the universe but itself a wandering star. Only mathematicians would, however, be clearly able to recognize this truth. 'Though these views are difficult, contrary to expectation, and certainly unusual,' wrote Copernicus in his *de Revolutionibus* (1543), 'yet in the sequel we shall, God willing, make them abundantly clear at least to mathematicians.'[14] Let us examine some of the arguments of Copernicus and the nature of the harmony he was able to demonstrate by postulating the earth to move.*

Some advantages of the Copernican system

(i) We have already seen that if the earth is assumed to rotate daily about its axis, then the daily rotation of all celestial bodies around the earth becomes only an apparent motion.

(ii) If the sun is assumed to be the centre of the cosmos and the (rotating) earth a wandering star orbiting the sun once a year, then the order of the planets is given unambiguously. Since Mercury and Venus never stray far from the sun, then it follows that they lie between the earth and the sun; since Mars, Jupiter and Saturn are sometimes opposite the sun in the heavens then they clearly orbit the sun at distances greater than the radius of the earth's

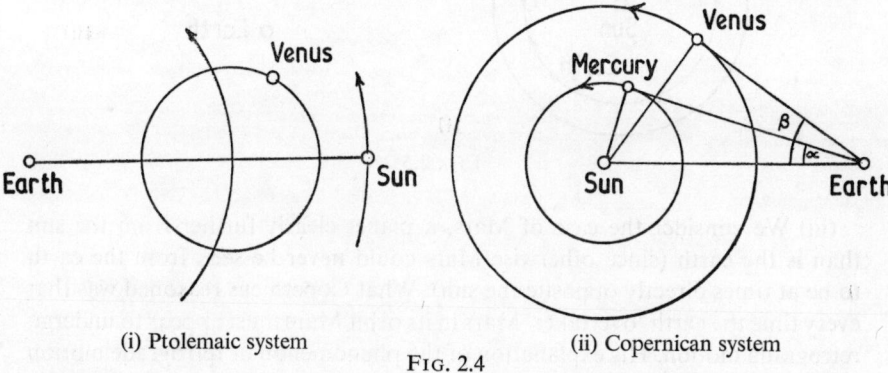

(i) Ptolemaic system   (ii) Copernican system

FIG. 2.4

orbit. Since Venus strays further from the sun than Mercury does, clearly Venus' radius of orbit around the sun must be greater than that of Mercury's. Whereas Ptolemy had to tie the centres of the epicycles of Mercury and Venus to the line joining the earth and the sun—an *ad hoc* procedure to give the correct results—the fact that these planets never stray far from the sun is built into the geometry of the Copernican system. The reader is referred to the diagrams of Fig. 2.4. We immediately see that in terms of the earth's radius of

---

* In my discussion of the world system according to Copernicus I have used many of the examples and arguments given by Kuhn, although without completely agreeing with his overall interpretation.

orbit r_E, Mercury's distance from the sun is given by $r_M = r_E \sin \alpha$ and Venus' by $r_V = r_E \sin \beta$, where α and β are the observed maximum angles of elevation of these planets with respect to the sun. The distances of Mars, Jupiter and Saturn from the sun can also be determined in terms of the radius of the earth's orbit. Before, however, we show this we discuss another notable achievement of the Copernican system—a convincing (?) explanation of retrograde motion. Just like the daily rotation of the heavens, the retrograde motions of the planets turn out to be only apparent—provided, of course, the earth *truly* orbits the sun.

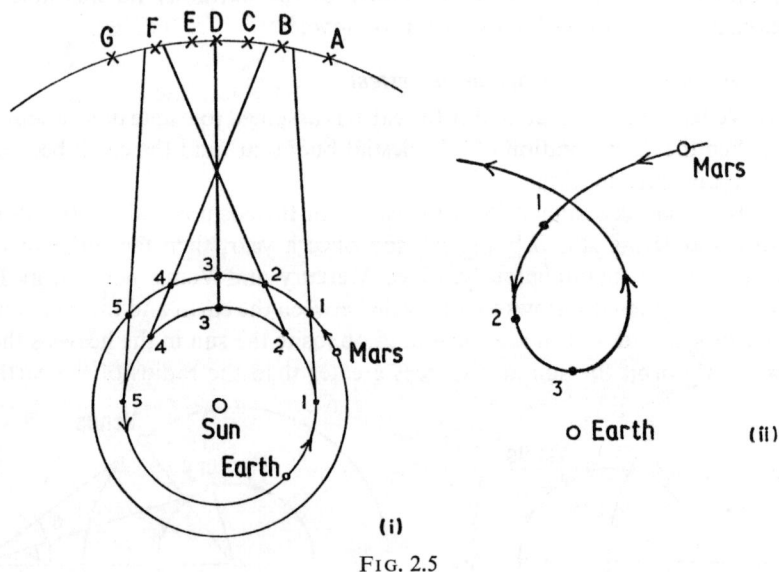

FIG. 2.5

(iii) We consider the case of Mars, a planet clearly further from the sun than is the earth (since otherwise Mars could never be seen from the earth to be at times directly opposite the sun). What Copernicus reasoned was that every time the earth 'overtakes' Mars in its orbit Mars must appear to undergo retrograde motion. His explanation of the phenomenon of retrograde motion is illustrated in Fig. 2.5(i). When the earth is at position 1, Mars is at position 1 in its orbit but appears from the earth to be between the fixed stars A and B. At a later time when the earth is at position 2, Mars has moved on to position 2 in its orbit and appears from the earth to be between stars E and F, still travelling in the 'forward' direction. As the earth begins to pass Mars, so Mars appears to stop and then change direction, shining at its brightest at the time of maximum retrograde motion when it appears from position 3 of the earth to be in line with star D. Once the earth has overtaken Mars, then the retrograde motion begins to slow down and by the time the earth is in position 5, then Mars has once more resumed its forward course and appears

THE COPERNICAN REVOLUTION 39

to lie between stars F and G. In Fig. 2.5(ii) the apparent motion of Mars is shown as viewed from the earth. The result is that strange motion most unacceptable to all astronomers who believed that the heavens should exhibit only harmonious behaviour. Copernicus' solution to the puzzle of retrograde motion was to propose that the motion is only apparent; the universe when looked at in the right way exhibits only harmony.

(iv) Furthermore, Copernicus was able to work out the periods of all the planets by observing the (mean synodic) times between two consecutive moments of maximum retrograde motion. Let S be this observed time for Mars (780 days) and T_M the period of Mars we wish to determine. T_E, the period of the earth's orbit, is, of course, approximately 365 days. Then T_M can be determined as follows. The time interval S is that time in which the earth performs exactly one more orbit than Mars in their journeys round the sun. Since in time S the earth makes S/T_E orbits while Mars makes S/T_M, we therefore have $S/T_E - S/T_M = 1$ and T_M can be determined since S and T_E are known. Likewise the periods for all the planets could be determined (maximum retrograde motion occurring for Mercury and Venus when they overtake the earth).

(v) We return now to the determination of the distances of Mars, Jupiter and Saturn from the sun in terms of r_E, the earth's distance from the sun. We illustrate the argument for the planet Mars in Fig. 2.6. When Mars and the

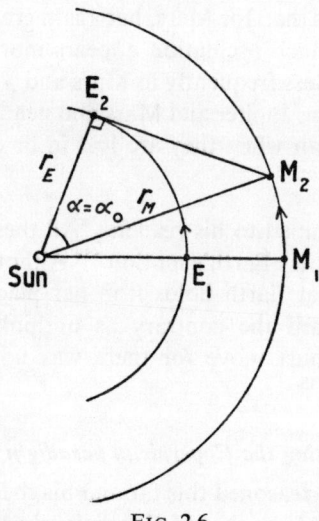

FIG. 2.6

sun subtend an angle of 90° at the earth (i.e. when the earth and Mars are at the positions E_2 and M_2 and subtend an angle $\alpha = \alpha_0$ at the sun), the required distance r_M is given by the relationship $r_E = r_M \cos \alpha_0$. The angle α_0 can be determined as follows. Since the earth and Mars are assumed to have constant velocities around the sun, then the angle α changes uniformly with respect to time and we can write $\alpha = ct$ where c is a constant to be determined. Thus when $t = 0$ then $\alpha = 0$ and the earth lies directly on the line joining the sun and Mars. However, after time S the earth once more lies in line with Mars, having swept through an angle of $\alpha = 360°$ with respect to Mars. Thus we have $360 = cS$ and therefore $c = 360/S$. We can now write that $\alpha = 360 \, t/S$ and by measuring the time $t = t^0$ it takes for the earth and Mars to move from positions E_1 and M_1 to positions E_2 and M_2 respectively (see Fig. 2.6), we can determine α and hence r_M. A similar argument can obviously be made for Jupiter and Saturn.

Thus Copernicus was able not only to order unambiguously the five planets with respect to the sun and the earth but also to determine their distances from the sun in terms of the earth's distance. His achievements were many but above all he had discovered harmony in the heavens. For in his system, and in his system alone, 'the orders and magnitudes of all stars and spheres, nay the heavens themselves, become so bound together that nothing in any part thereof could be moved from its place without producing confusion of all other parts and of the universe as a whole'.[12] Copernicus sums up:

> So we find underlying this ordination an admirable symmetry in the universe and a clear bond of harmony in the motion and magnitude of the orbits such as can be discovered in no other wise. For here we may observe why the progression and retrogression appear greater for Jupiter than Saturn, and less than for Mars, but again greater for Venus than for Mercury; and why such oscillation appears more frequently in Saturn than in Jupiter, but less frequently in Mars and Venus than in Mercury; moreover why Saturn, Jupiter and Mars and nearer to the earth at opposition to the Sun than when they are lost in or emerge from the Sun's rays.[13]

And as Copernicus explained to his readers: 'All these phenomena proceed from the same cause, namely Earth's motion.'[14] Although, wrote Copernicus, 'the authorities agree that Earth holds firm her place at the centre of the Universe, and they regard the contrary as unthinkable, nay as absurd', nevertheless the earth must move for there was no other way of finding harmony in the heavens.[15]

Some difficulties confronting the Copernican paradigm

Perhaps Aristarchus had reasoned this far and his theory of the universe had been rejected. Why, therefore, should Copernicus' proposals have met anything but a similar fate? And in addition, we have so far only stressed the geometrical simplicity and harmony inherent in his theory. However, Kuhn has argued that no paradigm is free of problems and certainly the Copernican system can be said to have been born in a sea of problems. Let us look at *some* of the difficulties it confronted.

THE COPERNICAN REVOLUTION

(i) Copernicus was able to explain the occurrence of the seasons (together with the fact that the North Star is stationary in the heavens*) only by supposing the earth to have a third motion, in addition to its daily rotation and annual journey around the sun. In the Ptolemaic theory the seasons were easily accounted for by supposing the plane of the sun's orbit around the earth to make a non-zero angle with the plane of the earth's equator (as shown in

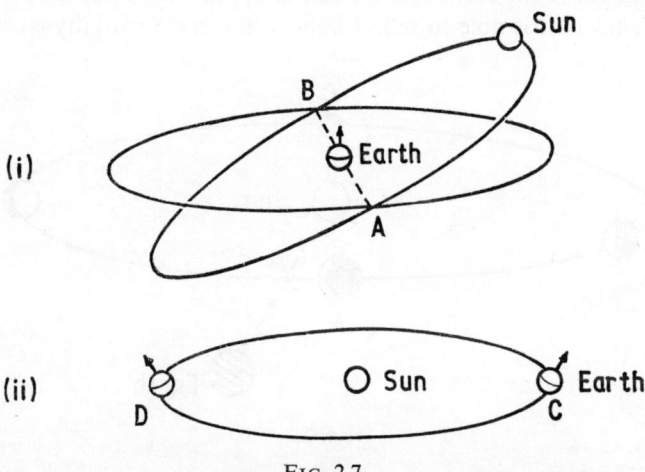

Fig. 2.7

Fig. 2.7(i)). When the sun is between A and B as shown in the diagram, the northern hemisphere is experiencing summer and the southern hemisphere winter; and conversely when the sun is between A and B in the other half of its orbit. Clearly in the Copernican system the axis of the earth's rotation cannot be perpendicular to the plane of the earth's orbit around the sun, otherwise no seasons will result. Hence the axis must have a tilt and must, of course, stay parallel with itself throughout its annual journey around the sun. Copernicus, however, believed that the natural motion of the earth's axis would be as that of a rod gripped in the hand and rotated through a full circle by turning one's body round, a motion that in six months would bring the earth's axis of rotation from position C to position D (as shown in Fig. 2.7(ii)). Thus Copernicus postulated that there must be a third motion of the earth that keeps the earth's axis of rotation always pointing in the same direction with respect to the sphere of the fixed stars. Needless to say, many of those philosophers and astronomers who were at least prepared to give the Copernican system some consideration could not bring themselves to accept three motions for the earth. One motion was manifestly absurd, but three!

(ii) Venus did not vary in brightness and size as predicted by the Copernican theory. Although it was, according to Copernicus, sometimes very close to the earth and sometimes far away, nevertheless its brightness and size

* We disregard here the additional complication of the precession of the equinoxes.

did not vary to the appropriate extent. The puzzle with respect to brightness, however, was resolved by supposing that Venus does not emit its own light but merely reflects the sun's light as does the moon. Thus when Venus is near the earth, at position A in Fig. 2.8, and almost in line with the sun, only a thin crescent of Venus will reflect light from the sun to the earth. When, however, Venus is far from the earth at position C, then nearly half its surface will reflect light to the earth and Venus will appear full. Thus the greater surface of Venus that is able to reflect light to the earth partially compensates

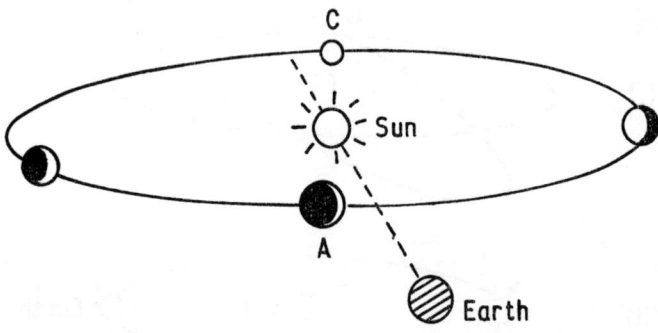

FIG. 2.8

for its greater distance from the earth. According to the Ptolemaic system, however (see the first diagram of Fig. 2.4), the phases of Venus should look quite different since Venus is always between the earth and the sun and therefore can never appear more than half-full. Unfortunately, however, this difference between the two theories appeared one that was capable of investigation only in principle and not in practice. Nevertheless, the followers of Copernicus had been able to solve this particular puzzle with which his theory had been confronted. But the phenomenon of the constant size of Venus remained an apparent refutation of the Copernican world system.

(iii) The absence of parallax motion had been regarded as disproof of the earth's motion. Copernicus, believing his theory correct, simply used this observational evidence to conclude that the universe was very much bigger than had been hitherto seriously considered: 'So great is this divine work of the Great and Noble Creator'.[14]

(iv) When Copernicus attempted to articulate his theory sufficiently to achieve the same degree of predictive accuracy as in the Ptolemaic system, he found that, despite earlier promises, he was not able to reduce the number of circles used by the Ptolemaic astronomers. This was mainly because Copernicus could not bring himself to make use of the equant position and therefore had to describe the planets' variable velocities around the sun by means of motion in circles upon circles. Thus the greater mathematical simplicity of the Copernican system lay in the number of phenomena explicable in terms of the earth's annual motion around the sun, not in terms of the number of circles Copernicus used compared with the number used by the Ptolemaics. Even so Copernicus was not able to achieve significantly better agreement with observation than the Ptolemaics had achieved. In particular, the orbit of Mars proved particularly troublesome.

(v) Finally, let us see how Copernicus attempted to meet the objections to the earth's supposed motion arising from Aristotelian physics. While he had negated fundamental principles underlying Aristotelian cosmology, nevertheless he still tried to make use of the Aristotelian categories of natural and forced motion in order to meet the objections that Ptolemy had made against moving-earth theories and that he knew would be similarly made against his. The earth's motion, he declared, is natural and not forced and this makes the crucial difference. 'Things subjected to any force, gradual or sudden, must be disintegrated, and cannot long exist,' Copernicus confirms. 'But natural processes being adapted to their purpose work smoothly.'[16] Hence, argues Copernicus, 'idle is the fear of Ptolemy that Earth and all thereon would be disintegrated by a natural rotation'. Indeed, argues Copernicus, if natural rotation produces disintegration then the tables can be turned against Ptolemy; for if the earth is at rest then 'should he not fear even more for the Universe, whose motion must be as much more rapid as the Heavens are greater than the Earth?' And as for objects on the earth's surface, they will share its motion since they share its nature. 'Perhaps the contiguous air contains an admixture of earthy or watery matter and so follows the same natural law as the Earth, or perhaps the air acquires motion from the perpetually rotating Earth by propinquity and absence of resistance.'[16]

Thus Copernicus defended his revolutionary theory. The earth was now afloat, orbiting the sun which 'sits as upon a royal throne ruling his children, the planets, which circle around him'. Aristotle's dichotomy of the cosmos had been challenged. A radically new world view was being proposed and in it the earth occupied no privileged position. And the sole defence of this revolutionary system of the world lay in its claim to mathematical harmony. It could not be expected that non-mathematicians would readily embrace it. Indeed, John Donne was of the opinion (1611):

> And New Philosophy calls all in doubt,
> The element of Fire is quite put out;
> The Sun is lost, and th' Earth, and no man's wit
> Can well direct him where to look for it.

And then followed his famous, anguished cry:

> 'Tis all in pieces, all coherence gone.

But this was not how it appeared to the mathematical, neo-Pythagorean, neo-Platonic astronomers for whom Copernicus had written his book. 'Mathematics is for Mathematicians' he had proclaimed. And for mathematicians alone coherence had been restored; for them alone, the pieces fitted together. Such a mathematical astronomer was the Englishman Thomas Digges, who, only 33 years after the publication of Copernicus' masterwork, himself published a popular defence of the Copernican system entitled *A Perfit Description of the Caelestiall Orbes according to the most aunciente doctrine of the Pythagoreans latelye reuiued by Copernicus and by Geometrical Demonstrations approued*. There were in fact no less than seven editions of his work (in 1576, 1578, 1583, 1585, 1592 and 1605). Nevertheless empirically-minded philosophers were not convinced; indeed, it is Burtt's opinion that 'contemporary empiricists, had they lived in the 16th century, would have been the first to scoff out of court the new philosophy of the Universe'.[17] Francis Bacon, for example, was positively hostile to the mathematical speculations announcing a new world system:

> In the system of Copernicus there are found many and great inconveniences; for both the loading of the earth with a triple motion is very incommodious and the separation of the Sun from the company of the planets with which it has so many passions in common is likewise a difficulty, and the introduction of so much immobility into nature by representing the Sun and the stars as immovable . . . all these are the speculations of one who cares not what fictions he introduces into nature, provided his calculations answer.[18]

The triple motion of the earth was felt to prove the inherent absurdity of the Copernican system. In 1581 Father Clavius, a Jesuit astronomer, published a textbook in which he commented:

> If the Copernican assumption implied nothing false or absurd, one might, so long as it were a question of preserving the appearances, be in doubt whether it is better to adhere to the opinion of Ptolemy or to that of Copernicus. But the Copernican theory contains many absurd or erroneous assertions: it assumes that the earth is not at the centre of the firmament; that it moves with a triple motion—a thing I find inconceivable, since, according to the philosophers, a single simple body has by rights a simple motion. . . . This is why it seems to us that Ptolemy's opinion should be given preference over the opinion of Copernicus.[19]

The Danish astronomer Tycho Brahe likewise rejected the radical nature of the Copernican system. 'But are not earthly things being confused with celestial things?', he complained. 'Is not the whole order of nature being turned upside down?'[20] The otherwise radical French thinker Jean Bodin, whose works were placed on the Index of forbidden books by the Catholic Church in 1628, ridiculed the new system of the universe:

> No one in his senses, or imbued with the slightest knowledge of physics, will ever think that the earth, heavy and unwieldly from its own weight

and mass, staggers up and down around its own centre and that of the sun; for at the slightest jar of the earth we would see cities and fortresses, towns and mountains thrown down. . . . For if the earth were to be moved, neither an arrow shot straight up nor a stone dropped from the top of a tower would fall perpendicularly but either ahead or behind. . . . Lastly, all things on finding places suitable to their natures, remain there, as Aristotle writes. Since, therefore, the earth has been allotted a place fitting to its nature, it cannot be whirled around by other motion than its own.[21]

3. The Articulation of the Copernican Paradigm

Galileo's contribution

Despite the opposition of the Aristotelian philosophers and the scepticism of 'reasonable', empirically-minded men, those thinkers who believed that the true system of the world must be mathematically harmonious committed themselves whole-heartedly to the defence and further articulation of the Copernican paradigm. Both Galileo and Kepler, for example, two of the most prominent defenders of the new philosophy, committed themselves at an early age to the truth of the Copernican world system. Of the two, however, only Kepler was from the outset prepared to defend the Copernican system publicly. In his exuberant way he explained the nature of his commitment: 'I certainly know that I owe it this duty, that as I have attested it as true in my deepest soul, and as I contemplate its beauty with incredible and ravishing delight, I should also publicly defend it to my readers with all the force at my command.'[22] Galileo similarly saw beauty in the Copernican system and hence truth as well. His reluctance, however, to publish in defence of Copernicus he explained to Kepler thus:

> I have written many arguments in support of him and in refutation of the opposite view—which, however, so far as I have not dared to bring into the public light, frightened by the fate of Copernicus himself, our teacher, who, though he acquired immortal fame with some, is yet to an infinite multitude of others (for such is the number of fools) an object of ridicule and derision. I would certainly dare to publish my reflections at once if more people like you existed; as they don't, I shall refrain from doing so.

Kepler, however, was not impressed and his rebuke to Galileo ended their incipient friendship:

> I would have wished, however, that you, possessed of such an excellent mind, took up a different position. With your clever secretive manner you underline, by your example, the warning that one should retreat before the ignorance of the world, and should not lightly provoke the fury of the ignorant professors; in this respect you follow Plato and

> Pythagoras, our true teachers. But considering that in our era, at first Copernicus himself and after him a multitude of learned mathematicians have set this immense enterprise going so that the motion of the Earth is no longer a novelty, it would be preferable that we help to push home by our common efforts this already moving carriage to its destination.[23]

Despite Galileo's early reluctance to publish, nevertheless Kepler and Galileo did both help to push home the 'already moving carriage to its destination', their combined efforts eventually leading to the complete rout of the Aristotelian philosophers. Let us look first at the nature of Galileo's commitment to the Copernican paradigm and of his contribution to its articulation.

In his *Dialogue Concerning the Two Chief World Systems* Galileo remarks that 'the true and the beautiful are the same, and so are the false and the ugly'.[24] Furthermore, it is his opinion that 'Nature does not act by means of many things when she can act by means of few'.[25] Already the reasons for his commitment to the Copernican system are clear. Only in this system is there mathematical harmony and therefore it is the only true system of the world. In a much quoted passage from *The Assayer*, Galileo states explicitly his belief in the Pythagorean conviction that the world is explicable only in mathematical terms:

> Philosophy is written in this grand book, the Universe, which stands continually open to our gaze. But the book cannot be understood unless one first learns to comprehend the language and read the letters in which it is composed. It is written in the language of mathematics, and its characters are triangles, circles and other geometric figures without which it is humanly impossible to understand a single word of it; without these, one wanders about in a dark labyrinth.[26]

When in the *Dialogue* Sagredo remarks how surprised he is that so few philosophers have believed the earth to move, Salviati, Galileo's mouthpiece, declares:

> No, Sagredo, my surprise is very different from yours. You wonder that there are so few followers of the Pythagorean opinion, whereas I am astonished that there have been any up to this day who have embraced and followed it. Nor can I ever sufficiently admire the outstanding acumen of those who have taken hold of this opinion and accepted it as true; they have through sheer force of intellect done such violence to their own senses as to prefer what reason told them over that which sensible experience plainly showed them to the contrary. For the arguments against the whirling of the earth which we have already examined are very plausible, as we have already seen; and the fact that the Ptolemaics and Aristotelians and all their disciples took them to be conclusive

THE COPERNICAN REVOLUTION 47

is indeed a stronger argument of their effectiveness. Indeed, the experiences which overtly contradict the annual movement are so much greater in their apparent force that, I repeat, there is no limit to my astonishment when I reflect that Aristarchus and Copernicus were able to make reason so conquer sense that, in defiance of the latter, the former became mistress of their belief.[27]

Despite his early reluctance to commit himself publicly, nevertheless it was Galileo's aim to prove, and prove publicly, that Aristarchus and Copernicus had been correct in their assertions that the earth *physically* moves around the sun. This was the point of issue. The Copernicans believed their system represented the true, physical reality, and it was Galileo's aim to prove that this was indeed so. In the *Dialogue* Salviati describes some of the problems confronting the Copernican paradigm and, as in the quotation above, praise Copernicus for believing his theory to be true while being unable to resolve these problems. It is, I think, worthwhile to quote 'Salviati' at further length: 'In this manner Copernicus pardoned Venus its unchanging shape, but he said nothing about its small variation in size; much less of the requirements of Mars. I believe this was because he was unable to rescue to his own satisfaction an appearance so contradictory to his view; yet being persuaded by so many other reasons, he maintained that view and held it to be true.'[28] There was another major difficulty: 'Besides these things, to have all the planets move around together with the earth, the sun being the centre of their rotations, then the moon alone disturbing this order and having its own motion around the earth . . . seems in some way to upset the whole order and to render it impossible and false.'[28] And so Salviati concludes: 'These are the difficulties which make me wonder at Aristarchus and Copernicus. They could not have helped noticing them, without having been able to resolve them; nevertheless they were confident of what their reason told them must be so in the light of many other remarkable observations. Thus they confidently affirmed that the structure of the Universe could have no other form than that which they had described.'[28]

Let us look at how Galileo attempted to resolve some of these difficulties and to *prove* the physical reality of the earth's motion. With respect to the infamous 'third motion' of the earth which, as he remarked, 'appeared to many a most improbable thing, and one that upset the whole Copernican system', Galileo convincingly demonstrated that it does not exist at all since the earth's axis of rotation stays *naturally* parallel to itself as the earth orbits the sun. Galileo's experiment was to place a ball in a bowl of water and then to pick up the bowl, hold it at arm's length, and rotate through 360°, so also rotating the bowl through a complete circle. 'It is certainly true that to the person holding the bowl such a ball appears to move with respect to himself and to the bowl, and to turn upon its axis. But with respect to the wall . . . the ball does not turn at all, and does not change its tilt, and any point upon it will continue to point toward the same distant object.'[29] Thus, said Galileo cunningly (for he had by this time been forbidden by the Church to argue that the earth *physically* moves), 'Copernicus had spoken falsely when he attributed his "third motion" to the earth.'

Furthermore, Galileo rolled balls down an inclined plane, allowing them to roll up a second inclined plane as shown in Fig. 2.9. The more polished the planes the more nearly a ball rolled up to the original height h from which it had started its descent, independently of the angle of elevation of the second plane. Galileo's hypothesis was that, could friction and air resistance be

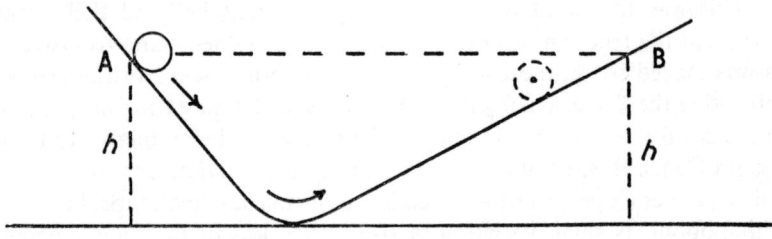

Fig. 2.9

eliminated entirely, the ball would roll up to exactly its original height. What, however, would the ball do if the second plane were horizontal? Then, Galileo hypothesized, it would continue rolling for ever were there no friction or air resistance at all. But, argued Galileo, a horizontal plane on the earth's surface is in reality a large circle. And hence Galileo's triumphant conclusion: all natural motion is circular. Thus wrote Galileo in the *Dialogue*: 'One may reasonably conclude that for the maintenance of perfect order among the parts of the Universe, it is necessary to say that movable points are movable only circularly; if there are any that do not move circularly, these are necessarily immovable, nothing but rest and circular motion being suitable to the preservation of order.'[30] Even Aristotle receives a rare commendation:

> Besides, straight motion being by nature infinite (because a straight line is infinite and indeterminate), it is impossible that anything should have by nature the principle of moving in a straight line; or, in other words, towards a place where it is impossible to arrive, there being no finite end. For nature, as Aristotle well says himself, never undertakes to do that which cannot be done, nor endeavours to move whither it is impossible to arrive.[31]

Nevertheless, the Aristotelian world-view is crumbling: the earth is one of the heavenly bodies, *all* natural motion is circular, and Aristotle's dichotomy of the Universe cannot be maintained. Even the values inherent in the Aristotelian universe are themselves under attack—immutability is no longer regarded as divine-like. For 'what greater stupidity can be imagined than that of calling jewels, silver and gold "precious", and earth and soil "base"?' asks Galileo.

> People who do this ought to remember that if there were as great a scarcity of soil as of jewels or precious metals, there would not be a prince who would not spend a bushel of diamonds and rubies and a cart-load of gold just to have enough earth to plant a jasmine in a little pot, or to sow an orange seed and watch it sprout, grow and produce its handsome leaves, its fragrant flowers, and fine fruit.[32]

Twenty years earlier Galileo had proclaimed in similar vein what was to be his life-long commitment: 'We shall prove the earth to be a wandering body surpassing the moon in splendour, and not the sink of all dull refuse of the universe.'[33]

Galileo has, however, not yet achieved his objective of proving the Aristotelian-Ptolemaic theory false and the Copernican theory true. Let us now consider some of his telescopic discoveries to see if these constitute the decisive proof for which he was searching.

The discovery of Jupiter's moons certainly showed that not all heavenly bodies revolve directly around the earth, an argument therefore against the Aristotelian world view. Conversely, Jupiter's moons showed that the earth is in no way privileged in having its own moon and this therefore constituted an argument in favour of the Copernican system. But no proof either way. That many more stars could be seen through the telescope than with the naked eye made the much larger universe postulated by Copernicus so much the more plausible but it did not prove that the earth moves. And while the roughness of the moon's surface again demonstrated the implausibility of the Aristotelian belief that the heavens are made of a fifth pure element, it nevertheless again did not prove the Copernican system true. And the same remark is valid with respect to the discovery of sun-spots. But, on the other hand, the phases of Venus did appear to constitute very definite disproof of the Aristotelian-Ptolemaic belief that Venus circles the earth always lying between the earth and the sun. For Galileo had observed that the phases of Venus are similar to those of the moon and this meant that when Venus is visible as an almost circular planet then it must lie on the far side of the sun with respect to the earth. And therefore Venus must circle the sun. Furthermore, through the telescope the size of Venus was seen to vary as predicted by the Copernican system. Thus the telescope had turned an apparent refutation of the Copernican system into an outstanding triumph—provided, of course, one was prepared to believe that the naked eye deceived but that the telescope did not!

Perhaps we should at least mention at this point the very great difficulties many of Galileo's contemporaries initially experienced in using the telescope —it was no easy matter to focus, for example, on the moons of Jupiter. Furthermore, Galileo had no theory as to how the telescope worked.

Moreover, the Aristotelians were able to claim that since the heavens were qualitatively different from matter in the sub-lunary sphere one could not be sure that what one was seeing through the telescope actually corresponded to the reality of the heavens. It seemed there could be no certainty. A theory of how the telescope worked would have been useful but it would not have compelled belief, for example, in the existence of Jupiter's moons.

The Tychonic 'compromise' paradigm

Nevertheless, the Aristotelian-Ptolemaic paradigm was now in great trouble in the minds of those astronomers who could manipulate the telescope successfully and who were prepared to believe in the reality of the evidence it presented. However, this did not mean that the only choice lay between the Ptolemaic paradigm and the radically new world system. On the contrary, it

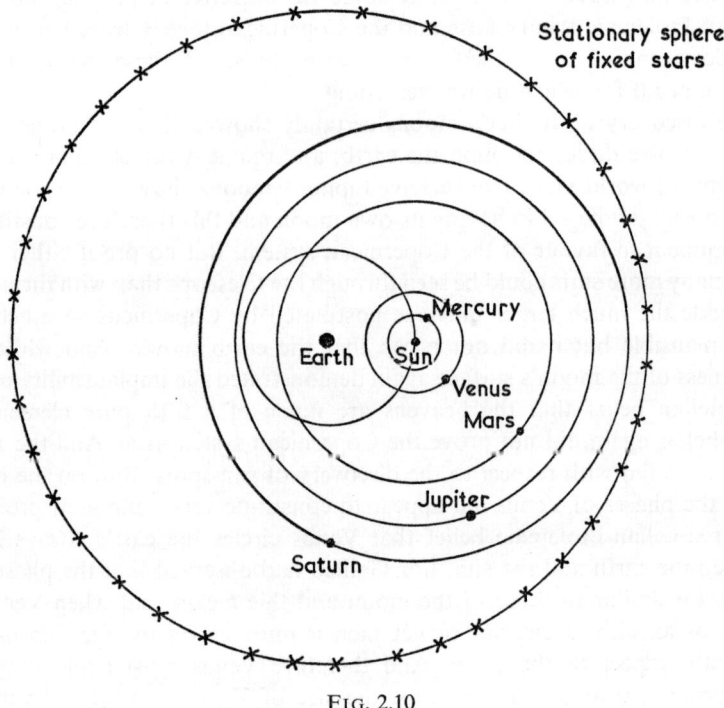

FIG. 2.10

is always possible to attempt to modify an existing paradigm so as to meet new difficulties rather than adopt a paradigm that means revolutionary changes. Thus the nobleman Tycho Brahe proposed a major modification of the Aristotelian-Ptolemaic world system, but which kept one of its essential features intact: the central position and immobility of the earth. His world system is that shown in Fig. 2.10. The earth is in the centre of a finite universe about which daily revolve the celestial bodies. In addition to this daily rotation, the sun orbits the earth once a year while the remaining planets orbit the sun and therefore, together with the sun, the earth as well. Observationally this world system is equivalent to the Copernican, except that it does not predict parallax motion (and no parallax motion had been observed) and does not suppose anything so contrary to common-sense as the motion of the earth. Galileo would, however, have nothing to do with this compromise solution. The title of Galileo's polemical work speaks for itself: there were only *two* great world systems to be considered. The Tychonic compromise, although as observationally accurate as the Copernican, was nevertheless beneath contempt. And the reason lies in the extreme lack of mathematical harmony that Galileo and Kepler regarded as characterizing Tycho's world system. Kepler's comments are revealing. 'Tycho,' he wrote, 'possesses the best observations and, consequently, as it were, the material for the erection of a new structure; he has also workers and everything else which one might desire. He lacks only the architect who uses all this according to a plan.'[34] Nevertheless, architecturally sound or not, the Tychonic compromise solution meant that the phases of Venus could not be regarded as conclusive evidence for the earth's motion and this left Galileo with the last but most important task still to perform—to produce concrete evidence that the earth moves. In the *Dialogue* he attempted to prove the reality of the earth's motion by asserting that the occurrence of tides is caused solely by the double motion of the earth, an argument perhaps presented only in desperation. Galileo therefore failed to prove his case beyond dispute and, tragically, was forced by the Inquisition to renounce the Copernican theory, his public commitment to prove the physical reality of the Copernican system thus ending in failure. To be sure, the choice appeared to lie only between the Tychonic compromise and the revolutionary Copernican world system but the reality of the earth's motion yet remained to be proved. And apart from this so elusive 'final proof' Galileo knew that many puzzles remained: 'How each planet governs itself in its particular revolutions,' he admitted, '... we cannot yet undoubtedly resolve. Mars that has so puzzled our modern astronomers is a proof of this.'[35] Some of these puzzles had, however, already been resolved by Galileo's great contemporary, Johannes Kepler.

Kepler's commitment and contributions to the Copernican paradigm

'Why waste words?', Kepler asked his readers. 'Geometry existed before the Creation, is co-eternal with the mind of God, *is God Himself*...; Geometry provided God with a model for the Creation and was implanted into man, together with God's own likeness and not conveyed to his mind through the eyes.'[36] And since the Copernican system was geometrically harmonious, this was for Kepler the one and true system of the world, no matter how much the eyes apparently received information to the contrary. Like Galileo then, Kepler knew that the Copernican system was true and he, like Galileo, devoted his life to its further articulation—an endeavour in which he was singularly successful. 'Holy Lactantius,' Kepler wrote, 'who denied that the earth is spherical; holy Augustine who acknowledged the sphericity of the earth, but denied the existence of antipodes; holy the Officium that recognized the antipodes, but rejects the motion of the earth . . . but holier yet is to me Truth, which reveals that the earth is a small sphere, that antipodes exist, and that the earth is moving.'[37]

Kepler's aim was no less than to show that each planet revolves about the sun in a circle but with the sun displaced somewhat from the centre of each circle. To this end he re-introduced the use of the equant position (which Copernicus had rejected) and was in this way quickly able to achieve his aim for all planets with the one exception of Mars. But Mars, Kepler decided, was to be made to succumb. Five years later and after some seventy tedious trial and error calculations, the weary but jubilant Kepler believed he had achieved the success for which he had dedicated his life: he had found a circular orbit for Mars which fitted Tycho's observational data—or, rather, which fitted those of Tycho's data he had selected to use. However, the cautious Kepler then checked with two more of the accurate observations that Tycho Brahe had made and found disagreement of 8 minutes of arc between his postulated circular orbit for Mars and Tycho's two observations. Now whereas Copernicus' data had been at most accurate only to 10 minutes of arc (and therefore such disagreement with respect to his data would have been perfectly acceptable) Tycho's data were accurate to 4 minutes of arc. There was therefore an undeniable difference between theory and observation amounting to 4 minutes of arc. Nevertheless, since Kepler had all but accomplished a two thousand-year-old dream, he perhaps could have felt himself justified in ignoring such a tiny 'puzzle', waiting possibly for someone else to clear up the small discrepancy. It is therefore of interest to follow Kepler's own thoughts on the problem of achieving a good fit with observational data.

As had Galileo, Kepler praised Copernicus for ignoring minor disagreement with observation: 'He sets an example for others,' Kepler wrote, 'by his contempt for the small blemishes in expounding his wonderful discoveries. If this had not been always the usage, then Ptolemy would never have been able to publish his *Almagest*, Copernicus his *Revolutions* and Rheinhold his *Prutenian Tables*.' And yet Kepler did not grant himself the privilege of ignoring small blemishes in his own work. On the contrary:

> But, but for us, who by divine kindness were given an accurate observer such as Tycho Brahe, for us it is fitting that we should acknowledge this divine gift and put it to use. . . . Henceforth I shall lead the way towards that goal according to my own ideas. For if I had believed that we could ignore those eight minutes, I would have patched up my hypothesis accordingly. But since it was not possible to ignore them, those eight minutes point the road to a complete reformation of astronomy: they have become the building material for a large part of this work.[38]

Kepler's momentous step was to decide to abandon the hypothesis of circular motion and therefore to investigate the possibility of the planets' describing closed but non-circular orbits. But if the planets did not describe circular motion at uniform speed why should the earth? How in fact does the earth move? This was the first problem Kepler posed for himself.

His method of solution was ingenious.[39] First of all he assumed that the orbit of Mars is closed and calculated the period T_M of the orbit (see the method previously explained). Kepler then observed the direction of Mars at one of the times when the earth lies on a straight line joining the sun and Mars. Let this direction—see Fig. 2.11—point to the fixed star F. At a time

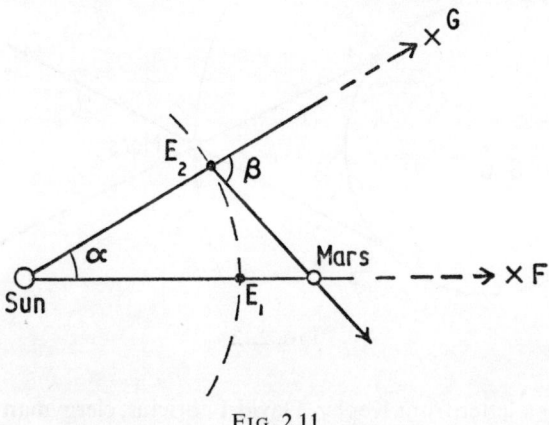

Fig. 2.11

T_M later Mars is, by hypothesis, back at the same point in space. The earth is, however, now at a different point E_2 in its orbit which can be determined by measuring the angle α the stars F and G subtend at the sun, and the angle β Mars and the star G subtend at the earth. Repeating this observation at intervals of T_M, Kepler was able to determine a sufficient number of points on the earth's orbit to allow him to hypothesize that the earth's orbit is approximately circular, with the sun displaced from the centre of the orbit, and that the earth's velocity is such that the radius vector from the sun sweeps out equal areas in equal times. (The latter he regarded as a very satisfactory formulation of the velocity 'law' since the velocity of the earth was now referred to the sun rather than to an empty point in space.) Making use of these hypotheses Kepler was then able to determine the earth's position in its orbit at any time and therefore, by observing Mars at intervals of T_M, to determine a number of positions of Mars on its orbit. Kepler's method is as shown in Fig. 2.12. After much trial and error, Kepler eventually hypothesized that the determined positions could all be regarded as lying on an ellipse with the sun at one focus. He had solved the problem of the orbit of Mars and in doing so had formulated his famous first law: the planetary orbits are elliptical.

Kepler was of course jubilant at his achievement. Nevertheless a nagging question remained—why are the orbits elliptical? The orbits could not be 'naturally' elliptical and Kepler felt that some explanation had to be found to account for the departure from circular motion, a departure, let us remember, that Galileo never even considered. He, Kepler, had striven to show that the system of the world was mathematically harmonious; he had eliminated the circles upon circles of Ptolemy and Copernicus; all was now harmonious in the heavens except for the very slight ellipticity of the orbits to which Kepler referred, in his own inimitable language, as the 'cart-load of dung' still

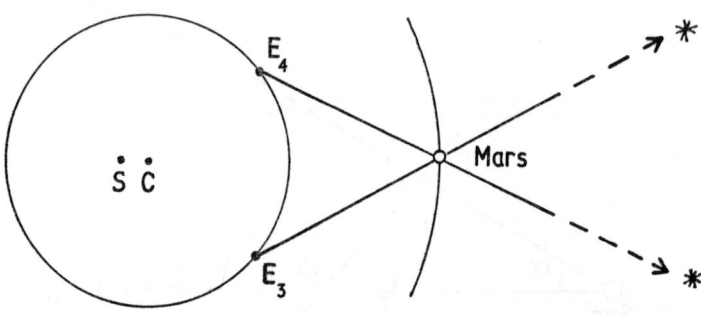

FIG. 2.12

remaining. To a letter from Kepler, David Fabricius, clergyman and amateur astronomer, replied thus: 'With your ellipse you abolish the circularity and uniformity of the motions, which appears to me the more absurd the more profoundly I think about it. . . . If you could only preserve the perfect circular orbit, and justify your elliptic orbit by another little epicycle, it would be much better.'[40] Kepler thought otherwise! The sun was responsible for the motions of the planets but an inherent 'laziness' in the planets prevented their describing perfect circular orbits. 'My aim,' he declared, 'is to show that the heavenly machine is *not* a kind of divine, live being but a kind of clockwork . . . in so far as nearly all the manifold motions are caused by a most simple, magnetic and material force, just as all motions of the clock are caused by a simple weight.'[41] This, then, was Kepler's research programme but it was one he was not able to complete.

At the start of this chapter we asked why the earth's motion should ever have been conceived at all. We have now reached a stage where in the decades following the deaths of Kepler and Galileo the earth's motion is held to be self-evident by all mathematically-minded astronomers. The telescopic discoveries of Galileo, combined with the unprecedented accuracy of Kepler's predictions of planetary motions, indicated that the true reality of the world system was now at last known. Nevertheless, there were still problems. The reality of the earth's motion had not yet been 'conclusively' demonstrated,

THE COPERNICAN REVOLUTION 55

the reasons for the ellipticity of the orbits remained unknown, and also the articulation of the Copernican paradigm had led to the bewildering new problem of the nature of infinity. For Kepler, the infinite was unthinkable: 'This very cogitation,' he wrote, 'carries with it I don't know what secret, hidden horror; indeed, one finds oneself wandering in this immensity, to which are denied limits and centre and therefore also all determinate places.'[42]

Nevertheless, the commitment to prove that the earth moves had been all but won. Isaac Newton, born in 1642, the year of Galileo's death, was to bring the 'already moving carriage to its destination'. Using his laws of dynamics and the law of universal gravitation, Newton showed that indeed the moon falls to the earth as does the apple; on the other hand, the sun is not falling constantly to the earth as is the moon but, on the contrary, the earth is falling always towards the sun as are all the other planets. As Cotes wrote in the preface to the second edition of the *Principia*: 'He has so clearly laid open and set before our eyes the most beautiful frame of the System of the World, that if King Alphonso were now alive, he would not complain for want of the graces either of simplicity or of harmony in it.'[43] There were, of course, many problems in the Newtonian paradigm still to be overcome, some of which we have already mentioned in the first chapter, but at any rate in England the men of the new science were convinced beyond doubt that the earth orbits the sun once a year and not vice versa.

4. Some Concluding Comments

It cannot be stressed too strongly that we have not attempted to discuss the origins of the 17th-century scientific revolution but limited ourselves to a much narrower objective, namely to explore the reasons why men conceived of the motion of the earth and attempted to prove its reality. Needless to say, even with this limited objective our discussion has been greatly simplified. We did not explore the differences between Aristotelian cosmology and Ptolemaic astronomy, or discuss the effects of Humanism in the Renaissance, the 'sun-worship' of the humanists and neo-Platonists, the related Hermetic tradition, the geographical discoveries which showed that Ptolemy as cartographer had certainly been wrong, the rise of trade and commerce in the Italian cities, the development of quite sophisticated machines, the difficulties with the calendar (which the Copernican system did not resolve), or the new ideas concerning motion developed by, for example, Buridan and Oresme. Nevertheless, we have explored at some length one extremely important aspect of the scientific revolution, namely the aesthetic commitment to the idea that nature possesses a mathematical simplicity and beauty if only looked at in the right way. And this commitment, as we argue in the next chapter, has not been confined to only the 17th-century scientific revolution but has been and still is an important factor motivating and directing research in the physical sciences. Basically men conceived of the idea that the earth moves because in this way the universe appeared to them to be mathematically simpler, more harmonious and more beautiful than when viewed from a stationary earth. No matter how difficult it may be to give operational definitions of such concepts as simplicity, harmony and beauty (or, for that matter, such concepts as love and justice) nevertheless it appears undeniable that the search for simplicity, harmony and beauty was an extremely important, perhaps essential, factor in the rise of modern science. Indeed, if men had not committed themselves to this quest perhaps we would still be puzzling over the epicycles of the Ptolemaic planetary system. If there is one outstanding moral to this story it is perhaps that concepts resisting operational definition should not be lightly discarded as 'unscientific'. For such an attitude would betray not only a misunderstanding of the nature of the scientific quest but would probably successfully prevent the development of new and revolutionary scientific paradigms.

Our sketch in the history of astronomy lends confirmation not only to the belief of Kuhn and Lakatos that theories are 'refuted' all the time but also to their belief that 'refutation' does not and must not mean *automatic* rejection. It was necessary to articulate the Ptolemaic paradigm to that point where it became possible for *some* mathematical astronomers to despair of finding solutions within the framework of circular motion around a stationary earth. Similarly it was necessary to articulate the Newtonian paradigm to that point where further progress could only come from the articulation of a radically new paradigm. Thus I cannot agree with the opinion of Needham, for example, who, after writing that 'Chinese astronomy is often reproached for its overwhelming observational bias, but the lack of theory was an inevitable result of the lack of deductive geometry,' goes on to add that 'it might well be argued that the Greeks had had too much, since the apparent beauty of "cycle on cycle, orb on orb" came eventually to constitute a strait-jacket posing unnecessary difficulties for a Tycho, a Copernicus and a Galileo.'[44] On the contrary, it was, I believe, the attempt to force planetary motions into the mathematical harmony of 'cycle on cycle, orb on orb' that led eventually to a Copernicus, a Tycho and a Galileo.

The system of the world proposed by Copernicus and developed by Kepler and Galileo was revolutionary in so far as it negated a basic feature of the Aristotelian-Ptolemaic paradigm, the immovability of the earth. Nevertheless, the Copernican and Ptolemaic paradigms are not 'incommensurable'; they can be compared, and compared against phenomena problematic for both paradigms, for example, the retrogade motion of the planets. However, since the proponents of each paradigm provided a solution to this 'puzzle', there were no 'objective' grounds in the 16th century for preferring the Copernican solution to the Ptolemaic. To be sure, the explanation proposed by Copernicus served to correlate together many of the phenomena associated with retrograde motion. On the other hand, Copernicus could only grope for answers to problems existing only within his paradigm, e.g. the problem of explaining how loose objects remain on the surface of a swiftly moving earth and indeed of explaining the motion of the earth itself. At this stage in the crisis state those astronomers who committed themselves to the articulation of the Copernican system did so because of their belief that the heavens must be mathematically harmonious; they were convinced that the many puzzles confronting the Copernican paradigm could be successfully met, whereas those confronting the Ptolemaic had not been and could not be met.

The phases of Venus were, for example, another phenomenon recognized by the proponents of both paradigms as needing explanation. However, although the Copernican paradigm offered an explanation that its rival could not match (indeed the Ptolemaic paradigm predicted a totally different sequence of phases), we remarked on the fact that Galileo could provide no satisfactory explanation as to how the telescope worked. Even so, it might reasonably be argued that at this point the Ptolemaic system had been objectively falsified. Yet the Copernican system had not been 'proved' to be physically true. In addition there was the Tychonic paradigm to contend with. However, for the same reasons that mathematical astronomers rejected the Ptolemaic paradigm in favour of the Copernican, so they rejected Tycho's compromise solution. But their choice was subjective; there were at the time no 'objective' grounds *compelling* rejection of Tycho's theory.

Gradually, however, the very great predictive accuracy resulting from the use of Kepler's laws convinced astronomers that the planetary motions were at last correctly known, while the hard-won successes of the Newtonian paradigm convinced *some* men of the 'new science' that a dynamical understanding of these motions had finally been achieved.

For let us note that Kepler's 'already moving carriage' had by this time arrived at a destination that many men of the new science objected to very strongly. In particular since the aim of the 'mechanical philosophy' had been to banish all non-material forces from nature, Newton's failure to achieve this led Huygens and other contemporaries of Newton to accuse him of reintroducing occult qualities into science, to declare his universal law of gravitation to be a retrogressive step. Newton himself was certainly worried over the nature of the gravitational force, writing in a famous letter to Bishop Bentley:

> It is conceivable that inanimate brute matter should, without the mediation of something else which is not material, operate upon, and affect other matter without mutual contact; as it must do if gravitation ... be essential and inherent in it. And this is one reason why I desired you would not ascribe innate gravity to me. That gravity should be innate, inherent and essential to matter, so that one body may act upon another at a distance through a vacuum without the mediation of anything else, by and through which their action and force may be conveyed from one to another, is to me so great an absurdity that I believe that no man who has in philosophical matters a competent faculty of thinking can ever fall into it.[45]

Huygens was amazed that Newton should ever have initiated such a research programme in the first place. For as he commented:

> I am astonished that Mr Newton has taken the trouble to construct, upon such an improbable and audacious hypothesis, so many Theorems and as it were a complete theory of the actions of the heavenly bodies. I mean his hypothesis that all[6]the little particles of diverse bodies attract one another, in reciprocal squared ratio of the distances.[46]

Here, then, was cause for concern and reason perhaps for abandonment of the Newtonian paradigm. That this did not happen was because the proponents of the Newtonian paradigm were able to resolve one by one the puzzles with which the paradigm had been initially confronted. Not only had the Ptolemaic paradigm been aesthetically unattractive to the scientific revolutionaries but it had also failed to predict the planetary motions with great accuracy. In the latter respect, however, the Newtonian paradigm was immensely successful and, for this reason, men slowly began to accept as natural and innate to matter a force which could exert its effect through empty space. 'There is but one universe,' wrote Lagrange in praise of Newton, 'and it can happen to but one man in the world's history to be the interpreter of its laws.'[47] A century later, however, and the intellectual edifice of the Newtonian paradigm lay in ruins.

Chapter 3

THE BREAKDOWN OF THE NEWTONIAN PARADIGM: A STUDY OF CRISIS

> The relativity theory arose from necessity, from serious and deep contradictions in the old theory from which there seemed no escape. The strength of the new theory lies in the consistency and simplicity with which it solves all these difficulties, using only a few very convincing assumptions.
>
> A. EINSTEIN and L. INFELD
> *The Evolution of Physics*

FOR our future use, and to throw further light on the Popper versus Kuhn debate, we sketch in this chapter a historical outline of the dramatic series of events which brought about a transformation of Newtonian physics so radical as to be almost comparable in impact with that produced by the Copernican revolution. For although physicists of the 19th century may well have thought, and very reasonably so, that the fundamental principles on which the whole of physics could be built had been discovered and had been proved to be true, the 20th century was to shatter this confidence and to plunge the physics community into the kind of crisis state Kuhn has discussed so lucidly in his 1962 essay. As in the case of the Copernican revolution we shall see that a well-established paradigm was articulated into crisis, followed by competition between rival paradigms with ultimate success going to the paradigm which, although the most outrageous from a common-sense standpoint, outrivalled its competitors in its unifying power and mathematical simplicity and beauty. We shall then briefly sketch the implications of the underlying analysis for the Popper-Kuhn-Lakatos debate before closing the chapter with some remarks on the very great importance of aesthetic criteria in the choice of commitment to a research programme. In a later chapter we shall compare the nature of objectivity and commitment in the physical sciences with that in the social sciences.

1. The Newtonian Paradigm

Newton had shown without any doubt (or so it seemed) that it is the earth which moves around the sun and not vice versa. For if his laws were true and if there existed a body in the solar system whose mass is very much greater than the masses of all other bodies then the latter bodies would perform elliptical motion about a common focus in the massive body (to a very good approximation). Clearly there was one, and only one, body around which all other bodies of the solar system moved *as predicted* by Newton's laws—and this body was the sun. Thus this fact, it was thought, not only showed Newton's laws to be true but also that the earth physically moves, that truly the sun 'sits as upon a royal throne ruling his children, the planets, which circle around him'.

Absolute space

But with respect to what do the planets move? For if the motion of the earth is referred to a reference frame whose centre is imagined to lie in the sun and one of whose (rigid) arms is fixed in the earth, then *with respect to this particular frame the earth is at rest*. But since the earth is being constantly pulled towards the sun (according to Newton's law of universal gravitation) then an observer stationary in this frame apparently has no answer to the question, 'Why does the earth not fall into the sun?' The answer, however, given by Newton is that this observer too is moving, together with his frame—really, truly, physically moving, just as the scientific revolutionaries had supposed the earth to be. And, in particular, moving with respect to *absolute space*, that space with respect to which all motion must be referred. 'Absolute space,' Newton wrote in the *Principia*, 'in its own nature, without relation to anything external, remains always similar and immovable'.[1] An extremely good approximation to a reference frame at rest in absolute space must be, Newton reasoned, one whose centre lies in the sun and whose arms do not rotate relative to the fixed stars. For as predicted by Newton's laws, it is with respect to such a frame that the earth physically orbits the sun once a year, Mars once every two years and the other planets in their respective periods.

The Newtonian principle of relativity

There was, however, one curious and very important aspect to motion through absolute space that Newton discussed at some length. For while the theory predicts that it is possible to detect accelerated motion of any kind through absolute space (e.g. an observer in an accelerating rocket experiences so-called fictitious forces acting on him which he cannot attribute to physical agencies acting within the rocket) it also predicts that straight line motion with a constant velocity through absolute space (which we shall call inertial motion) is indistinguishable from a state of rest in absolute space. Certainly, *according to Newton's laws, no mechanical experiment of any kind can determine whether or not a reference system is stationary in absolute space or undergoing inertial motion.* (For example, the occupants of a smoothly flying jet airliner can play a game of billiards using the same techniques as when stationary on the earth's surface.)

For convenience let us agree to give the name of 'inertial system' to any reference system that is either stationary in absolute space or undergoing inertial motion. Then we note that the earth itself can be regarded, for small-scale experiments of short duration, as a good approximation to an inertial system. But as soon as the motions of any of the other planets are considered,

BREAKDOWN OF THE NEWTONIAN PARADIGM 61

then it becomes clear that the earth is physically orbiting the sun once a year (in addition to rotating on its axis once a day) and therefore is not an inertial system. However a reference system which does not accelerate with respect to the fixed stars is an excellent approximation to an inertial system.

Absolute time

Finally, but of great importance, we must look at Newton's famous conception of time, his belief that time is absolute, that time runs at the same rate for all observers. Newton of course agreed that it would be difficult to make clocks which would not sometimes lose or gain time: 'It may be,' he wrote, 'that there is no such thing as an equable motion whereby time may be accurately measured.' But nevertheless, it was his opinion that 'the flowing of absolute time is not liable to any change'. 'Absolute, true and mathematical time, of itself and from its own nature, flows equably without relation to anything external.' Likewise 'all things are placed in time as to order of succession'.[2]

The Galilean addition of velocities and absolute time

Although stated as baldly as this Newton's concept of absolute time appears somewhat abstract, nevertheless this concept is basic to our intuitive notions of how velocities add up. For example, suppose a red car travelling at 30 m.p.h. along a road is passed by a blue car travelling at 50 m.p.h., both velocities measured with respect to the road. *Relative* to the red car, the blue car, we would all say, has a velocity of 20 m.p.h. and accurate measurements carried out in the red car as the blue car passed it would give this velocity of 20 m.p.h. Although this 'common-sense' result (the so-called Galilean addition of velocities) follows from Newton's concept of absolute time, it is exactly this result which is challenged by Einsteinian theory. The result according to Einstein would not be 20 m.p.h. but slightly more—to be sure, not, in this case, measurably more using our ordinary clocks but nevertheless more. If, however, the two velocities in question were almost as large as the velocity of light then the 'common-sense' Newtonian result would be catastrophically wrong and very measurably so. It is obvious therefore that we should look in more detail at Newton's concept of absolute time.

The synchronization of clocks

In order to be able to measure velocities in an inertial system we need to synchronize identically made clocks which are located at different places in the system. We first note that we would obviously not be justified in viewing a standard clock through a telescope and then synchronizing all clocks to the time shown on the standard clock without allowing for the time taken for the light to arrive from the standard clock. But since this would mean already knowing the velocity of light, before having measured it, a vicious circle is evident. Neither do we want at this stage to synchronize clocks by physically taking them to the standard clock, synchronizing them and then taking them back to their original positions. For this would mean making the assumption, as implicitly made by Newton, that the rate of a moving clock is independent of the velocity of its motion with respect to an inertial system. We therefore describe a method that is in principle acceptable no matter whether we live in a Newtonian or Einsteinian world.

Two identical guns are taken and it is checked that when both guns, lying side by side and pointing in the same direction, are fired simultaneously then both bullets travel together, i.e. they have the same velocity. In order to synchronize two of the identically constructed clocks A and B, separated by a carefully measured distance x, one of the guns is taken to clock A and the other to clock B. At the instant the gun at A is fired in the direction of the clock at B, clock A is set going at time t_0. At the instant the bullet arrives at clock B a mechanism fires the second gun at B in the direction of A, the bullet arriving at A at time t_1 say. Then the velocity of the bullet is, by definition,

$$v = 2x/(t_1 - t_0).$$

Knowing the velocity of the bullet, it is then possible to synchronize B with A and indeed any other clock with A. For instance, a bullet from A is fired at time T_0 to clock B. At the instant of arrival clock B is set going at time $T_0 + x/v$. Similarly with all other clocks.

Suppose now a clock C is synchronized with clock A and then taken from A to the clock B already synchronized with A. Then, as we have already stated, Newton's hypothesis that time is absolute means that when clock C arrives at B it will show the same time as does clock B. Expressed this way Newton's concept of absolute time appears very reasonable—which it is. Clearly no one seeks to re-synchronize his watch with Big Ben simply because he has just finished a bus ride. Hence the elaborate procedure described for synchronizing clocks appears unnecessarily complicated in view of the empirical evidence that clocks remain synchronized after periods of relative motion. Instead, therefore, of using such an elaborate procedure it appears that we could simply bring all clocks together, synchronize them with each other, and then take them apart to their respective positions.

Having discussed these important details we return to the development of the Newtonian paradigm.

The concept of the ether

Newton was much concerned to find a mechanism by which the force of gravity could be transmitted through apparently empty space and tried to develop the idea that space is in fact not empty but is filled with a medium, the ether, of sufficiently low density to permit the almost frictionless passage of bodies through it but sufficiently dense to be able to act as the agency by which gravitation is transmitted between bodies. Although he was unsuccessful in his research programme, nevertheless the idea persisted. When in the

BREAKDOWN OF THE NEWTONIAN PARADIGM   63

19th century it appeared increasingly clear that light is a wave motion (and not a stream of particles) then the ether was automatically taken to be that medium whose physical vibrations are none other than light waves. Indeed as early as 1760 Euler had written very confidently that 'light is nothing but an agitation or disturbance caused in the particles of the ether', adding that 'the parallel between light and sound is in this respect so well established that we can boldly maintain that if the air became as subtle and at the same time as elastic as the ether, the velocity of sound would also become as rapid as that of light [186 000 miles per second]'.[3] In 1864 Maxwell predicted that light waves are electromagnetic in origin, so unifying two branches of physics previously thought quite distinct. Clearly just as water waves cannot exist without water, or air waves (sound) without air, so light waves, it was reasoned, could not exist without the 'light medium' pervading all space—the ether. Thus according to a very prominent 19th-century physicist Lord Kelvin: 'This thing we call the luminiferous ether . . . is the only substance we are confident of in dynamics. One thing we are sure of, and that is the reality and substantiality of the luminiferous ether.'[4] Indeed Maxwell himself in an article entitled 'The Ether', written just before his death in 1879 (the year of Einstein's birth), expressed the conviction that 'whatever difficulties we may have in framing a consistent idea of the constitution of the ether, there can be no doubt that the interplanetary and interstellar spaces are not empty but are occupied by a material substance or body which is certainly the largest and probably the most uniform body of which we have knowledge'.[5]

So far nothing dramatically untoward had happened. To be sure, there had been constant paradigm articulation. So much so that even a brief glance at the overall physics paradigm of the 18th and 19th centuries would reveal the existence of several competing sub-paradigms. *But all these shared the basic conceptual framework of Newtonian space and time.* To this extent there had been no revolution in physics throughout the 18th and 19th centuries, no major paradigm change worthy of the name. But it was to come.

On the velocity of the earth through the ether

Maxwell himself had suggested how the earth's velocity through the ether was in principle quite directly measurable by determining the velocity of ether waves (i.e. light) both in the direction of the earth's supposed motion and in the opposite direction. Only if the earth were at rest in the ether (which it was not, since it orbited the sun once a year) would the two velocities be the same. For example, the ripples spreading out from a rocking boat will, relative to the boat, be the same in all directions only if the boat is stationary in the water; otherwise the velocity of the water waves will be least in the direction of the boat's motion (relative to the boat) and greatest in the direction opposite to the boat's motion. *For the most characteristic property of wave motion is that the velocity of the wave motion is a constant relative to the medium; it is independent of the velocity of the source giving rise to the wave motion.* Thus we see that if the earth's velocity is v through the ether and the velocity of light waves through the ether is c then light waves will appear to an observer on the earth to have a velocity of $(c - v)$ in the direction of the earth's velocity and $(c + v)$ in the opposite direction.

Note, however, one remarkable fact that will later be of great relevance. According to Newton's laws, as we have remarked, *inertial* motion through absolute space can never be detected by any kind of mechanical experiment. However, according to Maxwell's equations, inertial motion through the all-pervasive ether is in principle easily detectable by a straightforward optical experiment. *Hence there existed a very definite and very remarkable asymmetry between mechanical and optical phenomena as conceived by Newton and Maxwell respectively.* We shall see that this asymmetry was the starting point of Einstein's attack on the established paradigm. However, asymmetry or no, since Maxwell's equations predicted that the earth's velocity through the ether is directly measurable by an *optical* experiment—an experiment which would surely have warmed the heart of Galileo—an American physicist Michelson set out to devise an apparatus which would be sensitive enough to detect such a motion. The experiment has, of course, become a classic. Below we give a *greatly simplified* description of Michelson's apparatus (which, nevertheless, may be safely omitted by the non-physicist reader).

Light from a source S is partly transmitted and partly reflected by a mirror M as shown in Fig 3.1. The transmitted light is reflected back to M by the

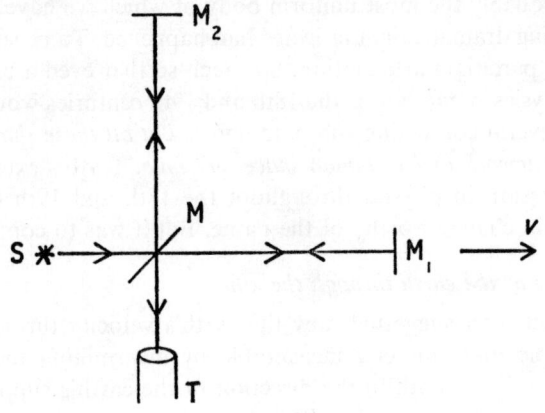

FIG. 3.1

mirror M_1 as is the reflected light by the mirror M_2. The two light waves recombine in the telescope T. Suppose now the interferometer is moving through the ether with velocity v in the direction of the arm MM_1. Then the time taken for the light ray to traverse the distance MM_1 and back will be

$$t_1 = \frac{l}{c-v} + \frac{l}{c+v} = 2\frac{l}{c} \frac{1}{1 - \frac{v^2}{c^2}}$$

where l is the length of the arm MM_1. On the other hand the time taken for the light to traverse the distance MM_2 (also of length l) and back will be

$$t_2 = 2\frac{l}{c}\frac{1}{\sqrt{1-\frac{v^2}{c^2}}}.$$

Since v^2/c^2 is very small, the optical path difference between the two recombining light waves is

$$c(t_1 - t_2) = c\frac{2l}{c}\left(\frac{v^2}{c^2} - \frac{1}{2}\frac{v^2}{c^2}\right) = l\frac{v^2}{c^2}.$$

When the apparatus is rotated through 90°, arms MM_1 and MM_2 interchange roles and therefore the interference fringes should shift by an amount corresponding to an optical path difference of $2\,l\,v^2/c^2$. Less then one-fortieth of the expected fringe shift was observed when, with great experimental skill, the experiment was finally performed in 1887 to the satisfaction of Michelson and his collaborator Morley.

The paradigm is articulated

Thus Michelson and Morley concluded from the result of their experiment that 'the relative velocity of the earth and the ether is probably less than one-sixth of the earth's orbital velocity, and certainly less than one-fourth'.[6] What should physicists have done? Should they, as good Popperians, have leapt for joy and abandoned all existing theories? We already know that physicists do not and, I believe, should not behave in this way; if they did there would be no physics. Clearly physicists took the null result of the Michelson-Morley experiment as yet another puzzle to be resolved— scarcely a unique occurrence in the history of physics from the publication of *de Revolutionibus* to Michelson's experiment!

First of all, the experiment was an extremely complicated one and difficult to interpret. The famous Dutch physicist Lorentz had previously found an error in the theoretical interpretation of a similar experiment performed in 1881 and had shown that Michelson had overestimated the expected fringe shift by a factor of two. This necessitated repeating the experiment with a more sensitive apparatus, an experiment performed in 1887 with the result described. This time Lorentz was baffled. The simplest explanation, namely that the ether at the earth's surface was carried along with the earth, was incompatible with other experiments (e.g. on the aberration of stars) and with Fresnel's explanation of them. As Lorentz later wrote in 1892 to the English physicist Lord Rayleigh: 'I am utterly at a loss to clear away this contradiction, and yet I believe if we were to abandon Fresnel theory [of the ether], we should have no adequate theory at all.'[7] 'Can there be,' he asked Rayleigh, 'some point in the theory of Mr Michelson's experiment which has as yet been overlooked?' He could not find it. Lorentz then proposed a hypothesis which has been much criticized as being *ad hoc* and has indeed often been taken as an outstanding example of incorrent scientific procedure, especially by Popper. The proposal was that the arm MM_1 contracts from length l to length $l\sqrt{1-v^2/c^2}$ (as does any body in the direction of its motion through the ether) thus rendering the optical path difference zero with an expected null result. According to Lakatos, however, this is most legitimate scientific practice corresponding in his terminology to a progressive problem shift. And to be sure the Lorentz hypothesis gave rise to the possibility of other experimental tests, such as that proposed by Lord Rayleigh in 1902 and tested in 1904 but with a result less than 1% that expected on the assumption the Lorentz contraction takes place. Other experiments were done which not only indicated that motion through the ether was undetectable but whose null results could not be accounted for even by invoking the Lorentz contraction hypothesis. Lorentz then began to develop a rather complicated theory in which motion of a body through the ether would set up forces which would not only produce contraction but would also slow down clocks in such a way that the motion would turn out to be undetectable. It was the only way Lorentz could see of solving the puzzle which had by this time turned into a counter-example challenging the ingenuity of the entire physics community.

2. The Einsteinian Paradigm

A new paradigm is proposed

In 1905 a paper modestly entitled 'On the Electrodynamics of Moving Bodies' appeared in the German journal *Annalen der Physik*. Its 26-year-old author, Albert Einstein, not a member of the physics community but an employee of the Swiss Patent Office at Berne, was nevertheless to see his paper produce the second of the great turning points in physics. Unlike Copernicus, however, Einstein was able to see the revolution through to its successful victory over all rival paradigms.

It is very interesting to note, as Holton has emphasized[8], that Einstein did not begin his paper by pointing to disagreement between theory and experiment, but rather to a feature of Maxwell's equations which was, to him, aesthetically unappealing. In the first sentence of his paper he states: 'It is known that Maxwell's electrodynamics—as usually understood at the present time—when applied to moving bodies, leads to asymmetries which do not appear to be inherent in the phenomena.'[9] The example he then gives (that the interaction between a conductor and a magnet depends only on the relative motion between the two, not on the separate motion of either) although familiar to Einstein's contemporaries, had not been previously thought of as a major problem in any way. Einstein then goes on: 'Examples of this sort, together with the unsuccessful attempts to discover any motion of the earth relatively to the "light medium", suggest that the phenomena of electrodynamics as of mechanics possess no properties corresponding to the idea of absolute rest.'[10]

BREAKDOWN OF THE NEWTONIAN PARADIGM 67

Like Copernicus, therefore, Einstein was dissatisfied with the existing paradigm primarily on aesthetic grounds. The lack of complete agreement with experiment, while relevant, was not the strongest motivating factor. For physical reality, Einstein held, was an indissoluble unity. Thus it was totally unacceptable to Einstein that while, according to Newton's laws, inertial motion could *not* be detected by a *mechanical* experiment, according to Maxwell's equations, the same inertial motion *could* be detected by an *optical* experiment. And thus when, to within their experimental accuracy, optical experiments failed to indicate the reality of inertial motion with respect to the ether, Einstein felt able to commit himself to taking a radically new path: no matter what the difficulties (to be overcome) en route, the complete relativity of inertial motion must be built into the very heart of physical theory, and not 'added in' as Lorentz was then attempting to do. Although Lorentz's work, Einstein generously wrote, was 'of such consistency, lucidity and beauty as has only rarely been attained in an empirical science', nevertheless 'to him [Lorentz], Maxwell's equations held only for a particular coordinate system distinguished from all other coordinate systems by its state of rest'.[11] And this, Einstein added, 'was a truly paradoxical situation because the theory seemed to restrict the inertial system more strongly than did classical mechanics'. 'This circumstance,' concluded Einstein, 'which from the empirical point of view appeared completely unmotivated, was bound to lead to the theory of special relativity.' Like Copernicus, therefore, Einstein decided against further articulation of the established paradigm. Relativity of inertial motion was to become the new foundation stone of physics. Accordingly Einstein proposed in his paper two postulates: (1) the special principle of relativity, namely that every inertial observer is entitled to regard himself as at rest, i.e. absolute rest is in principle undetectable by any kind of experiment, and (2) the very famous postulate that 'light is always propagated in empty space with a definite velocity c which is independent of the state of motion of the emitting body'.[10]

It is not my intention here to pursue the consequences of these two postulates except to show that however innocent each postulate appears in itself the two together necessitate the abandonment of the Newtonian space-time framework that for more than two centuries had been the foundation stone for the entire edifice of physical theory articulated since the publication of Newton's *Principia*. In thus striking at the very core of the old paradigm Einstein was proposing nothing less than revolutionary change.

The non-Galilean addition of velocities

That the Einsteinian postulates violate the 'common-sense' world of Newtonian physics can easily be shown. For suppose we have two inertial observers moving with a very great velocity (say $c/2$) with respect to each other and suppose a ray of light emitted by a light source passes each observer. By postulate (1) each observer can consider himself at rest and the other moving.

To be sure, the velocity of the light *source* is measured differently by the two observers. But since, by postulate (2), the velocity of the light ray, as measured by an inertial observer in empty space, is independent of the velocity of the source it follows that each observer must measure the velocity of the light ray to be the same, namely c. Viewed theoretically from within the Newtonian space-time framework the result appears absurd. For how can it be that while one observer can declare that a light ray passes him with velocity c, a second observer moving with velocity $c/2$ with respect to the first, can declare that the same light ray also passes him with the same velocity c? It is only possible, declared Einstein, because (among other things) time is not absolute. On the contrary, each inertial observer has his own proper time. The Newtonian space-time framework must be abandoned.* We proceed to demonstrate explicitly by means of a simple example that time can no longer be considered to run at the same rate for all observers if we assume Einstein's two postulates to be correct.

The relativity of time

Let us suppose (i) that an inertial observer O equipped with a clock C is passed by another inertial observer O' equipped with an identically constructed clock C' and (ii) that, at the moment of passing, the clock C' is

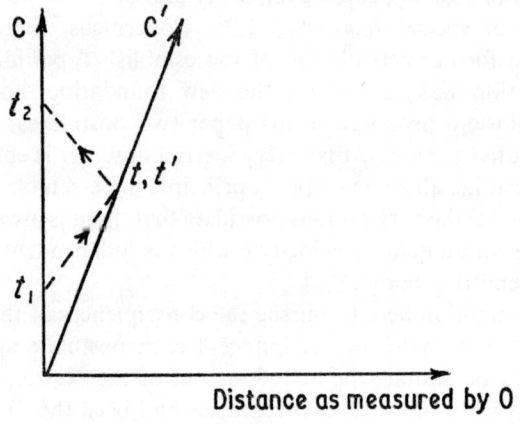

FIG. 3.2

synchronized to read the same time as the clock C, which for convenience we shall take to be zero. Let us further suppose (iii) that at time t_1 on clock C a light signal is emitted from O to O' which arrives at clock C' at time t' as measured on clock C'. Using Einstein's two postulates, our problem is to

* I.e. from the conceptual point of view. To be sure, the Newtonian *equations* still give experimentally acceptable results in those domains where all relative velocities of interest are very small compared with c. The point is that as these velocities increase with respect to c so the differences between the predictions of Newtonian and Einsteinian physics become progressively greater.

BREAKDOWN OF THE NEWTONIAN PARADIGM

find out the time of arrival t of the light ray at C' as would be registered by a clock C_2 stationary with respect to clock C, but coincident with clock C' at the instant of reception of the light signal. (We are, of course, assuming that the clock C_2 has previously been synchronized with clock C using the method described earlier.) It follows that if t is not equal to t', then Newton's hypothesis that time runs at the same rate for all observers is no longer valid and this foundation stone of physics is no more. Let us show that this is indeed the case.

Let us suppose that the clock C', at the instant it receives the light signal from C, emits a light signal back to C which arrives when clock C reads t_2. The situation is as shown in Fig. 3.2.

Let us write that $t' = kt_1$. Then we can use the first postulate to deduce that $t_2 = kt'$. The reasoning is as follows. According to O clock C' is leaving O with, say, velocity v; clock C emits a light signal at t_1 which is received by C' at time t'. Similarly according to O' clock C is leaving O' with velocity v; clock C' emits a light signal at t' which is received by C at time t_2. Hence by the special principle of relativity t' is to t_1 as t_2 is to t'. Hence if $t' = kt_1$ it follows that $t_2 = kt'$ and hence

$$t_2 = kt' = k^2 t_1.$$

By the second postulate we have that the velocity of light leaving C at time t_1 is the same as that arriving at C at time t_2. But the light ray leaving C travels the same distance in its journey to C' as the light ray emitted by C' travels in the 'return' journey back to C. Hence the time t must be exactly halfway between the two times t_1 and t_2 i.e.

$$t = \frac{t_1 + t_2}{2} = \frac{k^2 + 1}{2} t_1 = \frac{k^2 + 1}{2k} t'.$$

Unless $k = 1$ we will have t is unequal to t' and therefore the revolutionary result that Newton's hypothesis that time is absolute is incompatible with Einstein's two postulates. We can, however, quickly find k since the distance the light travels in time $(t - t_1)$ with velocity c is equal to the distance O' travels with velocity v in time t. Hence

$$c(t - t_1) = v t$$

i.e.
$$c \frac{(k^2 - 1)}{2} t_1 = v \frac{(k^2 + 1)}{2} t_1$$

giving
$$k = \sqrt{\frac{1 + v/c}{1 - v/c}}.$$

Therefore, we have

$$t = \frac{1}{\sqrt{1 - v^2/c^2}} t'.$$

Manifestly, therefore, t is not equal to t' and we have the revolutionary result derived by Einstein that the clock C′, moving in relation to the stationary clocks C and C_2, records a time interval between the two instants of passing C and C_2 less than the time interval recorded by the two clocks themselves. Time is not absolute but is relative to each inertial frame of reference. Each observer can be said to have his own proper time. A foundation stone of physics has therefore collapsed and the whole edifice must be rebuilt. Let us note now how Max Born would later describe his reaction to Einstein's paper: 'Einstein's reasoning was a revelation to me . . . For me—and for many others—the exciting feature of this paper was not so much its simplicity and completeness, but the audacity of challenging Isaac Newton's established philosophy, the traditional concepts of space and time.'[12] We shall in a later section discuss some very different reactions of other members of the physics community to Einstein's revolutionary proposals.

We might, however, at this point just mention another revolutionary consequence of Einstein's two postulates: namely, the denial of the ether's existence. The first postulate, we recall, declares all inertial frames to be equivalent in every way; the second postulate, combined with the first, gives the result that the velocity of light is the same in all inertial frames. It follows, therefore, that *either* each inertial system must be considered to have its own ether *or* that there is no ether at all. The first alternative, however, demands the existence of an infinity of ethers all moving through each other without any kind of interference. Einstein therefore accepted the second alternative, and declared light to be a wave motion in its own right—a wave motion in empty space. Kelvin's 'only substance we are confident of in dynamics' had ceased to exist! But to say that light is a wave motion without a medium appeared to many physicists equivalent to saying that sound waves could exist without air, water waves without water. It appeared completely nonsensical. The English physicist J. J. Thomson was therefore not exaggerating when he declared in 1909 that 'the aether is not a fantastic creation of the speculative philosopher; it is as essential to us as the air we breathe'.[13] But Einstein, and the articulators of the Einsteinian paradigm, thought differently.

3. Competition between Paradigms

According to both Kuhn and Lakatos a new paradigm is born in an ocean of counter-examples. If there are any exceptions to this rule the birth of Einstein's special theory of relativity is not among them. The following year after the publication of Einstein's paper in the *Annalen der Physik* the same journal published the first paper to discuss Einstein's theory. Its author, the eminent German experimental physicist Kaufmann, began his paper by starkly announcing that his results refuted both the theories of Lorentz and Einstein but were compatible with two competing theories being developed by Abraham and Bucherer. Kaufmann did not beat about the bush: 'I anticipate right here the general result of the measurements to be described in the following: *the measurement results are not compatible with the Lorentz-Einsteinian fundamental assumption.*' After a year's silence (Holton tells us) Einstein replied that although he could find no error in Kaufmann's work, nevertheless, 'whether there is an unsuspected systematic error or whether the foundations of relativity theory do not correspond with the facts one will be able to decide with certainty only if a great variety of observational material is at hand'.[14] Furthermore, although Einstein agreed that the rival theories of Abraham and Bucherer gave predictions much closer to Kaufmann's results than were the predictions of his own theory, he announced his intention to remain committed to his own research programme. 'In my opinion,' he wrote, 'both [rival] theories have a rather small probability, because their fundamental assumptions concerning the mass of moving electrons are not explainable in terms of theoretical systems which embrace a greater complex of phenomena.'[15] He considered his own theory to be more sweeping, more powerful than the rival theories.* Therefore, although the latter were in better agreement with the available experimental 'facts', Einstein was convinced that his own theory of relativity would win through in the end. Nine years later in 1916 Guye and Lavanchy established that Kaufmann's apparatus had been faulty.

This is typical of Einstein's attitude and the strength of his commitment to his research programme. Another example will be given later of Einstein's refusal to allow a few adverse 'facts' to shake his faith in the theory of relativity.

The competition between the Einsteinian and Lorentzian research programmes

In 1909 Lorentz described with appreciation the manner in which Einstein had made the special principle of relativity the foundation stone of his theory. Nevertheless, Lorentz was not convinced that this was the right approach:

> Yet, I think, something may also be claimed in favour of the form in which I have presented the theory. I cannot but regard the ether, which can be the seat of an electromagnetic field with its energy and its vibrations, as endowed with a certain degree of substantiality, however different it may be from all ordinary matter. In this line of thought it seems natural not to assume *at starting* that it can never make any difference whether a body moves through the ether or not [emphasis added].[16]

But this was not how all physicists regarded the two approaches. For some at least the greater simplicity, elegance and universality of Einstein's theory of

---

* Abraham likewise remained committed to his own research programme and never accepted Einstein's theory. Born and Von Laue were later to write of Abraham that 'he loved his absolute ether, his field equations, as a youth loves his first flame, whose memory no later experience can extinguish'. In 1914 Abraham explained in a popular article how 'the physicists of the older generation . . . regarded with scepticism the bold young men who undertook to overthrow the trusted foundations of all physical measurement on the basis of a few experiments which were still under discussion by the experts'.[15n]

relativity won their allegiance. In 1909 the German physicist Wien, after admitting the absence of clear-cut experimental evidence, pronounced in favour of Einstein's research programme: 'What speaks for it most of all, however, is the inner consistency which makes it possible to lay a foundation having no self-contradictions, one that applies to the totality of physical appearances, although thereby the customary conceptions experience a transformation.'[17] Two years later the German physicist von Laue felt able to write:

> A really experimental decision between the theory of Lorentz and the theory of relativity is indeed not to be gained; and that the former, in spite of this, has receded into the background, is chiefly due to the fact that, close as it comes to the theory of relativity, it still lacks the great simple universal principle, the possession of which lends the theory of relativity from the start an imposing appearance.[18]

In 1915, after the battle had been won by the Einsteinian research programme, Lorentz summed up the deciding factor: 'Einstein's theory ... gains a simplicity that I had not been able to attain.'[19]

In this competition between the Einsteinian and Lorentzian paradigms, we can draw an analogy with the competition, such as it was, between the Copernican and Tychonic theories of the universe. Although there was no clear-cut observational evidence in favour of the Copernican theory (and the absence of parallax motion favoured its competitor), nevertheless its greater simplicity and 'harmonies' won the allegiance of mathematically minded astronomers. The Copernican theory, not the Tychonic, was capable of greater articulation and out of the Copernican theory was to emerge Kepler's laws and following them the Newtonian synthesis and research programme. Similarly out of the Einsteinian programme was to emerge the General Theory of Relativity and the application of relativity theory to the domain of quantum mechanics. Einstein's theory appeared to physicists, because of its greater beauty, simplicity and universality, to be inherently more correct, more powerful than its most serious rival. It was therefore Einstein's research programme which eventually won their allegiance.

The reaction against the Einsteinian paradigm

Yet, like the progress of the Copernican theory, the progress of Einstein's theory was achieved in the teeth of much initial opposition from the community of scientists. Just as the Aristotelians demanded an explanation from the Copernicans as to how the earth could be considered to move, and just as Newton was criticized by Huygens for making a retrograde step in introducing action at a distance, so the classical physicists demanded from the Einsteinians an explanation as to how wave motion could exist without a medium. Thus in 1911 the professor of physics at Princeton University in his presidential address before the American Physical Society ventured the opinion that 'the abandonment of the hypothesis of an ether at the present time is a great and serious retrograde step in the development of speculative physics'.[20] He admitted that the principle of relativity, by turning the problem into an axiom, thereby accounts for the negative result of the Michelson-Morley experiment—although only at the cost, he emphasized, of creating an entirely new problem:

> But without an ether how do we account for the interference phenomena which made that experiment possible? There are only two ways yet thought of to account for the passage of light through space. Are the supporters of the theory of relativity going to return to the corpuscles of Newton? Are they willing to explain the colours of thin plates by invoking 'the fits of easy reflection and of easy transmission'?

So challenged Professor Magie.

In reporting this critical speech in *The Nation* the following year More, another physicist, similarly made many attacks on Einstein's theory (and the newly emerging quantum theory). Indeed his comments illustrate so well the crisis state in physics at the time that we quote his article at some length. First of all he described very approvingly the paradigm articulation of the preceding two centuries:

> While physicists feel that their subject has always shown a healthy growth, yet, as a rule, new discoveries have been made slowly enough to be fitted into theory without causing serious trouble. . . . Since the beginning of modern physics, from the days, that is, of Galileo and Newton, physicists have been building their laws and their theories on the same primary mechanical concepts of space, time, and mass. Through all this time, the first two have evoked little discussion, and differences of opinion about the concept of matter have been, for the most part, merely a question of precedence regarding mass, force, and energy. . . .[21]
>
> Now, in the past, as new phenomena were discovered, theories were advanced to explain them in terms of these primary mechanical concepts and if discrepancies remained between the theory and the phenomena, the theory was abandoned or allowed to lie dormant, but the concepts were not questioned. This may be called the classic attitude; but a new scientific method, which may be called the school of transcendental symbolism, has been lately evolved by German physicists. . . .[22]

More then described how Einstein, according to his new theory, 'draws the conclusion that we must radically alter our concepts of space and time, and abandon our concept of mass'. In highly critical vein, and referring also to the emerging quantum theory, he continued:

> Both Professor Einstein's theory of Relativity and Professor Planck's theory of Quanta are proclaimed somewhat noisily to be the greatest

revolutions in scientific method since the time of Newton. That they are revolutionary there can be no doubt, in so far as they substitute mathematical symbols as the basis of science and deny that any concrete experience underlies these symbols, thus replacing an objective by a subjective universe. The question remains whether this change is a step forward or backward, into light or into obscurity. It is held, and apparently rightly, that the revolution effected by Galileo and Newton was to replace the metaphysical methods of the schoolman by the experimental methods of the scientist. Now the new methods might seem to be just the reversal of that step, so that, if there is here any revolution in thought, it is in reality a return to the scholastic methods of the Middle Ages.[23]

It is ironic that More apparently did not understand the basis of Galileo's commitment to the Copernican system of the world (that 'the true and the beautiful are the same, and so are the false and the ugly') and did not realize how difficult it was for the Copernican revolutionaries to convince their conservative colleagues that the 'massive and unwieldly earth' really moves. The latter had been asked to believe the unbelievable because of the mathematical simplicity and harmony of the Copernican system. However, in all innocence More continued:

They [the principal advocates of the theory of relativity] may stagger us when they require us to believe that the length of a body becomes less if it is put in motion, and that clocks run slower when they move than when they are at rest; but on the other hand, they offer the most alluring seduction to the mind, when, by the simplest kind of mathematics, they appear to subdue the whole universe to their ideas. Professor Magie points out that the chief incentive to the development of the theory of relativity is the desire to express all natural phenomena by a set of simple equations; and he is right when he objects to making the demand for simplicity the chief purpose of a scientific theory. It is better to keep science in homely contact with our sensations at the expense of unity than to build a universe on a simplified scheme of abstract equations.[24]

The Copernican revolutionaries had been given exceedingly similar advice!

An outstanding puzzle — the Miller experiments

Finally I shall describe just one more example (given by Holton)[8] to illustrate Einstein's strength of commitment to his own research programme. As has been stressed, one foundation stone of Einstein's theory (his first postulate) was that no experiment could detect inertial motion through space. However, on Christmas Day 1925 Einstein received news that a repetition of the Michelson-Morley experiment had been performed by Miller which had yielded a small but positive result. Although it was only one-thirteenth of the fringe shift expected by Michelson, nevertheless a very definite fringe shift had been observed. It was an experimental 'fact'. The implication was that the earth is moving through the ether.

Holton relates that on the very same day that Einstein received news of this apparent experimental falsification of his theory he wrote to a friend: 'I think that the Miller experiments rest on an error in temperature. I have not taken them seriously for a minute.' Three months later Einstein wrote to a physicist: 'I believe that in the case of Miller, the whole spook is caused by temperature influences (air).'[25] Nevertheless, the puzzle could not be resolved. Clearly in view of its undoubted successes the research paradigm would not be abandoned merely because this particular puzzle struck a blow at the very heart of the theory. Physicists felt sure that something had been wrong with either the experiment or its interpretation by Miller but they were unable to identify the trouble. In 1950 Einstein told Shankland that both he and Lorentz considered Miller an excellent experimenter and thought his data must be good. Accordingly, Einstein told Shankland, Lorentz had studied Miller's work for many years (now, ironically enough, trying to explain away the *presence* of a fringe shift instead of the *absence* of one!) but had not been able to solve the puzzle. Shankland, too, was perplexed: 'I told him (Einstein) that it had always been a great puzzle to me why Miller's data seemed to yield this small positive result and that I had concluded that it might be due to his method of treating the data.'[26] After a most thorough analysis Shankland and his colleagues were able to report in 1955 that Miller's apparently positive results 'were in fact due to the greatly differing temperature conditions in the basement laboratory at Case and at Mt. Wilson'.[27] Although it was agreeable to have the puzzle resolved the physics community, to say the least, had not been seriously concerned over this particular puzzle confronting Einstein's theory.

The role of the Michelson-Morley experiment

As we have already mentioned, Einstein's basic motivation in rejecting the Newtonian space-time framework was based on aesthetic dissatisfaction with the asymmetry between Newton's laws and Maxwell's equations (the former satisfying the special principle of relativity and the latter not) rather than on specific disagreement between theory and experiment such as indicated by the Michelson-Morley experiment. Indeed, it has even been argued recently by Holton that 'the role of the Michelson experiment in the genesis of Einstein's theory appears to have been so small and indirect that one may speculate that it would have made no difference to Einstein's work if the experiment had never been made at all'.[28] The basic asymmetry already described was the sufficiently powerful motivating force driving Einstein on in the development of his relativity theory. Be this as it may, however, it is very important to clarify one very misunderstood feature of the revolution, namely that in no way did the null result of the Michelson-Morley experiment *demand* the rejection of the Newtonian space-time framework in favour of the Einsteinian.

Apart from the Lorentzian theory, there were other possibilities. Unfortunately this seems to have been insufficiently understood by many physicists and philosophers of science who consequently regard the Michelson-Morley experiment as a crucial one, unambiguously demanding rejection of the Newtonian space-time framework.

For example, Herbert Feigl, in an article entitled 'The Origin and Spirit of Logical Positivism', has made the following interesting comments:

> The ether theory, in its final stage of defence, appeared to us [of the Vienna Circle] to be a typical instance of a theory made proof against disproof. The hypotheses of Lorentz and Fitzgerald regarding contraction and 'local time' made the ether hypothesis immune to any conceivable test. . . . The problem that confronted the physicists at the time was not to be solved by further *ad hoc* hypotheses but, as Einstein's genius perceived it, by a critical revision of the concept of simultaneity and, in consequence of it, a revision of the concepts of distance and duration.[29]

With some qualifications so far so good. But then comes a most misleading statement: 'Einstein solved the problem by transforming the puzzle of the constancy of the speed of light into a postulate of the theory.' To be sure, it follows from Einstein's two postulates that the velocity of light is a constant for all inertial observers. But the null result of the Michelson-Morley experiment was open to several different interpretations. It was a puzzle because on the basis of the Newtonian space-time framework and the ether interpretation of Maxwell's equations a positive not a zero fringe shift should have been observed. Hence either (1) the Newton-Maxwell-ether theory had to be articulated (Lorentz's programme); or (2) the Newtonian space-time framework was invalid but Maxwell's equations were correct (Einstein's programme); or (3) Maxwell's equations were incorrect as a description of optical and eletromagnetic phenomena but the Newtonian space-time framework was still valid. Feigl, however, in not questioning the validity of Maxwell's equations, assumes the puzzle to be that of 'the constancy of the speed of light' and therefore that the choice lay only between the first two programmes. Similarly, in the same volume of essays, Hanson in his article 'The Interpretation of Scientific Theories' reinforces Feigl's misjudgement by explicitly writing that the negative result of the Michelson-Morley experiment 'conflicts with the Galilean transformation of velocities'.[30] Again this is true only if Maxwell's equations are unproblematically accepted as valid. The possibility that perhaps the Michelson-Morley experiment demonstrated the invalidity of Maxwell's equations rather than the invalidity of the Newtonian space-time framework appears to have received insufficient recognition from the physics community. And this is all the more surprising since a determined effort was made by several physicists, and in particular by Ritz, to provide an alternative and less radical explanation of the null result of the Michelson-

BREAKDOWN OF THE NEWTONIAN PARADIGM   77

Morley and similar experiments than that proposed by Einstein. We concentrate our attention on the programme proposed by Ritz.

4. The Ritzian Paradigm

Unconvinced by Einstein's research programme the Swiss physicist Ritz suggested in 1908 that Maxwell's equations were invalid and that the Newtonian space-time framework could be retained. In particular he suggested that light was not a wave motion but consisted of a stream of particles whose velocity is a constant relative to the source (and remains constant relative to the source after reflection at a mirror moving relative to that source). This, of course, is in direct conflict with Einstein's second postulate that the velocity of light is *independent* of the velocity of the source. However, experiments by Thomson (published in 1910), Stevant (1911) and Majorana (1918 and 1919) while disagreeing with other so-called emission theories were not in disagreement with predictions based on Ritz's theory. Clearly the Michelson-Morley experiment presented no problems for Ritz since all mirrors in this experiment remain at rest relative to each other. In fact, it was not until 1913 when an analysis was performed on the motions of eclipsing binary stars that evidence was thought to have been found against the Ritzian theory. For if the velocity of light is a constant relative to the source, then the velocity of light reaching the earth from the approaching star will be greater than that approaching the earth from the receding star and this should produce an observable effect as calculated using the laws of mechanics. Since the effect predicted by the Dutch physicist de Sitter was not observed it was concluded that Ritz's emission theory is incorrect. Indeed this is the only evidence against emission theories offered by Einstein in 1921 in his popular exposition *Relativity, the Special and General Theory*. In 1924, however, a second experiment involving an extra-terrestrial source of light (this time sunlight) was performed, the analysis of which also appeared inconsistent with Ritz's theory.

In view of the paucity of experimental evidence against the Ritzian theory, and in addition evidence very obviously requiring theoretical interpretation (general points so strongly stressed by Lakatos), it is surprising that Ritz's theory was not given more attention by physicists. Indeed the following are the views of one physicist who in 1938 undertook single-handed, but unsuccessfully, to try to rehabilitate the Ritzian research programme: 'And when, in spite of his acknowledged researches in spectroscopy and elasticity, the Swiss physicist, Walther Ritz, expressed heterodox views on electromagnetics in 1908, shortly before his death, his ideas were received with a chill silence and have ever since been systematically boycotted.'[31] Ritz himself felt that his research programme was not meeting with the attention it deserved, writing in a letter in 1908:

> I am now going to return to the optics of bodies in motion, to satisfy my conscience but without enthusiasm. I cannot indeed doubt that people

> will approach my ideas, whatever be the perfection I give them, only with extreme misgiving; a conversation with X after many other conversations has convinced me of this. Nobody can give me a valid objection, and I have silenced X himself. But that makes no difference—they find my ideas monstrous.[32]

But Ritz's death in 1909 prevented him from articulating his paradigm in the hope of gaining converts. Thus the de Sitter analysis was not considered a puzzle for the programme to solve, but as clear-cut falsification. It is ironical that years later the observational evidence was reinterpreted in a way which made the de Sitter analysis plus the new physical hypothesis in complete accord with the Ritzian theory. Indeed in a 1965 article reviewing evidence against emission theories the author declares that 'it is a curious historical fact that the recent experiments with moving sources and even the first data on time dilation [i.e. the first 'clear-cut' evidence against the Ritzian theory] were not obtained until long after special relativity had completely displaced the emission theory in physics'.[33]

On the corpuscular properties of light

It is even more surprising that physicists were not more sympathetic to the Ritzian emission theory when one considers that in the same year in which Einstein published his revolutionary theory, based on Maxwell's wave equations and a new space-time framework, he also published a paper in which he declared that new experimental results, *in contradiction to Maxwell's wave theory*, 'can be better understood on the assumption that the energy of light is distributed discontinuously in space, [and that light] consists of a finite number of energy quanta localized in space, which move without being divided'.[34] Thus even in 1905 Einstein had himself suggested that light in some cases seems to behave more like a stream of particles than like a wave motion. In 1909, the year of Ritz's death, he confirmed this very explicitly:

> It is undeniable that there is an extensive group of data concerning radiation which shows that light has certain fundamental properties that can be understood much more readily from the standpoint of the Newton emission theory than from the standpoint of the wave theory. It is my opinion, therefore, that the next phase of the development of theoretical physics will bring us a theory of light that can be interpreted as a kind of fusion of the wave and emission theories.[35]

There thus seems to be no reason based on experiment why Einstein should have rejected the Ritzian research programme in favour of a commitment to a new space-time framework in which Maxwell's equations would satisfy the special principle of relativity. For Einstein clearly believed Maxwell's equa-

BREAKDOWN OF THE NEWTONIAN PARADIGM

tions to have only limited validity. Indeed in 1955 he summed up the significance of his early work as follows:

> The new feature of it was the realisation of the fact that the bearing of the Lorentz transformation transcended its connection with Maxwell's equations and was concerned with the nature of space and time in general ... *This was for me of particular importance because I had already previously found that Maxwell's theory did not account for the micro-structure of radiation and could therefore have no general validity* [emphasis added].[36]

His commitment therefore to the new space-time framework was one of those many acts in science which Kuhn has so rightly described as based on faith. Eventually the power and the beauty of the theory developed by Einstein was to win many converts, physicists as devoted to the Einsteinian research programme as Einstein was himself, their goal being no less than to see the new space-time framework embrace all of physics, including whatever equations would eventually be found to describe the corpuscular properties of light. Since the Ritzian programme, however, was based on the old space-time framework, it clearly stood no chance of gaining the sympathetic attention of the revolutionary physicists. Yet, in rejecting Maxwell's equations and the ether, it also failed to gain the support of the conservatives. Like the Tychonic compromise solution it fell between two stools and its neglect, as in the case of the Tychonic theory, was considered to be amply justified by the triumphant progress of the revolutionary theory.

5. The Popper-Kuhn-Lakatos Debate

Although the development of the special theory of relativity has been presented in such a way as to throw light on the debate between Popper, Kuhn and Lakatos, nevertheless it may still be useful to present some of the implications a little more explicitly.

Clearly the sketch we have given does not support the Popperian injunction to be ruthless in rejecting theories. Indeed it is difficult to see how the classical paradigm could have been articulated into that crisis state out of which the relativity theory was eventually to emerge if physicists in the 18th and 19th centuries had abandoned Newtonian physics when confronted with puzzles. Furthermore, it is difficult to see how Einstein's theory could have emerged successfully if Einstein had not been content either to ignore or to leave for later consideration experimental evidence which apparently falsified his theory. The sketch therefore supports the broad outlines of Kuhn's and Lakatos' interpretation of the nature of scientific progress. For it was the triumphs of classical physics and the triumphs of Einstein's theory which justified commitment to these theories in the face of constant difficulties. Neither Kuhn nor Lakatos would condemn Lorentz for being 'non-scientific' in attempting to resolve the puzzle of the Michelson-Morley experiment within the framework of the ether theory. Indeed Lorentz, Poincaré and others succeeded in pushing the ether theory to that stage where Einstein could perform a 'gestalt-switch' in making a puzzling set of transformation equations the physical space-time framework of a new theory.

During the revolutionary period it was certainly not possible for proponents of the competing paradigms to argue on 'objective' grounds that their respective paradigms were superior to their rivals. The historical sketch clearly illustrates that commitment in the revolutionary stage is based to a large extent on 'physical intuition' as to the nature of reality, with aesthetic criteria playing a guiding if not leading role. Were the competing paradigms, however, 'incommensurable'? The answer must be in the negative for the proponents of all three paradigms (the Lorentzian, Einsteinian and Ritzian) *ultimately* acknowledged the importance of the Michelson-Morley experiment, agreeing on the interpretation of the experiment within the framework of the old ether theory, although claiming to have solved the problem differently within the frameworks of their respective theories. Because, however, the Ritzian programme failed to gain converts, the puzzles confronting it, easily resolvable by use of the 'extinction theorem', were interpreted as falsifying experiments. As Lakatos has so rightly stressed, it is not the case that theory is compared against experiment but rather theory against experiment itself needing theoretical interpretation. Furthermore, it is only recently that 'clear-cut' experimental evidence has refuted the Ritzian theory in favour of the Einsteinian. Why then did the Einsteinian triumph over its two main rivals? For two reasons: (1) because of its great mathematical appeal; and (2) because it was an immensely powerful theory embracing and unifying nearly all of physics and successfully applicable where its rivals were not. Although the 'classical' physicists could not accept a theory that was so physically absurd (wave motion without a medium) nevertheless in the end the power, the beauty and the triumphs of Einstein's theory left it the undisputed victor over all rival theories. Thus in not agreeing with the 1962 Kuhn that paradigms across the revolutionary divide are incommensurable, we agree in part with Lakatos' viewpoint: for the paradigms did have problems in common and the Einsteinian research programme eventually emerged as the most successful in solving those problems, in making new predictions experimentally confirmed, and in generating new research programmes.

However, Lakatos, as we pointed out earlier, is surely unjustified in claiming to have established objective criteria by which commitment to a research programme can be declared rational or irrational. According to Lakatos a scientist is acting irrationally if, when his own programme is faced with a successful rival, he remains committed to his own programme even after sustained theoretical and experimental effort has failed to solve the problems already successfully solved or successfully by-passed by the rival programme. While this definition of rational action is at first glance attractive the weakness, of course, lies in the vagueness of the word 'sustained'. For how is one

BREAKDOWN OF THE NEWTONIAN PARADIGM

to know that one more effort will not resolve the most serious of the problems confronting the programme? According to the 1962 Kuhn the most that one can say about a man who hangs on after all his colleagues have been converted is that *ipso facto* he has ceased to be a scientist, not that he is being illogical or unscientific. It is ironical that Einstein himself might justifiably be considered an outstanding example of such a man.

For the fate of Einstein with respect to the quantum theory is a most striking example of the predicament of a scientist who is unable to accept a once revolutionary paradigm which has become the new orthodoxy. Although it will not be possible to give any details of the theoretical problems involved, nevertheless, since the attitude of Einstein throws considerable light on the difference of view between Lakatos and Kuhn, we briefly discuss this most interesting of paradigm debates between Einstein and his critics.

Quantum Mechanics and Einstein's apostasy

In the 19th century T. H. Huxley made explicit an important feature of the scientific world view being developed. According to this spokesman for science:

> The fundamental axiom of scientific thought is that there is not, never has been and never will be, any disorder in nature. The admission of the occurrence of any event which was not the logical consequence of the immediate antecedent events, according to these definite, ascertained or unascertained rules which we call the 'laws of nature', would be an act of self-destruction on the part of science.[37]

The universe was considered to be completely deterministic in nature. Indeed earlier in the century Laplace had declared that given the forces acting between all particles, and given the velocity and position of each particle at any instant of time, then it would in principle be possible for a super-intelligence to predict the entire future of the universe. More humbly, given the forces acting on any one particle and its position and velocity at any one instant, then it would be possible to predict at any future time both the position and velocity of the particle. Whatever, then, the formulation given for classical physics, one thing was considered to be clear: the behaviour of inanimate matter is *in principle* completely predictable. Since a particle has at any instant of time both a definite position and velocity its trajectory through space and time is *in principle* completely predictable.

The revolutionary change

It is exactly this most fundamental of propositions which the proponents of quantum mechanics assert to be false. In principle, they claim, nature is indeterministic; identical systems can and do behave differently. The exact behaviour of individual atomic systems is not in principle predictable but is in principle *unpredictable*.

Einstein, who had done so much to develop the quantum theory, refused to accept such conclusions as final. Until his death in 1955 he remained an opponent of the research programme based on quantum theory and the interpretation given it by Bohr, Heisenberg and Born in the late 1920s. His aim was to show that the statistical feature of quantum theory does not represent an inherent indeterminism in nature but rather a wrong approach to the problems of atomic phenomena.

In 1926 he wrote to his friend Max Born (who was the originator of the interpretation with which he so strongly disagreed): 'Quantum mechanics is most awe-inspiring. But an inner voice tells me that this is not the real thing after all. The theory gives much, but it scarcely brings us nearer to the secret of the Old Man. In any case I am convinced that He doesn't play dice.'[38] And in a letter in 1928 to Erwin Schrödinger, originator of the famous equation in quantum mechanics which bears his name (but who also, like Einstein, was deeply sceptical of the interpretation being given his equation), Einstein wrote somewhat bitterly: 'The Heisenberg-Bohr tranquillizing philosophy—or religion?—is so delicately contrived that, for the time being, it provides a gentle pillow for the true believer from which he cannot very easily be aroused. So let him lie there.'[39] In 1944 Einstein wrote again to Born that 'the great initial success of the quantum theory cannot convert me to believe in that fundamental game of dice'.[40]

Was Einstein being 'unreasonable' but perfectly scientific or was he being irrational, unscientific? The questions Einstein was asking were now regarded by the physics community as unscientific. When will a particular radioactive atom decay? A non-scientific question, the quantum theorists answer. In principle it is possible to predict only the behaviour of a very large number of such atoms. In 1936 Einstein wrote that 'to believe this is logically possible without contradiction, but it is so very contrary to my scientific instinct that I cannot forego the search for a more complete conception'.[41] Commenting on this statement of faith Einstein's principal adversary Niels Bohr declared that 'even if such an attitude might seem well balanced in itself, it nevertheless implies a rejection of the whole argumentation . . . aiming to show that in quantum mechanics we are not dealing with an arbitrary renunciation of a more detailed analysis of atomic phenomena, but with a recognition that such an analysis is *in principle* excluded' (Bohr's emphasis).[42] Indeed a most interesting statement. Bohr had not only used his own paradigm to argue in that paradigm's defence but believed that he had been able to show that the paradigm advocated by Einstein was *in principle* impossible. Einstein was certainly unimpressed. In his reply he stated again that he was 'firmly convinced that the essentially statistical character of contemporary quantum theory is solely to be ascribed to the fact that this [theory] operates with an incomplete description of physical reality'.[43] And what was the basis of Einstein's conviction? Two years earlier he had written to Born: 'I cannot provide logical arguments for my conviction, but can only call on my little

BREAKDOWN OF THE NEWTONIAN PARADIGM

finger as a witness, which cannot claim any authority to be respected outside my own skin.'[44]

And so Einstein had become a heretic. His point of view was not regarded as scientific. In 1939 he confided to Schrödinger (a fellow heretic of different persuasion): 'I write to you not in the illusion that I will convince you but solely with the intention that you may understand my viewpoint which has led me into deep loneliness.'[45] And in a 1955 article, Schrödinger as if in public answer to Einstein's plea for understanding commented thus: 'Scientists are inclined to take their own outlook for the natural way of looking at things while the outlook of others, inasmuch as they differ from theirs, are adulterated by preconceived and unwarranted philosophical tenets, which unprejudiced science must avoid'.[46] The statement was strongly felt since only in the previous year he had seen Born reply to one of his several attacks on the new orthodoxy by quoting the following remarks written to him (Born) by one of the outstanding guardians of the new paradigm, Wolfgang Pauli:

> Against all retrograde efforts (Schrödinger, Bohm, etc., and in a certain sense, also Einstein) I am certain that the statistical character ... of the laws of nature—which you have, right from the beginning, strongly stressed in opposition to Schrödinger—will determine the style of the laws for at least some centuries. It is possible that later, for example in connection with the processes of life, something entirely new may be found, but to dream of a way back, back to the classical style of Newton-Maxwell (and it is nothing but dreams which these gentlemen indulge in), that seems to me hopeless, off the way, bad taste. And we could add 'it is not even a lovely dream'.[47]

To what extent then was Einstein, not to mention fellow heretics like Schrödinger, Bohm and others, acting irrationally, unscientifically? Einstein did not deny that quantum mechanics was an immensely powerful and successful theory. On the contrary, he acknowledged the many and impressive achievements of the quantum theorists. But what he wanted to do was to develop a 'complete physical description' to which quantum mechanics would be a limiting case. 'I am firmly convinced,' he wrote in 1949 after twenty years of fruitless effort, 'that the development of theoretical physics will be of this type; but the path will be lengthy and difficult.'[48] Of course, as Einstein recognized, his colleagues were equally convinced that such a research programme would be fruitless, that what Einstein wished to do could not be done. Despite his overall failure, Einstein's constant criticism of the foundations of quantum mechanics did, however, lead to clarification of the theory's fundamental principles and did keep alive the seed of dissent within the physics community. Indeed within the past decade it has become almost fashionable to query the basic principles of quantum mechanics. Just as Einstein felt so dissatisfied with the asymmetry between the Newtonian and

Maxwellian theories, so the new rebels feel dissatisfied with a theory which once again, they believe, destroys the unity of physics.

On the necessity of tolerance

It is surely never possible to be absolutely certain that one's own research programme is the most fruitful possible. For this reason it would seem only prudent for scientists to encourage 'responsible' dissenters, men who for aesthetic reasons cannot accept the fundamental principles of the community's paradigm *but who recognize its achievements* and wish to see them preserved in any future paradigm. Indeed Feynman in his Nobel Prize address of 1966 has appealed for a diversity of viewpoints with respect to any one theory. For 'theories of the known', he declared, 'which are described by different physical ideas may be equivalent in all their predictions, and hence are scientifically indistinguishable. However,' he continued, 'they are not psychologically identical when trying to move from that base into the unknown. For different views suggest different kinds of modifications which might be made, and hence are not equivalent in the hypotheses one generates from them in one's attempt to understand what is not yet understood.' The danger of total conformity is obvious and Feynman asks the relevant question: 'If, on the off-hand chance, [the truth] is in another direction—a direction obvious from an unfashionable point of view . . . who will find it?'[49]

Einstein, the heretic, was seeking 'the truth' but from an unfashionable viewpoint. That he was not acting irrationally can be argued from the fact that not only did he acknowledge the achievements of the paradigm to which he was opposed but he also understood and clarified its fundamental principles. The heretic who demonstrates understanding of the current orthodoxy and acknowledges its achievements but who chooses to pursue a different research programme cannot reasonably be accused of acting irrationally. At most he can be accused of knowledgeable but dogmatic commitment to a programme yet to prove its worth. However such knowledgeable dogmatism can surely do little harm in the world of physics. For the lives and well-being of physicists are not threatened by such stubborn dissenters who will at worst experience somewhat lonely, 'unsuccessful' lives although always with the hope of success in the research programme and triumphant recognition.

It is necessary to stress that without personal commitment based on 'physical intuition' and aesthetic criteria, the nature of which cannot be objectively justified, physics would come to a halt. If harmony lies in the eye of the beholder this is not to be regretted—at least not in physics. For the existence of different views on what constitutes harmony ensures that many alternative paths will be explored, each path hopefully leading to harmony. Einstein and Infeld express this belief in harmony so well:

> Without the belief that it is possible to grasp the reality with our theoretical constructions, without the belief in the inner harmony of our

BREAKDOWN OF THE NEWTONIAN PARADIGM

world, there could be no science. This belief is and always will remain the fundamental motive for all scientific creation. Throughout all our efforts, in every dramatic struggle between old and new views, we recognize the eternal longing for understanding, the ever-firm belief in the harmony of our world, continually strengthened by the increasing obstacles to comprehension.[50]

But harmony, as we have said, lies in the eye of the beholder. When in 1926 Einstein challenged Heisenberg (as Heisenberg reconstructs the conversation) with the question, 'How can you really have so much faith in your theory when so many crucial problems remain unsolved?', the reply was the following:

> If nature leads us to mathematical forms of great simplicity and beauty ... that no one has previously encountered, we cannot help thinking that they are 'true', that they reveal a genuine feature of nature. ... You may object that by speaking of simplicity and beauty I am introducing aesthetic criteria of truth, and I frankly admit that I am strongly attracted by the simplicity and beauty of the mathematical schemes with which nature presents us.[51]

Einstein, however, saw simplicity and beauty elsewhere!

Just as Einstein committed himself to a research programme opposed to that of the quantum theorists, so Dirac committed himself not only to the development of quantum mechanics but also to the attempt to ensure a successful marriage between Einstein's special theory of relativity and the new quantum mechanics. His successes are well known. But what in particular interests us here is that yet again we find a commitment to the belief that nature, if looked at in the right way, is mathematically simple and beautiful. 'It is more important,' declared Dirac, 'to have beauty in one's equations than to have them fit experiment.' Again physical intuition and aesthetic criteria are declared essential guides:

> It seems that if one is working from the point of view of getting beauty in one's equations, and one really has a sound insight, one is on a sure line of progress. If there is not complete agreement between the results of one's work and experiment, one should not allow oneself to be too discouraged, because the discrepancy may well be due to minor features that are not properly taken into account and that will get cleared up with further developments of the theory.[52]

Dirac is adamant that mathematical beauty is the only sure guide:

> With all the violent changes to which physical theory is subjected in modern times there is just one rock which weathers every storm, to which one can always hold fast—the assumption that the fundamental

laws of nature correspond to a beautiful mathematical theory based on simple mathematical concepts that fit together in an elegant way, so that one has pleasure in working with it.[53]

Physicists, to be sure, seek to understand nature and of physics it may well be said that 'the way to Truth lies through the realm of the Beautiful'. And beautiful mathematical theories give great pleasure to those who use them. If, indeed, the aim of human beings is to build a beautiful world, one giving pleasure and creative joy to its inhabitants, then science must surely be an integral component of such a quest. For Schrödinger, 'the chief and lofty aim of science today, as in every other age, [lies] in the fact that it enhances the general joy of living'.[54] In a civilized world physics would certainly be a delightful play, bringing at peak moments beauty, pleasure and joy into the lives of those who choose this way to dialogue with nature. But this is not a civilized world and the beautiful play called physics has given birth to monstrosities.

Chapter 4

POLITICAL COMMITMENTS IN THE AGES OF THE SCIENTIFIC AND INDUSTRIAL REVOLUTIONS

> I am persuaded that till property is taken away there can be no equitable or just distribution of things, nor can the world be happily governed . . . I confess without taking it quite away, those pressures that lie on a great part of mankind may be made lighter; but they can never be quite removed. For if laws were made to determine at how great an extent in soil, and at how much money every man must stop . . . these laws, I say, might have such effects, as good diet and care might have on a sick man, whose recovery is desperate: they might allay and mitigate the disease, but it could never be quite healed, nor the body politic be brought again to a good habit, as long as property remains; and it will fall out as in a complication of diseases that by applying a remedy to one sore, you will provoke another; and that which removes the one ill symptom produces others, while the strengthening one part of the body weakens the rest.
>
> THOMAS MORE *Utopia*

IF concern for the beautiful has been an important factor in the development of physics, it has been no less an important factor in the development of mankind. For men have always dreamed of journeying to a beautiful world, a world in which they might rest safe from all peril, living in joyful brotherhood with their fellow-men. Perhaps in the distant past (the dream went) such a Golden Age had once existed on the earth.* But the harsh reality of modern conditions meant that such a good life was no longer attainable in this world. In the *real* world life was characterized by pain and disease, natural disasters, scarcity of the basic necessities of life, and competition between men for power, for command over the available wealth, for the right of entry into a *future* heaven. If utopia was a good place, it could be no earthly place.

But the 17th century was to see the completion of the scientific revolution, an event that in the opinion of the historian Herbert Butterfield 'outshines everything since the rise of Christianity'. And with good reason. For with the scientific revolution there came the realization that mankind could make

---

* This is Ovid's description of a once earthly paradise, for ever in the past: 'In the beginning was the Golden Age, when men of their own accord, without threat of punishment, without laws, maintained good faith and did what was right. . . . The peoples of the world, untroubled by any fears, enjoyed a leisurely and peaceful existence, and had no use for soldiers. The earth itself, without compulsion, untouched by the hoe, unfurrowed by any share, produced all things spontaneously, and men were content with food that grew without cultivation. . . . It was a season of everlasting spring . . .'[1n]

progress, that the good life did not have to be of necessity a life after death but could be achieved on this earth. The utopian writers of the early 17th century were convinced that wise use of the new knowledge would lead mankind to the realization of societies free from disease, violence and brute labour, where human fulfilment would consist of peaceful and brotherly cultivation of intellectual, sensual and spiritual delights. Science would give mankind an understanding of nature and thereby provide the technical means for realizing the good life on earth.

Motivated therefore by their desire to find beauty in nature, scientists would provide mankind with the techniques necessary for the construction of a beautiful world. The beautiful path of science would lead to a world that was itself beautiful.

Our quest in this book is to try to find out why the path taken by Western society has led to a world that in far too many aspects is so ugly and brutal. To do so we look not only at the path of science (in chapter 10) but at those other paths along which Western man has travelled—in particular, the path of capitalist society. For the Hermetic utopians had declared that a just and humane society could not be realized until private property had been abolished and all goods held in common. The social consequences of the development of English society during the 18th and 19th centuries—the society that had witnessed the apotheosis of the Copernican revolution and that was to put science to such great industrial use—appeared to vindicate the Hermeticists' beliefs. But men were not sure. Some claimed that the initially stony path of capitalist society would become progressively less stony. Others that the path would, on the contrary, become progressively stonier but would then abruptly turn and lead to another path of altogether different texture. There was 'paradigm' commitment and 'paradigm' debate. In this chapter we briefly follow aspects of this debate; and in so doing we set the stage for a discussion and analysis of the important paradigm debate generated by the universal crisis of the 1930s—that social abyss into which the capitalist path was to lead its hitherto prosperous followers. This will then set the stage for identification and examination not only of the path actually taken by Western societies as they eventually emerged on the other side of this abyss but also of the alternative paths they might have taken and might still have the opportunity to take should the present path not end in irremediable disaster.

1. Utopian Visions in the Age of the Scientific Revolution

In the Renaissance a new sensibility had emerged. Man had become proud of his body and mind. The nude human body no longer represented sin but goodness and beauty. Self-flagellation and mortification of the flesh were renounced; the body was to bring pleasure and joy to life. To the neo-Platonists, Humanists and Hermeticists, the sun symbolized God himself and for good reason. 'Like a caress,' declared Ficino, translator of the sacred

Hermetic texts,* the light from the sun 'penetrates all things harmlessly and most gently' just as 'the heat which accompanies it fosters and nourishes all things'.² Ficino told his contemporaries how 'similarly the Good is itself spread everywhere, and it soothes and entices all things. It does not work by compulsion, but through the love which accompanies it.' (The use of force was abhorrent to the Hermetic philosophers.) 'Just look at the skies, citizens of heavenly fatherland,' he appealed; 'the sun can signify God himself to you and who shall dare to say the sun is false.'

Ficino's attitude is typical of the Hermetic approach to nature and society. Although, as we shall discuss in more detail in chapters 10 and 12, the Hermeticists espoused a magical-animistic view of nature, they nevertheless believed that knowledge of the secrets of nature could be gained through careful experimental study and that these secrets could and would be used for the good of man and in the service of Love. For (of great importance) the Hermeticists believed that man was not a passive victim of forces beyond his control but a creature who had the power to understand these forces and thereby to achieve well-being for himself and his fellow men.

To the new Hermetic sensibility the central position of the sun was, in general, totally acceptable and indeed Copernicus' references to the sun have much in common with Ficino's. Even, however, before the publication of *de Revolutionibus*, the belief had been rejected that man occupied the central position in the Universe because of his essential degeneracy and sinfulness. In Pico della Mirandola's *Oration on the Dignity of Man*, God tells Adam that 'I have placed you in the midst of the world so that from there you might better see what is in the world. I have made you neither heavenly nor earthly, neither mortal nor immortal in order that, like a free and sovereign artifice, you may mould and sculpt yourself into that form you will have chosen for yourself.'³ A new spirit of enquiry was all-pervasive, a new attitude to the world that was to result in both the geographical discoveries of the 15th and 16th centuries and the scientific revolution of the following century.

Despite the new spirit, however, the harshness of life remained inescapable. Poverty and brute labour remained the lot of the majority while disease struck at rich and poor alike. Man was still at the mercy of the elements. Nevertheless, the new geographical discoveries suggested that the feudal social order was in no way sacrosanct and that qualitatively different societies in which wealth and duties were equitably shared were at least conceivable in the imagination, even if they could never be achieved in reality. More's *Utopia*, published in 1516, describes such a society.⁴ The Portuguese explorer, Raphael Hythloday, who discovers the island of Utopia, argues that in England

---

* Texts written in the 2nd and 3rd centuries A.D. but believed in the 16th century to have been written during or before the time of Moses. These Hermetic texts, supposedly based on the writings of the 'thrice-blessed' Egyptian 'Magus' Hermes Trismegistus, were believed to consist of a Divine revelation of the secrets of the natural world.

the majority of people live in poverty in order that a few can live in great wealth. It is his opinion, after living in Utopia for some time, that all other governments are merely 'a conspiracy of the rich who on pretence of managing the public only pursue their private ends. Where possessions are private, where money is the measure of all things', says Hythloday, 'it is hard and almost impossible that the commonwealth should have just government and enjoy prosperity.' If, therefore, man was to a large extent at the mercy of the *natural* environment, More nevertheless believed that at least much *social* injustice could be eliminated by the abolition of private property. Thus, as exemplified by the quotation given at the beginning of this chapter, More argued (let us ascribe the views of Hythloday to More) that while the ills of society could perhaps be *partly* remedied by laws restricting the amount of property that an individual could own, nevertheless the basic ills would remain and could not be removed without radically reorganizing society. In More's highly regimented society, all would work but for only six hours a day. Nevertheless, this would suffice to produce such an abundance of goods that quite probably, More thought, the citizens of Utopia would need to work even fewer hours per day than the allotted six. For, declared More, 'the magistrates never engage the people in unnecessary labour since the chief end of the constitution is to regulate labour by the necessities of the public, and to allow all the people as much time as is necessary for the improvement of their minds, in which they think the happiness of life consists.' To this end public lectures are given every morning for the enjoyment and enlightenment of the citizens of Utopia. After supper, when the six hours of work are over, the inhabitants of Utopia relax in their summer gardens, More tells us, while in the winter they entertain each other in the halls where they eat 'with music and discourse'. Thus are men made rich, declared More, for although no citizen of Utopia possesses any material wealth a man is indeed rich who leads 'a serene and cheerful life, free from anxieties, neither apprehending want himself, nor vexed with the endless complaints of his wife'! And 'so easy a thing would it be to supply all the necessities of life', argued More, 'if that blessed thing called money, which is pretended to be invented for procuring them, was not really the only thing that obstructed their being procured.' More subscribed strongly to the Renaissance humanistic position. The citizens of Utopia sought pleasure in life, 'as the end of all they do', but not the false pleasures of an avaricious society. For those 'who think themselves really the better for having fine clothes . . . are doubly mistaken', More tells us, 'both in the opinion that they have of their clothes and in that they have of themselves'. Of the true pleasures, More argued, some belong to the body and others to the mind. 'The pleasures of the mind lie in knowledge and in that delight which the contemplation of truth carries with it' while the pleasures of the body lie in giving our senses delight as 'that which arises from satisfying the appetite which Nature has wisely given to lead us to the propagation of the species . . . the pleasure that arises from music,' and that bodily pleasure 'which results from an undisputed and vigorous constitution of body, when life and active spirits seem to actuate every part.' The new spirit is here seen at its clearest. Medieval pessimism, Christian asceticism and mortification of the flesh are renounced. The body and the mind are both to be used for the attainment of pleasure. While Copernicus spoke of the 'unbelievable pleasure of the mind' that science gives—and not for him the ugly Ptolemaic system of the world—Francis Bacon's empiricism was clearly an expression of confidence in the body's sense organs. It can justifiably be argued that such prior confidence in mind and body was a necessary precursor to the scientific revolution of the 17th century.

And with the scientific revolution came a new hope for mankind. The emancipation of human imagination from the Aristotelian-Thomistic cosmology resulted not only in a scientific revolution but also in the awareness that the technical means were available or would soon become available for the construction of a society in which all citizens not only would be freed from pain, disease and back-breaking labour but would also be sufficiently educated to be able to enjoy the pleasures of body and mind described by More. The Hermetic utopians of the early 17th century believed that society could be and would be radically transformed so that men could live more human lives. To two of them, Andreae and Campanella, it appeared obvious that two prior conditions had to be met before a just and happy human society could exist, the first the abolition of private property and the second the education of all citizens. Scientific techniques (natural magic) would then enable the members of such a society to grow ample food for all, cure and prevent disease and otherwise enable every man to use and enjoy to the full the body and mind that a loving God had granted him. The utopian writers appealed to the hearts and minds of men; they did not, however, show how the (highly regimented) societies they desired could be constructed out of the rapidly changing feudal order in which they found themselves.

Kepler's friend Andreae, who had studied at Tübingen under Kepler's teacher and who had been strongly influenced by the Hermetic texts, believed as did Campanella and Bacon that knowledge of the secrets of Nature would make possible the construction of a just and Christian society. While Kepler wrote of a future journey to the moon, Andreae wrote of the future society, Christianopolis, that Kepler's astronauts would leave behind them, a society based on equality, the desire for peace, and the contempt for riches—'as the world is tortured primarily with the opposite of these', declared Andreae. Andreae's ideal society avidly sought after scientific knowledge to such an extent that its principal town, City of Peace, is described by Andreae as 'one single workshop ... of all sorts of crafts'. Manual and intellectual work are, however, not separated for the inhabitants believe that 'neither the subtleness of letters is such, nor yet the difficulty of work, that one man, if given enough, cannot master both'. The citizens also enjoy their 'very few working hours' each day, especially since 'they have been trained long before in an accurate knowledge of scientific matters, and find their delight in the inner parts of nature'.[5] It is science, believes Andreae, that will make a community possible in which God's love for men will become manifest.

Campanella, a man of immense will-power, who had spent twenty-seven years in prison because of his part in an attempt to realize an earthly utopia, eight of them in a dungeon, often tortured, nevertheless did not lose faith in the possibility of realizing a very different society from the one in which he suffered so much. As Andreae admired Kepler, so Campanella, although a Hermeticist, admired Galileo. While in prison in 1622 Campanella wrote a defence of Galileo, and when released in 1629 offered to defend Galileo before the Inquisition, a courageous gesture which necessitated flight to France. We find therefore that the chief ruler, Hoh, of Campanella's ideal society, is a man well versed in scientific techniques, a surprising departure from tradition which the inhabitants of Campanella's City of the Sun attempt to justify by comparing the merits of one 'who knows most of grammar, or logic, or of Aristotle or of any other author' against one who knows the ways in which God rules the universe, 'the laws and customs of Nature and the nations'.[6] In the former case 'much servile labour and memory work is required, so that a man is rendered unskilful'. Indeed, it is obvious that a man who 'has contemplated nothing but the words of books and has given his mind with useless result to the consideration of the dead signs of things' cannot be 'equal to our Hoh'. It is because of Hoh's knowledge of all the sciences, gained by learning and practical experience, that he is 'the most capable to rule ... never cruel, nor wicked, nor a tyrant, inasmuch as he possesses so much wisdom'. Campanella paints an idyllic picture of this Hermetic society. Because 'duty and work is distributed among all ... it only falls to each one to work for about four hours every day. The remaining hours are spent in learning joyously, in debating, in reading, in reciting, in writing, in walking, in exercising the mind and body, and with play.' Sanitation and medicine are highly developed, and as a result people live to the age of 100, some to 200. In the City of the Sun people are not rich because they own property, poor because they own none. On the contrary, there is no private property and therefore, Campanella argues, all people are both rich and poor. 'They are rich because they want nothing, poor because they possess nothing; and consequently they are not slaves to circumstances, but circumstances serve them.' The aim of the society was to promote joy and contentment. To this end, the inhabitants of Campanella's 'scientific' utopia 'repeat but one prayer, which asks for health of body and mind, and happiness for themselves and all people, and they conclude it with the petition "As it seems best to God".' For the God of the Hermetic utopians was one who no longer demanded mortification of the flesh but one who had put the sun in the centre of the universe so that its light could nourish all the inhabitants of the newly discovered family of planets.

Likewise, Comenius, the Czech educationist who had been strongly influenced by Andreae, stressed the desirability of peace and advocated the unity of mankind. He was opposed to the incipient European imperialism, arguing that 'these voyages of Europeans to foreign lands have brought evil to Europeans no less than to those peoples from which we obtain worldly goods'. The aim should not be exploitation but unity. 'We are all fellow citizens of one world, all of one blood, all of us human beings. Who shall prevent us from uniting in one republic? Before our eyes there is only one aim: the good of humanity, and we will put aside all considerations of self, of nationality, of sectarianism.'[7]

For Francis Bacon, propagandist for the scientific revolution in England, the purpose of science was none other than the relief of man's estate. Although Bacon did not see fit to advocate the abolition of private property, nevertheless no other result but the increased welfare of mankind could be expected from the progress of science. To be sure, Bacon occasionally tempered his optimism with words of warning. He recognized that the technological arts 'have an ambiguous or double use, and serve as well to promote as to prevent mischief and destruction, so that their virtue almost destroys or unwinds itself'. But it would not happen. 'Let none,' said Bacon, 'be alarmed at the objection of the arts and science becoming depraved to malevolent or luxurious purposes and the like.' People could trust in the scientists, *above all* in the scientists. For were they not dedicated to the betterment of human life, to the relief of man's estate? Not all scientific discoveries could therefore be made public or revealed to the State; the knowledge of those that could bring harm to mankind would be suppressed. The scientists of New Atlantis 'have consultations, which of the inventions and experiences which we have discovered shall be published, and which not; and take all an oath of secrecy, for the concealing of those which we think fit to keep secret'.[8] Privileged rulers or not, the scientists could be entrusted with the responsibility of knowledge and power: for the scientist was, above all else, a man devoted to the public good.

Such was *in part* the spirit of the scientific revolution. The 'new science' was to be pursued, wrote Descartes, 'principally because it brings about the preservation of health, which is without doubt the chief blessing and the foundation of all other blessings in this life'. And the blessings of life were no longer confined to the inhabitants of the planet earth. Huygens, in his book *The Celestial Worlds Discovered; or Conjectures Concerning the Inhabitants, Plants and Productions of the Worlds in the Planets*, believed that the inhabitants of other worlds would enjoy 'these blessings', namely 'the pleasures of eating and copulation', the 'smell in flowers and perfumes', 'the sight in the contemplation of beauteous shapes and colours', 'the hearing in the sweetness and harmony of sounds'. The purpose of life was to enjoy it. 'What an admirable providence it is,' declared Huygens, 'that there's such a thing as pleasure in the world.' Similarly in England the scientists subscribed to such an ethic.

Concerning Gresham College, until 1710 the home of the Royal Society, Glanvill wrote:

> The noble learned Corporation
> Not for itself in thus combined
> But for the public good of the Nation
> And general benefit of Mankind.

Sprat, for example, who in 1667 published an 'official' history of the Royal Society, defended the new scientists against 'the men . . . of severe devotion [who] tell us that we cannot have sufficient leisure to reflect on another life, while we are so taken up about the Curiosities of this'. On the contrary, Sprat argued, far from requiring men to 'cast away all the thoughts and desires of humanity . . . the Law of Reason intends the happiness and security of mankind in this life; and the Christian Religion pursues the same ends, both in this and a future life . . .'.[9]

Let us note in passing that the aim of the scientists to obtain mastery over nature did not meet with the approval of all. Indeed, it met with mocking ridicule from Jonathan Swift who argued that the scientists would not succeed in enriching society but, on the contrary, would succeed only in impoverishing it. In *Gulliver's Travels*, published 1726, Gulliver finds Lagado to be such a society. Forty years earlier (Gulliver learns) the Lagados had received a Royal Patent from Laputa to erect an Academy of Projectors so that the prosperity of the community could be increased. Unfortunately, however, because no project had yet been successfully accomplished the entire community was in a state of collapse: the houses were in ruins, and the people without food and clothes. The solution the scientists proposed was to redouble efforts to bring the projects to triumphant conclusion. The first scientist Gulliver met in the Academy had been working for eight years on an as yet unsuccessful project to extract sunbeams from cucumbers and store them for use on overcast summer days. Gulliver generously gave money to the fundless scientist for the purchase of a fresh supply of cucumbers. Other scientists were at work on such projects as turning ice into gunpowder, building houses from the roof downwards and softening marbles for pillows. In the School of Languages a project was under way to improve language through the elimination of all verbs and particles while more far-sighted scientists had plans to abolish language altogether.

2. The Emergence of Capitalist Society

If, however, the scientists of Lagado had brought their community to ruin, the scientists of England were achieving great successes in their aim of understanding and dominating nature. While no doubt there had been many failures, the overwhelming feeling was one of confidence in the 'new science'. The control of nature appeared to lie within the grasp of the scientists and of the society in which they worked. The dreams of More, Andreae, Campanella and Bacon could, it seemed, be realized, indeed improved upon. A society in which all citizens would be educated and in which all would have sufficient time to enjoy the many pleasures of body and mind was now, it seemed, thanks to the new science, a realizable possibility.

Andreae, Campanella and Bacon were utopians; each desired the realization of what he believed to be a just and humane society. However, although the three utopians each believed their respective utopias to be technically possible, they did not show how they could be achieved and perhaps even less did they analyse the direction in which their own and other societies were moving. Had it been possible to undertake such analyses, then it would have been evident that their prerequisites for the building of a just and humane society were indeed unrealizable ideals, at least in that historical period. More, Andreae and Campanella had wished to proscribe private property and profit but the guild system of the Middle Ages which had strictly limited competition and profits, which had forbidden advertising and had attempted to maintain high standards of production, was already fast declining. In its place a market society was beginning to develop in which worldly success would become socially acceptable and in which the making of money would be the overriding aim of the growing number of merchants, traders and producers.

In England the first enclosures, the Civil War, the slave trade and technical change all contributed towards the development of this new form of society. Towards the end of the 17th century the banking system was introduced; indeed 'the Bank of England by the 1750s had become a national institution as banker to the government and most of its departments'. Manufacturing production began to increase considerably and with it the number of wage labourers. The Lancashire cotton industry, one of the motor forces of the new society, expanded considerably. Between 1680 and 1760 the population of Liverpool, the major cotton port, increased ten times. Between 1750 and 1770 the export of cotton goods increased also by a factor of ten, and after 1770 British cotton goods began to dominate the European market.[10] A new society was emerging, a society which tended to divide all men into two classes, those who owned private property—in particular the means of production—and those who did not, and who, consequently, were obliged to sell their labour to the owners of the means of production.

The question was, could such a society—a society based on private property—provide its citizens with a constantly rising material standard of living? And would it be a society in which men could live together in creative comradeship? Adam Smith was in no doubt concerning the first part of the question. Capitalist society would necessarily bring material prosperity to all citizens, provided only that the government did nothing to hinder the development of free-trade and the free play of market forces. It was therefore a society in which the development of science and the mechanical arts would find application in the promotion of human welfare. It was therefore a society to be promoted.

3. The Commitment of Adam Smith: Promotion of the New Social Order

The principal function of the government, argued Smith, should be to ensure the preservation of private property. This belief was stated very emphatically. 'The acquisition of valuable and extensive property . . . necessarily requires the establishment of civil government,' a government which, 'so far as it is instituted for the security of property, is in reality instituted for the defence of the rich against the poor, or of those who have some property against those who have none at all'.[11] But the institution of private property was not to the disadvantage of the poor. On the contrary, thought Smith, it was to the advantage of the entire society, rich and poor alike. For provided individual capitalists enjoyed freedom from government obstructions, then each capitalist, striving only to 'find out the most advantageous employment for whatever his capital can command', would be 'led by an invisible hand to promote an end which has no part of his intention', namely, the public good. 'The study of his own advantage necessarily leads him to prefer what is most advantageous to the society.' Indeed, argued Smith, 'by pursuing his own interest he frequently promotes that of the society more effectually than when he really intends to promote it'. According, therefore, to Smith's economic analysis of capitalist society,* increasing material welfare for all would be ensured if competition and free trade could be guaranteed. In such a *laissez-faire* society 'the rich are led by an invisible hand to make nearly the same distribution of the necessities of life which would have been made had the earth been divided into equal proportions among all its inhabitants'. Smith therefore advocated the minimum of government intervention in the economy; the only public works he advised the government to undertake were ones in which he believed the profit motive would prove inadequate, such as road and canal construction.

Smith argued that one of the most important factors in the great improvement in the productive powers of labour lay in 'the effects of the division of labour'. His description of such effects in the pin-making trade is justly famous:

> A workman not educated to this business . . . nor acquainted with the use of the machinery employed in it . . . could scarce, perhaps, with his utmost industry, make one pin in a day, and certainly could not make twenty. But in the way in which this business is now carried on . . . one man draws out the wire, another straights it, a third cuts it, a fourth points it, a fifth grinds it . . . the important business of making a pin is, in this manner, divided into about eighteen distinct operations.[12]

* It is not our intention in this chapter to describe the economic theories of Smith, Mill, Marx and Marshall. However in chapter 5 we do discuss in considerable detail the economic theories of the chapter's principal protagonists.

Smith explains that he knows of ten workmen who in this way make upwards of 48 000 pins a day; but without division of labour he believes that they could not have made 200 a day between them, perhaps not even 10.

All, however, was not well. For although continuing adoption of such practice would certainly bring society great *material* wealth, it might also bring, Smith warned, very undesirable consequences of a *non-material* kind. In particular, widespread practice of the division of labour would, Smith believed, unless remedial action was taken by the government, necessarily lead to the emotional and intellectual impoverishment of working-class people. For:

> the man whose whole life is spent in performing a few simple operations . . . has no occasion to exert his understanding or to exercise his invention in finding out expedients for removing difficulties which never occur. He naturally loses, therefore, the habit of such exertion, and generally becomes as stupid and ignorant as it is possible for a human creature to become. The torpor of his mind renders him, not only incapable of relishing or bearing a part in any rational conversation, but of conceiving any generous, noble, or tender sentiment . . .[13]

Not only is such a man 'mutilated and deformed', but, Smith argues, 'he is incapable of defending his country in war'. And this, he points out, is very much the concern of the government. Since, therefore, a state of 'gross ignorance and stupidity' is one in which 'the labouring poor, that is, the great body of the people, must necessarily fall, unless government takes some pains to prevent it', Smith recommends the government of 'a civilized and commercial society' to make sure that 'the whole body of the people' receives 'those most essential parts of education'.

The Hermetic utopians had wanted all people educated so that each could enjoy the pleasures of body and mind described by More. Smith emphasizes a further reason for advocating universal education. Not only must a 'martial spirit' be inculcated into 'the great body of the people' but 'an instructed and intelligent people are always more decent and orderly than an ignorant and stupid one'. Nothing but the public good would result from universal education. For educated people, Smith explains,

> feel themselves, each individually more respectable, and more likely to obtain the respect of their lawful superiors, and they are therefore more disposed to respect those superiors. They are more disposed to examine, and more capable of seeing through, the interested complaints of faction and sedition, and they are, upon that account, less apt to be misled into any wanton or unnecessary opposition to the measures of government.[14]

The social and political commitment of Smith was therefore to promote competition and free trade. A perfectly competitive capitalist society would be a self-regulating social order in which the common good would automatically emerge out of individual selfishness. The principal functions of government ought to be to defend private property and to ensure universal education. To be sure, the capitalist social order would continue to have its problems but these, Smith believed, would automatically get resolved as capital accumulated and the society as a whole grew wealthier. There would therefore be no necessity for further qualitative social change—the existing social order would be able to ensure not only social stability but also a rising standard of living for all citizens.

But by 1776, the year of publication of Adam Smith's *The Wealth of Nations*, the first phase of English capitalism was coming to an end. The next phase, the industrial revolution, was not only to greatly develop the industrial capacity of England but to plunge that society into crisis and to give birth to a working-class movement that Marx and Engels believed would take up the challenge to private property laid down by More and the Hermetic utopians. In the next section we look briefly at some social conditions during the trauma of the industrial revolution.

4. The Industrial Revolution (1780–1850): Capitalism in Crisis

By the end of the 18th century the peasantry as a class had been virtually eliminated in England, unlike the situation in France and the rest of the Continent. English agriculture had become 'rationalized'. The common-field system, characterized by 'dispersed holdings, common grazing rights and sharing of the commons', had been inefficient and expensive. On the other hand, enclosure offered lower costs per acre and therefore greater profits on the larger areas under cultivation. By the end of the enclosure movement there existed a few thousand landowners leasing out land to tens of thousands of tenant farmers who in turn hired those peasants who had remained on the land instead of seeking work in nearby towns. According to Kransberg, 'Serfdom and the guild-system, two obstacles to the workers' freedom of movement and occupation, had been almost completely broken down in England by the beginning of the 18th century. . . . Hence,' declared Kransberg, 'English working-men, many of whom had been forced off the soil by the enclosure movement, were freer to move from farm to factory.'[15]

Thus the Enclosures Acts paved the way for the mechanization of agriculture and helped provide at the same time an increasing supply of labour for the growing number of factories. The population of England was also growing rapidly: from 6·5 million in 1730 to more than 9 million in 1800 and to about 16 million in 1840. This is the century referred to with much justification as the Age of Disaster for working-class people. Torn from the land the peasant could not easily adapt to the different and harsh conditions of the factory system. An alternative to factory employment was a life of crime. It is scarcely surprising, therefore, that the English penal code became increasingly severe in the 18th and early part of the 19th century. In 1700 there were some

50 crimes punishable by death. But by 1750 the number had trebled and by 1800 there were 220 capital offences. The vast majority of the new crimes punishable by hanging were those against wealth and property. Although hanging for witchcraft was abolished in 1736, the death penalty was introduced for stealing 40/- from a house, 5/- from a shop. In 1727 it became a capital offence to break the hated machines. According to Radzinowicz, because of 'the composite character of these enactments, each of which was so broadly framed as to allow for the infliction of the death penalty for a considerable number of variations of the same offence . . . the actual scope of the death penalty was therefore as much as 3 or 4 times as extensive as the number of capital provisions would seem to indicate'.[16] In 1810 Sir Samuel Romilly told Parliament that 'there is probably no other country in the world in which so many and so great a variety of human actions are punishable with loss of life as in England'. Capital crimes applied to men and women equally, although women were burnt, not hanged, for counterfeiting (and in 1779–89 were burnt for coining). 'Adolescents as well as children could be—and actually were—sentenced to death and even executed,' Radzinowicz writes. Even as late as 1814 a boy of 14 was hanged at Newport for stealing.

During this period there were attacks on the prevailing system of poor relief. Adam Smith had argued that the government was as mistaken in legislating to protect the poor as in allowing monopolistic privileges. For in his opinion the existing Poor Laws, which made residence within a particular parish a condition of eligibility for relief, were responsible for restricting the mobility of labour, and thereby unnecessarily restricted the rate of economic growth. Ricardo, Smith's intellectual and very influential successor, agreed with this verdict on the Poor Laws and argued of them (1821) that 'every friend to the poor must ardently wish for their abolition'. In less measured terms the Rev. J. Townsend, calling himself a 'well-wisher of mankind', protested that the Poor Law 'tends to destroy the harmony and beauty, the symmetry and order of that system which God and Nature have established in the world'.[17] Arguing that the social security offered by the Poor Law did not encourage the poor to work, Townsend expressed the opinion that:

> It seems to be a law of Nature that the poor should be to a certain degree improvident that there may always be some to fulfil the most servile, the most sordid, and the most ignoble offices in the community. The stock of human happiness is thereby much increased whilst the more delicate are not only relieved from drudgery . . . but are left at liberty without interruption to pursue those callings which are suited to their various dispositions.

First published in 1786 and then again in 1817, the Rev. Townsend's *Dissertation on the Poor Laws* was representative of the swing of opinion away from a relatively humane system of public relief in favour of a harsher, centralized system in the hands of elected guardians of the poor, a system finally institutionalized by the Poor Law Amendment Act of 1834.[18]

The promise of science betrayed?

The Hermetic utopians had wished to proscribe private property; capitalist society sanctioned it. The Hermetic utopians had declared that all goods should be held in common; capitalist society sanctified an unequal distribution of the national wealth. The Hermetic utopians had wished to provide universal education; this was not, however, a feature of capitalist society during the industrial revolution. The Hermetic utopians, intent on giving pleasure to men's hearts and minds, wished work hours to be reduced to only a few hours each day; this was scarcely a goal of factory owners. Even children, from the age of seven upwards, were required to work 6 days a week, 12 to 15 hours a day. And pleasure and understanding to men's hearts and minds? In a letter to the *Sheffield Courant* (1831) Thomas Arnold, headmaster of Rugby, declared: 'A man sets up a factory and *wants hands*. I beseech you, sir,' he appealed, 'to observe the very expressions that are used, for they are all significant. What he wants of his fellow creatures is the loan of their hands; of their heads and hearts he thinks nothing.'[19] And, of special relevance for the theme of this book, as science in the 18th and 19th centuries came to have increasing practical and technological relevance,* scientists and 'technologists' appeared to place themselves whole-heartedly at the service of private wealth and that commercial spirit which More and the Hermetic utopians, rightly or wrongly, had so greatly deplored.

In 1802 Humphry Davy, scientist and first Director of the Royal Institute, later President of the Royal Society, declared that 'the unequal division of property and labour, the difference of rank and condition amongst mankind, are the sources of power in civilized life, its moving causes, and even its very soul'.[20] And when in 1807 Samuel Whitbread introduced a bill proposing the education of all children—a measure the Hermetic utopians considered an essential prerequisite for the realization of a just and happy society—Giddy, patron of science and later President of the Royal Society, (in disagreeing with Adam Smith) answered thus:

> However specious in theory the project might be, of giving education to the labouring classes of the poor, it would in effect be found to be prejudicial to their morals and happiness; it would teach them to despise their lot in life, instead of making them good servants in agriculture, and other laborious employments to which their rank in society had destined

---

* 'Contrary to long-accepted ideas,' write Musson and Robinson, '[the industrial revolution] was not simply a product of illiterate practical craftsmen, devoid of scientific training. In the development of steam power, in the growth of the chemical industry, in bleaching, dyeing and calico-printing, in pottery, soap, and glass manufacture, and in various other industries, scientists made important contributions and industrialists with scientifically trained minds also utilized applied science in their manufactures.'[19n]

them; instead of teaching them subordination, it would render them factious and refractory, as was evident in the manufacturing countries; it would enable them to read seditious pamphlets, vicious books, and publications against Christianity; it would render them insolent to their superiors; and in a few years the result would be that the legislature would find it necessary to direct the strong arm of power towards them, and to furnish the executive magistrate with much more vigorous laws than were now in force.[21]

It was recognized by many that science had not led to the furthering of freedom for all. Andrew Ure, a spokesman for the new social order, welcomed the fact: 'When capitalism enlists science into her service, the refractory hand of labour will always be taught docility.'[22] Both Mill and Marx, whose views we discuss in the next two sections, regretted the fact. Both, however, held out hope for a future society in which science and industry would contribute to the emancipation of all mankind.

The industrial revolution was an age of turmoil and unrest. Between 1778 and 1830 there were constant revolts against the spreading machinery that threatened to force the village craftsmen and domestic producers of goods into the factory system. In 1811 at Lancaster and Chester 13 Luddites were publicly hanged while the following year at York a public hanging of 17 such offenders took place. In 1830 the unrest appeared all but revolutionary in character. During the years 1841–2 the Chartist agitation reached its peak. On a visit to England in 1845, an American, Colman, described his (exaggerated?) reaction to the condition of the English working class as he saw it: 'Wretched, defrauded, oppressed, crushed human nature lying in bleeding fragments all over the face of society. Every day that I live I thank Heaven that I am not a poor man with a family in England.'[23] Investigating the causes of crime in this period Tobias comes to the conclusion that 'crime, and especially juvenile crime, in the first half of the nineteenth century was the crime of a society in violent economic and social transition'.[24]

But in transition to what kind of society? This question leads us directly to a consideration of the competing views and political commitments of John Stuart Mill and Karl Marx, views and commitments that we look at in some detail. For they have lost none of their relevance in the second half of the 20th century.

5. The Commitment of John Stuart Mill (1806–73): Progress through Reform

Mill, with the experiences of the industrial revolution behind him and witnessing social convulsions on the Continent, found it difficult at the turn of the half-century to be as optimistic about the future of capitalism as Adam Smith had been. On the contrary, he believed himself at that time to be living in an 'age of transition', that the existing social institutions were 'merely provisional', that it was an age of crisis, an age, he wrote, 'when a general reconsideration of all first principles is felt to be inevitable'.[25]

We recall that Kuhn identifies a crisis state in a science by noting 'the expression of explicit discontent, the recourse to philosophy and to debate over fundamentals'. And in a crisis state essentially three different kinds of paradigm commitments are possible: to continue to have faith in the existing paradigm, or to advocate revolutionary changes, or to try to compromise between these two extremes. We shall see that in the social crisis in which Mill found himself he returned to 'debate over fundamentals' and in doing so considered the possibility of three very different kinds of social and political objectives. One of these was the preservation of the existing system of individual property, a system characterized by the institutionalization of competition and the existence of two principal classes, capitalists and labourers. The second objective Mill saw as the attainment of socialism, defined by him to be a system of 'associations of labourers' competing in 'friendly rivalry' among themselves. The third objective recognized by Mill was what he called 'communism', a social system he described as 'the extreme limit of socialism', in which 'not only the instruments of production, the land and capital, are the joint property of the community, but the produce is divided and the labourer apportioned, as far as possible, equally'.[26] In analogy with our classification of competing paradigms in the physical sciences, we shall refer to Mill's three social systems as the existing system (a class-structured, competitive society), the compromise system (a classless, competitive society), and the revolutionary system (a classless, cooperative society). Beginning with quotations from the third edition (1852) of Mill's major work *Principles of Political Economy* we attempt to give at least the flavour of this 'paradigm' debate that Mill undertook with himself and with his contemporaries.

Ethical evaluation and social goals

We take up first Mill's specific complaints against the society in which he lived, complaints which gave him cause to give sympathetic attention to the advocates of radically different societies.

In the first place industrial society had not, in Mill's opinion, markedly improved 'the universal lot' of mankind: the greater part of the population remained confined, he believed, to 'the same life of drudgery and imprisonment'. He was disappointed to find that the great wealth of mechanical inventions had not significantly 'lightened the day's toil of any human being'. For this to be possible, he stated—for 'the conquests made from the powers of nature by the intellect and energy of scientific discoverers' to produce 'those great changes in human destiny, which it is in their nature ... to accomplish' —then at least two conditions would have to be met: 'just institutions' would have to be established and the population growth rate would have to be placed 'under the deliberate guidance of judicious foresight'.[27]

POLITICAL COMMITMENTS

Not only, however, did Mill condemn the manifestly unjust distribution of work and wealth characteristic of the existing society—'that the produce of labour should be apportioned as we now see it, almost in an inverse ratio to the labour'—he also disapproved strongly of the ethos of the existing society. He could not bring himself to believe that 'the trampling, crushing, elbowing, and treading on each other's heels, which form the existing type of social life, are the most desirable lot of mankind'. On the contrary, they were, he believed, nothing but 'the disagreeable symptoms of one of the phases of industrial progress'. Mill passionately desired a stationary *material* state 'in which, while no one is poor, no one desires to be richer, nor has any reason to fear being thrust back by the efforts of others to push themselves forward'. The rich countries were, he believed, rich enough. 'It is only in the backward countries of the world', he declared, 'that increased production is still an important object: in those most advanced, what is economically needed is a better distribution, of which one indispensable means is a stricter restraint on population.'

Mill was insistent that the world population should be brought to a stationary state. Even if it were the case that the world could sustain a much greater population than the existing one, he saw 'very little reason for desiring it'. Mill's comments are of great relevance for the contemporary world and bear quotation at length:

> A world from which solitude is extirpated, is a very poor ideal. Solitude, in the sense of being alone, is essential to any depth of meditation or of character, and solitude in the presence of natural beauty and grandeur, is the cradle of thoughts and aspirations which are not only good for the individual, but which society could ill do without. Nor is there much satisfaction in contemplating the world with nothing left to the spontaneous activity of nature; with every rood of land brought into cultivation, which is capable of growing food for human beings; every flowery waste or natural pasture ploughed up, all quadrupeds or birds which are not domesticated for man's use exterminated as his rivals for food, every hedgerow or superfluous tree rooted out, and scarcely a place left where a wild shrub or flower could grow without being eradicated as a weed in the name of improved agriculture. If the earth must lose that greater portion of its pleasantness which it owes to things that the unlimited increase of wealth and population would extirpate from it, for the mere purpose of enabling it to support a larger, but not a better or a happier population, I sincerely hope, for the sake of posterity, that they will be content to be stationary, long before necessity compels them to it.[28]

Mill took pains to emphasize that a stationary *material* state did not imply a stationary intellectual or moral state. On the contrary, the former would, he believed, be more conducive than the existing system to the development of a diversity of tastes and talents, and of a stimulating variety of intellectual points of view. In no way, therefore, did Mill welcome stagnation of mental and moral life. Indeed, his strong wish to see a 'multiform development of human nature' was one of the principal reasons for his rejection of radical social change. 'No society,' he wrote, 'in which eccentricity is a matter of reproach, can be in a wholesome state.'[26] His fear was precisely that communist society would make eccentricity not merely a matter of reproach but an impossibility.

Mill's social goals were therefore the attainment of a just distribution of work and wealth, a stationary material state combined with a guarantee of personal freedom of intellectual and moral life, and an environment which allowed men the possibility of solitude and enjoyment of the beauty of an uncultivated nature. These were his goals. The failure of capitalist society to achieve these goals was apparent to Mill. Yet to what extent did Mill consider radical social change feasible, and, if feasible, desirable? To what extent did he expect either socialist of communist society to satisfy his criteria of the 'good life' more effectively than the existing system or possible improvements within this system?

On the possibility of change

Mill considered it unrealistic to suppose that ruling élites would voluntarily relinquish their privileges. In his opinion 'all privileged and powerful classes . . . have used their power in the interest of their own selfishness' and that this 'evil . . . cannot be eradicated, until the power itself is withdrawn'.[29]

But if the ruling élites of the existing social system could not be expected to voluntarily abdicate their power, they could certainly be expected to encounter increasing opposition to the use of that power—an opposition, thought Mill, that was gradually developing and would continue to develop within the ranks of 'the working classes'. Indeed, since 'ideas of equality' were spreading rapidly 'among the poorer classes' and since Mill believed that this process could not be stopped 'by anything short of the entire suppression of printed discussion and even of freedom of speech', Mill felt himself forced to conclude that 'the division of the human race into two hereditary classes, employers and employed', could not be permanently maintained. 'The relation of masters and workpeople,' he wrote, 'will be gradually superseded by partnership, in one of two forms: in some cases, association of the labourers with the capitalist; in others, and perhaps finally in all, associations of labourers among themselves . . . on terms of equality, collectively owning the capital with which they carry on their operations, and working under managers elected and removable by themselves'.[30]

The important point, however, was that such changes would be gradual, 'without violence or spoliation, or even any sudden disturbance of existing habits and expectations'. 'Eventually,' hoped Mill, 'and in perhaps a less remote future than may be supposed, we may . . . see our way to a change in society, which would combine the freedom and independence of the individual, with the moral, intellectual, and economical advantages of aggregate production,' a change that would result in 'the transformation of human life, from a conflict of classes struggling for opposite interests, to a friendly rivalry in the pursuit of a good common to all'.[31]

On the desirability of change

Mill was therefore not opposed in principle to the socialist goal as he conceived it to be—provided no sudden disturbances of the existing order were contemplated. Communism, however, could not be recommended. Indeed, Mill took great pains to make clear his very fundamental reservations with respect to the desirability of communist society, reservations which centred around two connected problem areas, the future of competition and the future of individual freedom. We consider in turn each of Mill's principal objections to the communist goal of a classless, cooperative society.

In so far as socialists desired to form associations of labourers Mill found himself, in principle, in agreement. But Mill believed that ideally such associations ought to compete among themselves in 'friendly rivalry' and therein lay, as he saw it, the source of a fundamental disagreement with communist thinkers. 'I utterly dissent,' he wrote, 'from the most conspicuous and vehement part of their teaching, their declamations against competition.' In words reminiscent of Adam Smith, Mill proclaimed that 'instead of looking upon competition as the baneful and anti-social principle which it is held to be by the generality of [communists], I conceive that, even in the present state of society and industry, every restriction of it is an evil, and every extension of it, even if for the time injuriously affecting some class of labourers, is always an ultimate good'. To be sure, Mill agreed that competition was 'a source of jealousy and hostility among those engaged in the same occupation'. But if this was an evil resulting from competition, greater evils were thereby prevented. The trouble with communists, Mill thought, was their tendency to overlook 'the natural indolence of mankind'. For if men were once to attain any state of existence they considered comfortable, then Mill thought that they would not only be in danger of stagnating intellectually and morally but would perhaps lose even the energy required to preserve themselves from deterioration. Competition was therefore a necessary evil; and to be protected against it was 'to be protected in idleness, in mental dullness; to be saved the necessity of being as active and as intelligent as other people'. The conclusion Mill drew was quite emphatic: 'Competition may not be the best conceivable stimulus, but it is at present a necessary one, and no one can foresee the time when it will not be indispensable to progress.'[32] And progress for Mill, let us recall, means not greater material wealth but more free time for the cultivation of mental life and for improving 'the Art of Living'.

Connected with Mill's fear of intellectual and moral stagnation was his fear that the realization of communist society would mean the impossibility of any individual freedom. 'The question is,' he emphasized, 'whether [in communist society] there would be any asylum left for individuality of character; whether public opinion would not be a tyrannical yoke; whether the absolute dependence of each on all, and surveillance of each by all, would not grind all down into a tame uniformity of thoughts, feelings and actions.' In the opinion of Mill this was already 'one of the glaring evils' of capitalist society, 'notwithstanding a much less absolute dependence of the individual on the mass than would exist in the communist régime'. And thus the threat to human liberty would be all the greater in communist society. To be sure, Mill was of the opinion that if the institution of private property necessarily led to the produce of labour being apportioned in an inverse ratio to the labour, then 'if this, or communism, were the alternative, all the difficulties, great or small, of communism would be but as dust in the balance'. But Mill declared that to make the comparison applicable communism at its best must be compared with 'the régime of individual property, not as it is, but as it might be made'. And Mill was definitely of the conviction that 'the principle of private property has never yet had a fair trial in any country'.[33]

Mill's research-action programme

Convinced, on the one hand, that capitalism had not yet been developed to its full extent and, on the other, that the labouring classes were as yet totally unfit 'for any order of things, which would make any considerable demand on either their intellect or their virtue',[34] Mill saw only one possible choice of action. The socialists ought to continue with their experiments 'on a moderate scale', thereby enabling experience to decide 'how far or how soon any one or more of the possible systems of community of property' could serve as a substitute for the existing system. And then comes the 'normal science' commitment:

> In the meantime we may . . . affirm that the political economist, for a considerable time to come, will be chiefly concerned with the conditions of existence and progress belonging to a society founded on private property and individual competition; and that the object to be principally arrived at in the present stage of human improvement, is not the subversion of the system of individual property, but the improvement of it, and the full participation of every member of the community in its benefits.[35]

In the remaining twenty years of his life, Mill's commitment to reform *within the existing capitalist system* strengthened considerably. In rough drafts of a planned book on socialism, written shortly before his death, Mill rebuked socialists for claiming that real wages were decreasing in capitalist countries. Such an assertion, he wrote, 'is in opposition to all accurate information'. Real wages were, if anything, he declared, increasing everywhere, not decreasing; and moreover, the rate of increase of real wages was itself increasing, not decreasing. He told socialists that the existing system was not 'hurrying us into a state of general indigence and slavery from which only

POLITICAL COMMITMENTS 107

socialism can save us'. He admitted that the existing evils were great 'but they are not increasing; on the contrary, the general tendency is toward their diminution'.

Once again, therefore, Mill finds himself justified in advising against a programme of revolutionary change. 'An entire renovation of the social fabric, such as is contemplated by [the revolutionary form of] socialism, establishing the economic constitution of society upon an entirely new basis, other than that of private property and competition, however valuable as an ideal, and even as a prophecy of ultimate possibilities, is not available as a present resource, since it requires from those who are to carry on the new order of things qualities both moral and intellectual, which require to be tested in all, and to be created in most.' Mill is in no doubt about the matter: 'For a long period to come the principle of individual property will be in possession of the field.'[36]

6. The Commitment of Marx and Engels: Progress through Revolution

Looking at the same social reality as John Stuart Mill, Marx and Engels came to very different conclusions. Unlike Mill, Marx and Engels believed that far from being able to solve its problems capitalism would inevitably lead to universal crisis—and that the latter would occur not in the remote but in the near future. Such a crisis was, however, to be welcomed. For it would make possible the construction of a qualitatively different society that would enable all men and women not only to enjoy the fruits of science and industry 'according to their needs' but to achieve the intellectual and spiritual diversity so desired by Mill. Whereas Mill feared that the effect of communist society on human freedom would be essentially negative, Marx and Engels saw (their version of) communist society as offering the only way of guaranteeing human dignity and freedom to each and to all.

Although Marx thought it wrong to class Mill, and men like him, 'with the herd of vulgar economic apologists', nevertheless he thought it useless to try, as Mill had tried, 'to harmonize the Political Economy of capital with the claims, no longer to be ignored, of the proletariat'. In the opinion of Marx, Mill was trying to 'reconcile irreconcilables' and had inevitably produced only 'a shallow syncretism'.[37] What was needed, and what would 'inevitably' occur, was *revolutionary* change. Marx and Engels committed their emotional energies and intellects to the development and promulgation of this belief.

Ethical evaluation and social goals

The scientific and technical achievements of the bourgeoisie were not overlooked by Marx and Engels. As they acknowledged:

> The bourgeoisie during its rule of scarce one hundred years, has created more massive and more colossal forces than have all preceding generations together. Subjection of Nature's forces to man, machinery,

application of chemistry to industry and agriculture, steam navigation, railways, electric telegraphs, clearing of whole continents for cultivation, canalization of rivers, whole populations conjured out of the ground — what earlier century had even a presentiment that such productive forces slumbered in the lap of social labour?[38]

The bourgeoisie's development of the productive forces was indeed greatly welcomed by Marx and Engels. For such a development, they believed, would provide the necessary material basis on which could subsequently be built a classless, and therefore non-exploitative, society.

Nevertheless, despite their conception of such a welcome historical role for capitalism, the writings of Marx and Engels constantly express anguish over the accompanying mutilation of the individual, temporary though they conceived it to be. However necessary a stage capitalism was in mankind's evolution, it was also, Marx and Engels lamented, a brutal and dehumanizing stage.

Thus in the first volume of *Capital* Marx agrees with Adam Smith when he declares that:

> [in capitalist society] all means for the development of production transform themselves into means of domination over, and exploitation of, the producers; they mutilate the labourer into a fragment of a man, degrade him to the level of an appendage of a machine, destroy every remnant of charm in his work and turn it into a hated toil; they estrange from him the intellectual potentialities of the labour-process in the same proportion as science is incorporated in it as an independent power; they distort the conditions under which he works, subject him during the labour-process to a despotism the more hateful for its meanness.

Thus Marx concludes that 'in proportion as capital accumulates, the lot of the labourer, be his payment high or low, must grow worse . . . Accumulation of wealth at one pole is, therefore, at the same time accumulation of misery, agony of toil, slavery, ignorance, brutality, mental degradation, at the opposite pole.'[39] Much more was therefore needed than an enforced increase of wages which would, Marx declared, 'be nothing but *better payment for the slave*, and would not win either for the worker or for labour their human status and dignity'.[40]

But it was not only the worker who was degraded by the capitalist mode of production; the capitalist himself was also unable to be a *human* being. It is true (Marx wrote) that the capitalist 'shares with the miser the passion for wealth as wealth. But that which in the miser is mere idiosyncrasy is, in the capitalist, the effect of the social mechanism, of which he is but one of the wheels'.[41] Under the system of private property, Marx explains, 'every person speculates on creating a *new* need in another, so as to drive him to a fresh sacrifice, to place him in a new dependence, and to seduce him into a new mode of gratification and therefore economic ruin'.[42] Thus for the capitalist 'every new product represents a new *possibility* of mutual swindling and mutual plundering'. The capitalist was therefore as emotionally exploited as the worker. For 'the less you think, love, theorize, sing, paint, fence, etc., the more you save', Marx tells the capitalist, 'the greater becomes your treasure which neither moths nor dust will devour—your capital. The less you *are*, the less you express your own life, the greater is your *alienated* life, the more you *have*, the greater is your store of estranged being. Everything which the political economist takes from you in life and in humanity, he replaces for you in *money* and in *wealth*; and all the things which you cannot do, your money can do'.[43] On the other hand (Marx continues), 'assume *man* to be *man* and his relationship to the world to be a human one: then you can exchange love only for love, trust for trust, etc. If you want to enjoy art, you must be an artistically cultivated person; if you want to exercise influence over other people, you must be a person with a stimulating and encouraging effect on other people'.[44] But in bourgeois society there is 'no other nexus between man and man than naked self-interest, than callous "cash-payment"'. Only 'capital is independent and has individuality, while the living person is dependent and has no individuality'. But, said Marx and Engels, after the inevitable revolution, 'in the place of the old bourgeois society, with its classes and class antagonisms, we shall have an association, in which the free development of each is the condition for the free development of all'.[45]

Such was Marx's utopian goal. This is how Fromm assesses it: 'Marx's aim was the spiritual emancipation of man, of his liberation from the chains of economic determinism, of restituting him in his human wholeness, of enabling him to find unity and harmony with his fellow man and with nature.'[46]

On the 'inevitability' of communist society according to Marx

Marx convinced himself that the capitalist social order was only a transient, although necessary, stage in the evolution of society towards the realm of human freedom. Of the dialectical method he used, Marx wrote:

> In its rational form dialectic is a scandal and abomination to the bourgeoisie and its doctrinaire professors because it includes in its comprehension and affirmative recognition of the existing state of things, *at the same time also*, the recognition of the negation of that state, of its inevitable breaking up; because it regards every historically developed social form as in fluid movement, and therefore takes into account its transient nature not less than its momentary existence; because it lets nothing impose upon it, and is in its essence critical and revolutionary.

There was nothing the bourgeoisie could do, Marx believed, to prevent the final crisis of capitalism; the contradictions inherent in the capitalist mode of production would automatically generate those social conditions which would lead to capitalism's negation and replacement. For 'the natural laws of capitalist production' were 'working with iron necessity towards inevitable results'.⁴⁷ Indeed, the bourgeoisie was very aware of such contradictions, declared Marx. In particular, 'the contradictions inherent in the movement of capitalist society impress themselves upon the practical bourgeois most strikingly in the changes of the periodic cycle, through which modern industry runs, and whose crowning point is the universal crisis. That crisis is once again approaching,' he wrote in 1873, 'although as yet but in its preliminary stage'.⁴⁸ According to Marx, increasing competition between capitalists would result in the control of the ever-developing forces of production by ever-fewer capitalists while at the same time the proletariat would grow in number, organizing power and militancy. Eventually a severe depression and large-scale unemployment would be followed not by an up-swing in the economy but by revolution and expropriation of the appropriators:

> The advance of industry, whose involuntary promoter is the bourgeoisie, replaces the isolation of the labourers, due to competition, by their revolutionary combination, due to association. The development of Modern Industry, therefore, cuts from under its feet the very foundation on which the bourgeoisie produces and appropriates products. What the bourgeoisie, therefore, produces, above all, is its own grave-diggers. Its fall and the victory of the proletariat are equally inevitable.⁴⁹

Inevitable?

On the inevitability of communist society according to Engels

In his *Socialism, Utopian and Scientific* Engels attempts to convince the reader of the eventual triumph of Communism. In doing so, however, he exposes serious weaknesses in his own argument. 'The final causes of all social changes and political revolutions,' Engels writes, 'are to be sought, not in men's brains, not in man's better insight into eternal truth and justice, but in changes in the modes of production and exchange.'⁵⁰ In this way Engels attempts to distinguish his and Marx's theories from those of 'utopian' socialists. Furthermore, the

> conflict between productive forces and modes of production . . . exists, in fact, objectively, outside us, independently of the will and actions even of the men that have brought it on. Modern Socialism is nothing but the reflex, in thought, of this conflict in fact; its ideal reflection in the minds, first, of the class directly suffering from it, the working class.⁵¹

Nothing but the reflex! Surely it is a mistake to describe the conception of a qualitatively different society in the minds of working men, and of appropriate action to achieve such a society, as nothing but the (automatic?) reflex of the conflict between the productive forces and the social relations of production. But even so, why should the proletariat necessarily act against the bourgeoisie? This is Engel's answer: 'Whilst the capitalist mode of production more and more completely transforms the great majority of the population into proletarians, it creates the power which, under penalty of its own destruction, is forced to accomplish this revolution.'⁵²

Under penalty of its own destruction! Thus there is no alternative for the workers. Either they set up a classless society or they will become totally degraded as human beings. They *must* therefore choose the former. Engels continues:

> To accomplish this act of universal emancipation is the historical mission of the modern proletariat. To thoroughly comprehend the historical conditions and thus the very nature of this act, to impart to the now oppressed proletarian class a full knowledge of the conditions and of the meaning of the momentous act it is called upon to accomplish, this is the task of the theoretical expression of the proletarian movement, scientific socialism.⁵³

But here, it would seem, is the fundamental difficulty: the proletariat is oppressed—and, furthermore, mentally degraded. And yet the proletariat must be receptive to the idea of a classless society and act accordingly. But how can there be any guarantee that this revolutionary awareness will develop? Engels does not discuss this problem except in arguing that after capitalistic competition has led to capitalistic monopoly the exploitation will become so apparent 'that it must break down'. For, according to Engels, 'no nation will put up with production conducted by trusts, with so barefaced an exploitation of the community by a small band of dividend-mongers'.⁵⁴ The point he stresses is that the new society is now realizable:

> The possibility of securing for every member of society, by means of socialized production, an existence not only fully sufficient materially and becoming day by day more full, but an existence guaranteeing to all the free development and exercise of their physical and mental faculties—this possibility is now for the first time here, but *it is here*.⁵⁵

And the possibility must be turned into successful action. Engels, after apparently downgrading the autonomy of the human will, now chooses to emphasize how important is the human will in changing the social reality:

> Active social forces work exactly like natural forces: blindly, forcibly, destructively, so long as we do not understand, and reckon with, them. But when once we understand them, when once we grasp their action, their direction, their effects, it depends only upon ourselves to subject them more and more to our own will, and by means of them to reach our own ends. And this holds especially for the mighty productive forces of today.⁵⁶

The end result would be a truly free society:

> The laws of his own social action, hitherto standing face to face with man as laws of Nature foreign to, and dominating, him will then be used with

full understanding, and so mastered by him. Man's own social organization, hitherto confronting him as a necessity imposed by Nature and history, now becomes the result of his own free action. The extraneous objective forces that have hitherto governed history, pass under the control of man himself. Only from that time will man himself, more and more consciously, make his own history—only from that time will the social causes set in motion by him have, in the main and in a constantly growing measure, the results intended by him. It is the ascent of man from the kingdom of necessity to the kingdom of freedom.[57]

7. The Evolution of English Capitalism: 1850–1900

So Mill, Marx and Engels came to their diametrically opposed commitments in relation to the development of capitalist society. Mill believed (although uneasily) that capitalism would survive for a long period to come and that it had yet to produce the benefits for man of which it was inherently capable. The duty of the scientist was to help improve the existing social system. Marx and Engels, on the other hand, believed capitalist society to be inherently unstable, that it necessarily deprived capitalist and worker alike of dignity and freedom, and that it would inevitably undergo universal crisis. Marx and Engels committed themselves to helping to ensure the triumph of the revolutionary class—the proletariat—in that future period of universal crisis.

How, then, did English capitalism develop during the second half of the 19th century?

The landed aristocracy and emerging bourgeois class weathered the storm of the 1830s and 1840s—the 'age of transition'. Not only was there to be no revolution but repression eased and the conditions of life for the proletariat became less severe. Although the Factory laws of 1819 had stipulated that no children under 9 were to be employed in factories, and no children under 16 to be employed more than 12 hours per day, the Justices of the Peace were themselves often millowners and consequently the laws were seldom enforced. By the early 1840s, however, there was better enforcement of the laws and by 1867 no women and children were allowed to be employed in factories for more than 10 hours a day. Admittedly the Poor Law Amendment Act made residence in workhouses to be as disagreeable as was consistent with health, but after about 1850 the unemployed and disabled received somewhat more sympathetic help. By 1837 the number of capital crimes had been reduced to 15 and by 1870 there were only two, for murder and high treason. English capitalist society seemed secure. The working class no longer possessed a revolutionary consciousness but rather the workers were determined to get what they considered to be a fairer share of what the social order, whose basic institutional framework they no longer seriously disputed, was able to produce. Thomas Cooper, a former Chartist, wrote sadly in 1872 of the change in attitude that he observed:

My sorrowful impressions were confirmed. In our old Chartist times, it is true, Lancashire working men were in rags by the thousands; and many of them lacked food. But their intelligence was demonstrated wherever you went. You would see them in groups discussing the great doctrine of political justice.... *Now* you will see no such groups in Lancashire. But you will hear well-dressed working men talking, as they walk with their hands in their pockets, of 'Co-ops' and their shares in them, or in building societies. And you will see others like idiots leading small greyhound dogs.[58]

It was the opinion even of Engels that 'the mass of the working people' in the 1880s 'looked up with respect and deference to what used to be designated as "their betters", the middle class. Indeed', wrote Engels in 1892, 'the British workman, some 15 years ago, was the model workman, whose respectful regard for the position of his master, and whose self-restraining modesty in claiming rights for himself' served as a source of consolation for certain German 'socialists' who deplored 'the incurable communistic and revolutionary tendencies of their own working men at home'. Although the optimistic Engels thought he saw signs of change he considered it only too apparent that in both France and Germany 'the working class movement is well ahead of England'.[59]

Clearly English capitalism still had its problems. But the problems to many no longer seemed to threaten the basic class structure of the society. Like a paradigm in the physical sciences, capitalism had been born 'refuted'; the men who committed themselves to the 'articulation' of this social 'paradigm' did so on faith alone—they enjoyed no 'objective' guarantee that the new social order would ultimately prevail and that they would become the ruling class in society. Nevertheless, one by one the problems confronting the bourgeois class were apparently resolved and by the latter half of the 19th century the new social order in England appeared to have triumphed as effectively as had its companion at birth, the Newtonian paradigm. To be sure, both the capitalist social order and the Newtonian paradigm faced unresolved problems but, it seemed to so many, all remaining problems would ultimately prove themselves soluble within the existing social and conceptual frameworks respectively. The new science and the new society would, it seemed, eventually produce material abundance for all mankind.

8. The Commitment of Alfred Marshall (1848–1924): Progress through Perpetuation of the Status Quo

This was certainly the way capitalist society was regarded by Alfred Marshall, England's most influential economist towards the end of the 19th century. Unlike Marx, Marshall was convinced that capitalist society was essentially stable, that it would provide employment for all workers (apart from temporary fluctuations of a minor nature), and that the further development of capitalism would result in a growing material standard of living for all citizens. His economic theory we sketch at some length in the following chapter. It suffices to state here that Marshall strongly advised against tampering with the 'delicate and complex' self-adjusting organism he conceived capitalist society to be. He was of the opinion 'that large ill-considered changes might result in grave disaster'. In any case, capitalism was working well—why tamper with a successful system in favour of one that existed only as an idea? And Marshall was not fond of the idea of socialism:

> The collective ownership of the means of production would deaden the energies of mankind, and arrest economic progress; unless before its introduction the whole people had acquired a power of unselfish devotion to the public good which is now relatively rare. And ... it might probably destroy much that is most beautiful and joyful in the private and domestic relations of life. These are the main reasons which cause patient students of economics generally to anticipate little good and much evil from schemes for sudden and violent reorganization of the economic, social and political conditions of life.[60]

Marx had predicted universal crisis and had advocated the revolutionary reorganization of society. Marshall had predicted the steady and uneventful development of the existing system and had strongly advised against large-scale changes. In the next chapter we sketch just what did happen when in the 1930s capitalist society plunged into its gravest-ever crisis, a crisis that was, to say the least, to call into question the ability of capitalist society to satisfy even the material requirements of mankind, let alone requirements of an intellectual, sensual and spiritual nature.

Chapter 5

CAPITALISM IN CRISIS AND THE KEYNESIAN REVOLUTION

> The completeness of the Ricardian victory is something of a curiosity and a mystery. It must have been due to a complex of suitabilities in the doctrine to the environment into which it was projected. That it reached conclusions quite different from what the ordinary uninstructed person would expect, added, I suppose, to its intellectual prestige. That its teaching, translated into practice, was austere and often unpalatable, lent it virtue. That it was adapted to carry a vast and consistent logical superstructure, gave it beauty. That it could explain much social injustice and apparent cruelty as an inevitable incident in the scheme of progress, and the attempt to change such things as likely on the whole to do more harm than good, commended it to authority. That it afforded a measure of justification to the free activities of the individual capitalist, attracted to it the support of the dominant social force behind authority.
>
> JOHN MAYNARD KEYNES
> *The General Theory of Employment, Interest & Money*

IN chapter 4 we outlined the evolution of English capitalism, 'the workshop of the world', the country in which the Copernican Revolution had achieved its apotheosis and which was to give rise to such a breathtaking development of the productive forces. We noted, however, that the new English society had caused great suffering for the working class it had brought into being, that competition and self-seeking behaviour had been institutionalized as a way of life. It was a society in which income was inequitably distributed and inevitably so.

Underlying our outline was the question underlying each chapter of this book: can such a society use its developing science and technology for human benefit and only for human benefit? We noted that Marx and Engels thought not. Nevertheless, capitalist society would, they believed, eventually undergo a period of universal crisis that would be resolved by the development of a qualitatively different society, one that would be able to use science for 'the relief of man's estate', one that would ask 'from each according to his ability' and give 'to each according to his needs'.

Yet there were to be no social revolutions in the advanced capitalist countries. Instead the followers of Marx seized power in a backward peasant country in only its initial stages of industrialization; intervention by the capitalist powers followed, together with a bloody civil war. The aftermath was to put not socialism on trial but the Bolshevik party. In a devastated country surrounded by powerful enemies, the world's first 'socialist' state was formed. But there could be no socialization of wealth, only 'socialization of poverty'. It was an ominous and tragic start.

However, some fifteen years after the Russian revolution the capitalist world found itself in the depths of a universal crisis of unprecedented magnitude, the kind of crisis that Marx had predicted would herald the dawn of a civilized world.

1. The Great Depression

In the United States of the 1920s, the world's now leading capitalist country, whose debtor status the First World War had changed into that of general creditor, the economy was going from strength to strength. Throughout the decade unemployment remained low while productivity increased, so much so that in the summer of 1928 President Hoover felt able to declare that 'We in America today are nearer to the final triumph over poverty than ever before in the history of any land.'[1] The claim seemed reasonable. Had the total national income been shared equally among America's 36 million families of 1929, the average yearly income would have been some $2 300, sufficient to take a family over the poverty line then estimated at $2 000 per annum. Of course the total income was not distributed equitably: the top 1% received nearly 19% of the total income while 60% of American families were living in poverty with incomes of less than $2 000. But the corner it seemed had been turned. Further investment and production would, even with the same inequitable distribution of income, take all American families over the poverty line and to affluence. The growth of the economy, it seemed, could not and would not stop. For, as Lipsey tells us in his *Positive Economics*, 'The human wants that can be satisfied by consuming goods and services may be regarded, for all practical purposes in today's world, as insatiable,' a fact which, he says, 'gives rise to one of the basic problems encountered in economics, the problem of *scarcity*' (Lipsey's emphasis).[2] Even, therefore, after poverty has been eliminated, it would seem that insatiable wants must compel the economy forwards to ever greater production of goods and services; resources will always remain scarce. But the American economy of 1929, still characterized by poverty yet on the threshold of affluence, did not continue its forward momentum. Four years later nearly thirteen million people—one quarter of the labour force—were unemployed. 'Although,' as Heilbroner writes, 'these unemployed men and women were eager to work, although empty factories were available for them to work in, despite the existence of pressing wants, somehow a terrible and mystifying breakdown short-circuited the production process, with the result that an entire third of our previous annual output of goods and services simply disappeared.'[3]

For ten years unemployment remained high; as late as 1939 over 9 million people, some 17% of the labour force, were looking for work. Between 1929 and 1939 net *per capita* growth was zero. Investment fell in housing, commercial building, and in manufacturing plant and equipment. Between 1929 and 1933 the output of investment goods shrank by 88% in real terms. In 1932 total investment expenditure was less than 15% of the amount needed merely to keep the country's stock of capital equipment intact. In fact for the five years 1931 to 1935 net investment remained negative so that in each year the capital stock decreased. After a slight improvement there was a new recession in 1937-8; in 1939 the economy faltered again. As Carl Becker aptly wrote in that year: 'A survey of human history will often disclose millions of men starving in time of famine: what we see now is something unprecedented—millions of men destitute in the midst of potential abundance.'[4] The economy was saved, however, by the outbreak of the Second World War; as arms orders increased, unemployment fell. Once America entered the war, production rapidly increased and in 1942 President Roosevelt set a target of 60 000 aircraft, 45 000 tanks, and 8 million tons of merchant shipping. The war revealed what had been obvious to all: the enormous hidden reserves of productive power within the American economy. Yet why had a war been necessary in order to eliminate unemployment? Why was it that only in the production and consumption of armaments had all Americans seeking work been able to find it? How, we must ask, had the economists and their theories measured up to this social crisis?

We shall return to discuss the so-called Great Depression, but before beginning a discussion of economic theory it is as well to point out that this prolonged traumatic experience seems at the very least to indicate that Lipsey's emphasis on the problem of perpetual scarcity is not the whole story in economics. Most economists, we should note, seem to place great stress on scarcity. In Samuelson's textbook, for example, we find that 'Economics is the study of how men and society *choose* ... to employ *scarce* productive resources ...' (his emphases).[5] Similarly Dernburg and McDougall inform their readers that 'economic activity involves the use of scarce resources (including time) in the provision of goods to satisfy unlimited wants'.[6] And yet for ten years these scarce productive resources were lying idle while men sought desperately to find work—idle machines, idle factories and 13 million unemployed people! Scarce resources, one might think, should be used to maximum capacity, not left lying idle. Furthermore, in 1933, at the midst of the crisis, American farmers were paid *not* to produce: the production of cotton, wheat, corn, hogs, tobacco, rye, flax, barley, peanuts, sugar beet, cane, beef and dairy cattle and other products was deliberately restricted in return for government subsidies; 10 million acres of cotton and 12 000 acres of tobacco were ploughed back into the soil, 6 million pigs were slaughtered, Californian fruit crops were left to rot.[7] In a country with millions hungry and destitute, a cynic might well be forgiven for remarking that this seems a strange way to treat 'scarce resources'. Indeed, it was nothing but a planned and systematic attempt to prevent abundance and create scarcity! Again we must ask, what were the reactions of economists to this sudden turn the economy had taken?

The depression of the 1930s was a serious counter-example to the existing economic paradigm but, as we have now many times noted and argued, such counter-examples do not, and should not, automatically result in rejection of the established paradigm. Indeed in 1932 the American economist Hansen did not altogether see the downturn as even a problem confronting the established paradigm:

> Cyclical fluctuations have the effect of giving the whole economic structure a good shake-up and keeping the system reasonably flexible and mobile. With the business cycle eliminated there would not be the periodic rubbing down, so to speak, which gives industry a fresh lease of life. Depression, like a cruel and heartless tyrant, clubs down the impossible demands made by the employed agents of production, until the earnings of the factors have again reached a point at which full employment becomes profitable.[8]

Nevertheless, as the unemployment continued to increase it became increasingly clear that the American economy was experiencing something out of the usual. However, although the 'classical' economists claimed that prolonged unemployment was impossible in a perfectly competitive society, its presence did not mean that the hard core principles of the existing paradigm had necessarily to be challenged and replaced. Three different choices of action were possible. Economists could, as scientists, retain their commitment to the existing paradigm, seeking to develop its positive heuristic in order to understand better why the unemployment had occurred, and thereby to suggest measures by which it could be eliminated. Or they could, as scientists, commit themselves to *revolutionary* paradigms with the intention *either* of investigating the possibility of building societies based on economic and social principles radically different from capitalist ones *or* of seeking to retain the essential features of capitalist society but challenging the hard core principles on which hitherto an explanation of capitalist economies had always been founded. A revolutionary economist of the politically conservative kind was John Maynard Keynes, politically committed to the preservation of capitalist society, but not committed to the theoretical principles underlying generally accepted explanations of capitalist economies.

2. The Theoretical and Political Commitments of John Maynard Keynes

Keynes was very explicit about the political commitments guiding his theoretical research. His rejection of the Marxist political programme was emphatic. 'How,' he asked in 1925, 'can I adopt a creed which, preferring the mud to the fish, exalts the boorish proletariat above the bourgeoisie and the intelligentsia, who, with all their faults, are the quality of life and surely carry the seeds of all human achievements?'[9] And yet, as he had to admit eleven years later, 'It is certain that the world will not much longer tolerate the unemployment which, apart from brief intervals of excitement is associated

—and in my opinion, inevitably associated—with present day capitalistic individualism.' Hence his commitment to a politically oriented programme of research and action in defence of capitalism: 'But it may be possible for a right analysis of the problem to cure the disease whilst preserving efficiency and freedom.'[10]

That the way he viewed the workings of the capitalist economy was revolutionary Keynes was in no doubt. In 1935 he wrote to George Bernard Shaw: 'I believe myself to be writing a book on economic theory which will largely revolutionize—not, I suppose, at once but in the course of the next ten years—the way the world thinks about economic problems.'[11] In fact, looking through the writings of economists of the time demonstrates all the ingredients Kuhn describes as those of a scientific revolution. Let us note some of them.

In his book *The Years of High Theory* Shackle describes the power of the paradigm which Keynes inherited at Cambridge:

> The forty years from 1870 saw the creation of a Great Theory or Grand System of Economics, in one sense complete and self-sufficient, able, on its own terms, to answer all questions which those terms allowed. . . . In its arresting beauty and completeness this theory seemed to need no corroborative evidence from observations. It seemed to derive from these aesthetic qualities its own stamp of authentication and an independent ascendancy over men's minds. The intellectual establishment were basically content, and therefore passive. Only a few questions, that lay outside the terms on which the Grand Theory allowed itself to be consulted, remained as scraps to satisfy the prowlers round the edge of the camp. The overwhelming concentration of intellectual power within the camp was such as to daunt any possible attacker, and the Great Theory, thus guarded, remained inviolate for two decades.[12]

In the quieter years of 1922 Keynes, himself, described perfectly the activity of normal science. With respect to the then existing and (as he saw it) unchallenged paradigm Keynes was of the opinion:

> Before Adam Smith, this apparatus of thought scarcely existed. Between his time and this it has been greatly enlarged and improved. . . . It is not complete yet, but important improvements in its elements are becoming rare. The main task of the professional economist now consists, either in obtaining a wide knowledge of *relevant* facts and exercising skill in the application of economics to them, or, in expounding the elements of his method in a lucid, accurate and illuminating way, so that, through his instruction, the number of those who can think for themselves may be increased.[13]

In other words, Keynes is recommending that activity which Kuhn calls paradigm articulation.

But when in 1936 Keynes published his attack against the accepted paradigm, he complained, exaggerating not a little, that 'Ricardo conquered England as completely as the Holy Inquisition conquered Spain. Not only was his theory accepted by the city, by statesmen and by the academic world. But controversy ceased; the other point of view completely disappeared; it ceased to be discussed.'[14] And he confessed to his readers: 'I myself held with conviction for many years the theories which I now attack and I am not, I think, ignorant of their strong points. ... The difficulty lies, not in the new ideas, but in escaping from the old ones, which ramify, for those brought up as most of us have been, into every corner of our minds.' And as for the contemporary thought of his colleagues, Keynes wrote that they could not maintain the old standpoint consistently: 'For their thought of today is too much permeated with the contrary tendency and with facts of experience too obviously inconsistent with their former view. But', Keynes criticized, 'they have not drawn sufficiently far-reaching consequences; and have not revised their fundamental theory.'[15]

Keynes' former teacher, Professor Pigou, was not impressed. In his 1936 review of Keynes' *General Theory*, he wrote scathingly that 'Einstein actually did for Physics what Mr. Keynes believes himself to have done for Economics. He developed a far-reaching generalization, under which Newton's results can be subsumed as a special case.'[16]

But ten years later, in 1946, the American economist Samuelson could write that:

> The General Theory caught most economists under the age of thirty-five with the unexpected virulence of a disease ... Economists beyond fifty turned out to be quite immune to the ailment. With time, most economists in-between began to run the fever. ...'[17] Gradually and against heavy resistance, the realization grew that the new analysis ... was not to prove such a passing fad, that here indeed was part of 'the wave of the future'.[18]

Samuelson's comments are interesting and relevant. Does refutation of a theory necessitate rejection? Samuelson answers:

> The nature of the world did not suddenly change on a black October day in 1929 so that a new theory became mandatory. ... The Great Depression of the thirties was not the first to reveal the untenability of the classical synthesis. The classical philosophy had always had its ups and downs along with the great swings of business activity. Each time it had come back. ... Theorists can always resist facts. ... Inevitably, at the earliest opportunity, the mind slips back into the old grooves of thought, since analysis is utterly impossible without a frame of reference, a way of thinking about things or, in short, a theory.[19]

Why, then, had rejection, not simply refutation, become a serious possibility? Samuelson answers: 'A new *system*, that is what requires emphasis. Classical economics could withstand isolated criticism. . . . But now for the first time, it was confronted by a competing system . . . and one which could swallow the classical system as a special case.'[20]

Hansen paints the same picture:*

> Facts upon facts were piled up . . . which often seemed to indicate that the conclusions of orthodox theory failed to conform to the real world. . . . [But] facts alone will not destroy a theory. . . . In the literature of 1900 to 1936 one finds numerous efforts . . . to challenge the prevailing orthodoxy. . . . But few took any notice. What was required was nothing less than a *general theory* sufficiently comprehensive to supplant the orthodox theory of automatic adjustment. This truly herculean task Keynes essayed in his *General Theory*.[21]

Let us now attempt to sketch the 'classical' theory,† that 'orthodox tradition' in which Samuelson wrote that he had been brought up, and compare it against what was for a time, until many of the classical paradigm's leading protagonists admitted defeat, its rival research programme, the Keynesian paradigm. This comparison between paradigms in the social sciences will later be useful in a comparison between the natural and the social sciences, and we shall see that an understanding of Keynesian theory is essential for an understanding of the nature of advanced capitalist society and of the role of scientists in such societies.

The Great Depression showed that, at least in the 1930s, capitalism could not satisfy man's basic material needs, let alone any spiritual ones. Yet the majority of economists remained committed, not only to the social and political principles of capitalism, but also to the hard core principles on which their theory of a capitalist economy was based. According to the classical economists, the Great Depression should never have happened. In a perfectly competitive capitalist economy prolonged unemployment, they declared, is impossible—an automatically adjusting mechanism ensures the

---

* Hansen's conversion to the Keynesian paradigm is particularly interesting. In 1932 he argued that it was better not to try to prevent the cycle from reaching its natural low mark and that government intervention was both unnecessary and undesirable. In 1936 he reviewed Keynes' 'herculean' effort in a rather less than enthusiastic tone: 'The book under review is not a landmark in the sense that it lays a foundation for a new economics. It . . . is more a symptom of economic trends than a foundation stone upon which a science can be built.'[20n] But in 1938 these sentences were omitted in a reprint of the review and by 1941 the conversion was complete, his book *Fiscal Policy and Business Cycles* being not only an exposition but also a plea on behalf of the Keynesian theory.

† Of course, no one theory, let alone a sketch of a theory, can do justice to the multiplicity of competing sub-paradigms sharing a common hard core which Keynes somewhat inappropriately called classical theory. Provided, however, it is understood that our sketch is a reconstruction based on a considerable number of overlapping theories, we will follow Keynes' terminology and refer to this reconstruction as the classical theory and to the men whom Keynes was trying to convert as the classical economists.

maximum utilization of all scarce resources, including labour. Their diagnosis of the social malady was that real wages were too high for full employment; their prescription—a cut in money wages. 'Until recently,' Pigou emphasized in 1937 (a year after the publication of Keynes' *General Theory*), 'no economist doubted that an all-round reduction in the rate of money wages might be expected to increase the volume of employment.'[22] Of course, the classical economists agreed that this seemed a bitter remedy in such troubled times but only with the subsequent fall in real wages would all workers seeking employment be able to find it. As human beings they sympathized with the resistance of trade unions to wage cuts but, as economists (given their political and theoretical commitments), they believed that this was the only way to restore full employment. Let us try to understand this argument.

3. The 'Classical' Paradigm

A model economy

In order to focus attention on salient features of the classical paradigm we consider a closed capitalist society (i.e. an economy with no external trade of any kind) and in which there is no government taxation or spending. Thus there are only firms and 'households' to be considered, the firms making either consumer goods or capital (investment) goods. In addition, we assume that the capitalist economy is in a stage of perfect competition, i.e. that no one firm is sufficiently large so as to be able to significantly alter the total supply of a commodity by increasing its own production of that commodity.

Let us now return to our original problem to investigate what exactly the classical economists meant by full employment. To do this let us call the general price level P and the money wage per period W. Then the real wage $w = W/P$ represents the physical volume of goods per period that a worker can buy with his money wage W. Should the average price of goods rise but W remains the same, then clearly the worker's real wage falls. Now the classical economists assumed that the number of workers willing to work at wage w increases as the real wage offered increases, i.e. the 'supply curve' of labour $w(n)$ looks qualitatively as shown in Fig. 5.1. If w is the wage offered by the owners of capital and n is the number of workers willing to accept this offer, then the portion of y, the national product per period, that labour can buy is nw and therefore the portion available to the capitalists is $(y - nw)$. The question we now wish to ask is, how will the equilibrium values of y_0 and w_0 be determined? The classical economists reasoned as follows.

Given a certain stock of physical equipment K then the more workers employed the greater will be the national product. But, beyond a certain point, the additional product each extra worker contributes decreases as the total number of workers employed increases—this is the famous law of diminishing returns. If the total stock of equipment remains constant, then, beyond a certain point, the efficiency of $(n + 1)$ workers is less than the efficiency of *n working with the same equipment*. If the product added by the last man employed (the marginal product) is greater than his real wage, then it will be to the capitalist's benefit to employ him; if less, then the capitalist will certainly not employ him. Thus individual capitalists will find it profitable to employ workers until the marginal product of the last worker has a value equal to the average wage level w. Hence the lower the real wage accepted by workers the greater will be the capitalists' demand for workers. Writing y as a function of the number of workers employed i.e. $y = y(n)$, then the marginal product we write as $dy(n)/dn$, a decreasing function of n by hypothesis. The equilibrium values of w and n will be given by the intersection of the two curves $dy(n)/dn$ and $w(n)$ as shown in Fig. 5.1. At O the marginal productivity

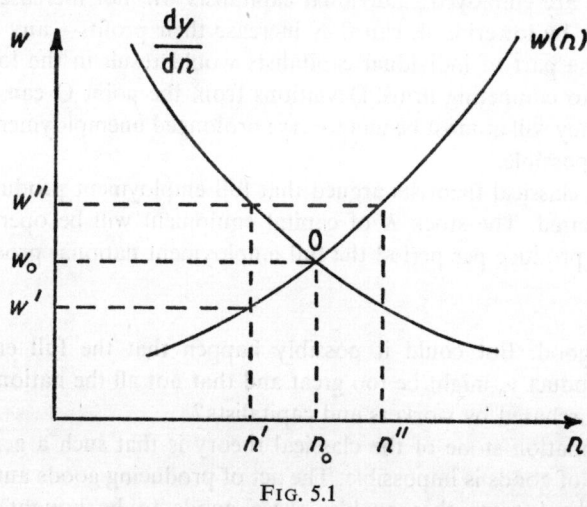

FIG. 5.1

of (n_0) workers is $(dy/dn) = w_0$. Since the going wage is w_0, industry maximizes its overall profit by employing n_0 workers, neither more nor less. On the other hand, not more than n_0 workers are prepared to work for this wage. Hence there is equilibrium, full employment equilibrium *by definition*, since all workers wishing to work at the wage w_0 are employed.

Note that in our perfectly competitive economy no one capitalist can offer a wage below the going one. If he did so, then workers would seek employment elsewhere. Of course, if there were only a few employers in our model economy then the story would be a very different one.

Suppose, however, that for some reason n' workers are employed at a wage w'' which is greater than the full employment wage w_0. The national product y' is then less than the full employment national product y_0, although the workers who *are* employed receive a wage higher than that they would get at full employment equilibrium. The situation is one, however, which cannot remain for any length of time, the classical economists argued. At the wage w'' there are n'' workers who wish to work but only n' of these are actually employed—hence there is unemployment of $(n'' - n')$ workers. Indeed, for any wage lower than w'' (but greater than w') there are more than n' workers who wish to work. Competition between the workers, employed and unemployed, will thus lower the real wage and hence, as it becomes more profitable for capitalists to employ workers, employment will continue to increase until full employment equilibrium is reached at the wage w_0, represented by the point O. There are now no unemployed workers seeking work at this wage w_0 and hence no competition between workers for jobs; furthermore, the marginal product is now such that it would be unprofitable for capitalists to attempt to employ additional men at the wage w_0. Thus the point O represents a stable equilibrium: all those workers wishing to work at the going wage are employed; individual capitalists will not increase this wage and neither, by lowering it, can they increase their profits—any attempt to do so on the part of individual capitalists would result in the loss of their employees to competing firms. Deviations from the point O can, of course, occur but they will at most be temporary; prolonged unemployment, it would seem, is impossible.

Thus the classical theorists argued that full employment production is on average assured. The stock K of capital equipment will be operated by n_0 workers to produce per period the full employment national product y_0.

Say's Law

Well and good. But could it possibly happen that the full employment national product y_0 might be too great and that not all the national product could be purchased by workers and capitalists?

The foundation stone of the classical theory is that such a general overproduction of goods is impossible. The act of producing goods automatically generates the income that enables these goods to be bought. Popularly expressed, Say's Law states that 'supply creates its own demand'. For the final price of a product is after all nothing but the cost of the raw materials with which it is made, the cost of the labour used in producing it, and the profit accruing to the producer. Hence the supplier of the raw materials, the labourer and capitalist have enough income between them to buy back the final product. The argument, it was claimed, holds for the complete economy. The production of the national product generates that national income which is sufficient for workers and capitalists together to buy back the total product. If there is unemployment, then the national product y will be less than the possible full-employment national product y_0. However, increased employment and production will generate that income which will be spent in the purchase of the added product. Since obviously no one will be content with anything less than the maximum possible national product, i.e. when there is full employment, additional production will take place until all workers seeking work are employed. Clearly individual firms could overproduce and find their produce unsold; but in this case the firms in question would either become bankrupt or begin to manufacture commodities that the general public wished to purchase—should they fail to do so other firms would step in, employing the redundant workers caused by the decline in business of the unlucky firms, and produce saleable products. While temporary unemployment was held to be obviously possible due to a changing pattern of consumer demand, general prolonged unemployment appeared an absurdity. 'From the time of Say and Ricardo,' Keynes wrote, 'the classical economists have taught that supply creates its own demand,' and 'Say's Law', as he explained, 'is equivalent to the proposition that there is no obstacle to full employment.'[23]

Again, well and good: the production of the national product generates the income which enables the public, if it wishes, to purchase that product. But will the public do so? Is not, after all, some income always saved? Surely not all profits, for example, are spent on the purchase of commodity goods? Of course they are not, but it is exactly this act of saving which, the classical economists declared, makes investment possible; the income saved is not hoarded (apart from the odd Silas Marner around) but is invested by capitalists on equipment, factories and similar investment goods. It is this act of saving (i.e. refraining from immediate consumption) that makes investment possible and therefore enables the economy to grow. Saving is to be welcomed. The more households save, the faster the economy grows.

It is important to look at this aspect of Say's Law in more detail. To do so, let us suppose our model economy to be in an equilibrium state such that the total number of goods produced in any given period is sold in that period and that therefore the inventory held by firms is neither growing nor shrinking in any period. Let us suppose that the capital goods industries pay out I in wages and dividends each period, produce goods to which they give a value I during the period and receive an income I each period from the sale of these capital goods to the consumer goods industries. The consumer goods industries produce and sell goods of value C each period and, let us suppose, collectively save an income S_F each period. (Thus each period the consumer goods firms pay out an income $(C - S_F)$ and receive an income C.) It follows that from the consumer and capital goods industries an income of $(I + C - S_F)$ flows each period into the households H who spend C of it (on consumer goods) and save an amount $S_H = I - S_F$ by, let us say, placing this part of their income in banks. Clearly since the capital goods industries receive income I each period the consumer goods industries order capital goods each period to the value of I which they pay for partly with their saved income S_F and partly with the saved income S_H which the consumer goods industries borrow from the banks. Since in each period the income I invested is equal to the income $(S_F + S_H)$ saved, the circular flow of income remains constant and the economy remains in equilibrium. A diagram schematically representing this circular flow of income is shown in Fig. 5.2.

In general the amount S_F saved by firms will be sufficient not only to cover the cost of replacing any capital equipment that wears out (depreciates) through the year but also to *increase* the stock of capital equipment. In any case the amount S_H specifically borrowed for investment purposes each period definitely leads to an increase in the stock of capital goods and hence to a growing national product. In what follows, however, we shall assume that I refers to net investment so that $(C + I)$ measures the value $Y (= yP)$ of the *net* national product, i.e. the total value of all final goods produced in the period *minus* the value of those capital goods that wear out during the period. To simplify the coming analysis we shall assume that firms do not save to

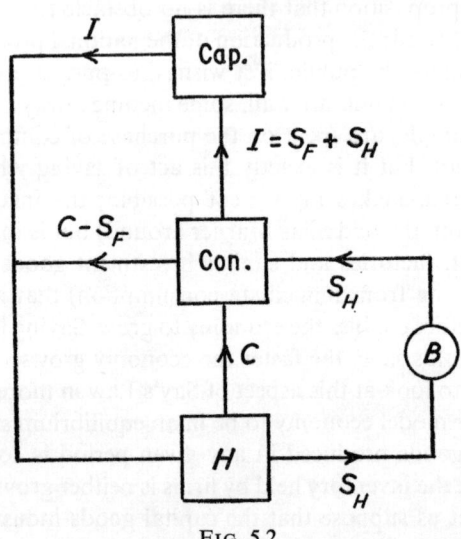

FIG. 5.2

increase the capital stock, i.e. $S_F = 0$, and that therefore all net saving, $S = S_H$, is done by households. The condition for inventory equilibrium remains the same, namely that income saved per period is equal to planned investment expenditure per period, i.e. $S = I$. In each period, of course, the consumer goods industries get progressively in debt by amount S. However, since their assets increase each period to match this increasing debt (each period the capital stock of buildings and equipment is correspondingly growing), all parties remain content. The interest on loans that is paid to banks, say rS each period, in part returns to industry through spending on consumer goods by those households who receive interest on their deposit accounts (or who are employed by banks), and in part is itself saved and is therefore part of the total saving by households for the period. For reasons of simplicity, however, these flows of income are not shown in the flow diagram of Fig. 5.2.

Say's Law at Work

Suppose, however, that in this equilibrium-state economy households suddenly increase the proportion of income they save, so that S become greater than I, the actual investment expenditure. There will no longer be equilibrium since inventories of consumer goods will begin to increase, i.e. *unintended investment* will occur, and it will begin to look as if a general *overproduction* of goods is occurring, contrary to Say's Law. However, such a situation cannot last for any appreciable time, the classical economists reasoned. For there is in the economy a self-adjusting mechanism—variations in the rate of interest—that ensures that the income saved by millions of households is eventually transformed into investment expenditure on the part of capitalists so restoring inventory equilibrium. For should income saved tend to increase faster than investment expenditure, then the rate of interest will automatically decline, making it therefore less profitable for households to save but, at the same time, more profitable for firms to invest (since the cost of borrowing money becomes less, i.e. the interest payable on loans decreases). Thus while the amount S of income that households desire to save is an *increasing* function of the rate of interest r, the intended investment expenditure I is on the other hand a decreasing function of r (as shown in Fig. 5.3). Where the two curves intersect is the equilibrium point ensuring

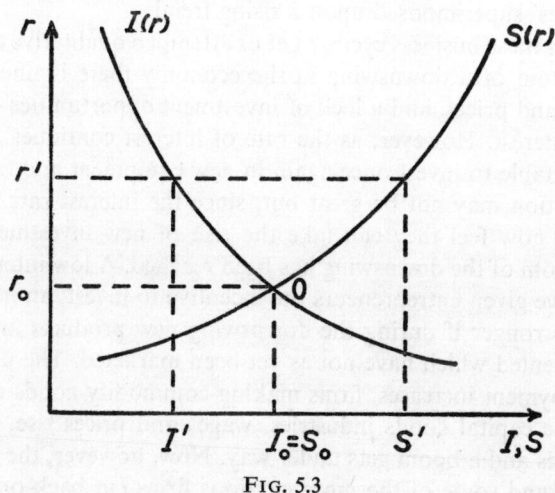

FIG. 5.3

the equality of intended investment and saving. At the equilibrium point O the rate of interest is r_0, thus ensuring that it is profitable for capitalists to invest the savings of the general public. Were the rate $r = r'$ greater than r_0, as shown in Fig. 5.3, then since S' is greater than I' (i.e. there is *unintended* investment of inventory of amount $S' - I'$) there would be disequilibrium in the economy. The rate of interest would then decrease and with it S until investment expenditure I is again equal to S at the equilibrium rate of interest r_0.

The money supply and the price level

To complete our sketch of the classical economic theory let us now note that, according to the theory, the price level P is determined by (i) the quantity of money M made available to the public by the monetary authorities and by (ii) the fraction l of income that is needed in the form of cash balances to satisfy the transactions demand for money (to a large extent, l is determined by the institutional structure of the economy, i.e. whether workers are paid daily, weekly, monthly, etc.). Clearly we have $M = lY$ and since $Y = Py$, we have, given M, l and y a relationship that determines the price level:

$$M = lPy, P = M/ly. \tag{5.1}$$

The money wage W is determined from the real wage w by

$$W = Pw. \tag{5.2}$$

The business cycle

According to our model, net positive investment is taking place each period and therefore the productive capacity of the economy must be growing; with full utilization of capital equipment y must be growing as well. Hopefully in a real economy y might be expected to grow smoothly over time. The reality, however, is different; the national product grows, to be sure, but with 'business cycles' superimposed upon a rising trend.

What causes these business cycles? Let us attempt a qualitative explanation. Near the bottom of a downswing in the economy there is unemployment, falling wages and prices, and a lack of investment opportunities—and also a low rate of interest. However, as the rate of interest continues to fall, so it becomes profitable to invest once again in new equipment and factories; the profit expectation may not be great but, since the interest rate is very low, entrepreneurs now feel they can take the risk of new investments. At this point the bottom of the downswing has been reached. A low interest rate and low wages have given entrepreneurs the incentive to invest, an incentive that will be even stronger if during the downswing new products and processes have been invented which have not as yet been marketed. The upswing then begins; employment increases, firms making commodity goods order equipment from the capital goods industries, wages and prices rise, demand for goods increases and a boom gets under way. Now, however, the interest rate begins to rise and some of the more cautious firms cut back on investment plans, or simply do not continue to maintain their investment expenditure once their initial plans have been realized. This has a very great effect on the capital goods industries: suppose a firm making consumer goods has ordered a machine each year from a capital goods firm in order to replace worn out equipment and then decides to expand its stock of 10 machines to a total of 11 in order to meet an envisaged 10% increase in the demand for the consumer goods it produces. That year the capital goods industry will receive not the usual order for one machine but for two—a 100% increase in the demand for capital equipment caused by only a 10% increase in the demand for the goods the machine is designed to produce. Thus when a boom is under way the capital goods industries tend to expand faster than the consumer goods industries. This means that as the upswing levels off, with full employment, high wages and prices, and a high interest rate, the capital goods industries will immediately feel the effect of reduced orders for equipment. Employers in these industries then dismiss surplus workers, thereby reducing demand for goods produced by the consumer goods industries. This in turn causes further unemployment. A slump gets under way. This continues until wages and prices have fallen appreciably, together with the interest rate, until it once again becomes profitable for business men to resume investment.[24]

Although the above sketch of a business cycle obviously means that the efficacy of Say's Law needs qualifications, nevertheless the classical economists still assumed that full employment was the norm and that unemployment represented only a *temporary* deviation from this norm. The tendency was to regard the business cycle as inevitable—that a boom period of sharply rising investment expenditure, prices and wages would inevitably be followed by a recession but which would in turn be *inevitably* followed by an upswing leading once again to the norm of full employment. Clearly there was much paradigm articulation to be done in order to produce a satisfactory theory of the business cycle, but there was no reason at all to suppose, thought the classical economists, that Say's Law needed to be challenged or that the economy could get stuck in the bottom of a slump. Although, of course, if the economy did get so stuck, then Say's Law and the classical paradigm could be used in order to advise the government, industry and labour on ways to get out of the depression. Clearly, however, according to the classical economists, *if* wages, prices and interest rates remained flexible, then the economy had built into it a self-regulating mechanism which would make prolonged unemployment impossible.

The classical economists and the Great Depression

This, then, was the trend of the advice offered by economists in the worst of the depression years. The severity of the depression, its duration and the excessive unemployment (up to 25% instead of the usual 6–10%), convinced most economists *not* that there was no self-adjusting mechanism, but rather that this delicate mechanism was being prevented from working properly. In a perfectly competitive, frictionless capitalist system there could be no persistent unemployment. Since there was, the cause lay in frictions, rigidities and obstructions, one of the major culprits being trade union obstruction to the downward movement of the real wage. Lord Robbins (then Professor Robbins), in his 1934 book *The Great Depression* declared that 'in general it is true to say that a greater flexibility of wage rates would considerably reduce unemployment'. Despite the fact, he regretted, that 'men of humanity, especially those who are not themselves of the wage-earning classes . . . feel a natural reluctance' to advocate that 'the position of others should be temporarily worsened', nevertheless, 'the true humanitarian', once he understands economic theory, 'will realize that a policy which holds wage rates rigid when the equilibrium rate has altered, is a policy which creates unemployment' and 'he will regard with some contempt the attitude of those who, unwilling to face the facts of poverty, content themselves with approving the enforcement of a wage higher than industry can bear and avert their gaze from the unemployment which they have thus created, or deceive themselves that it springs from other causes'.[25] Economic theory (or rather that version espoused by those economists committed politically to the perpetuation of capitalist society and theoretically to Say's Law) pointed an accusing finger unhesitatingly at the labouring class: if only those working would agree to eat less, their unemployed colleagues would later be able to join them in the canteen.

Not all economists, of course, supported a direct reduction in the real wage as a means of increasing employment. A few (Keynes and Robertson in England as early as the mid-1920s) suggested a massive programme of public works which would not only directly give work to the unemployed but would thereby enable much needed houses, schools and hospitals to be built. The classical economists regretfully pointed out that (i) additional government expenditure would result in a budget deficit—unsound at the best of times and possibly disastrous in a depression when it was of the utmost importance for the government to balance the budget—and (ii) that the saved income so used by the government for the construction of houses would thus not be available to industry for the construction of factories. And since the latter kind of investment by industry was one which raised both productive potential and employment, it was 'obviously' preferable to one which raised only the level of employment. Clearly government intervention, if not causing a final and catastrophic loss of business confidence and initiative, would certainly delay recovery. As Winch has written, 'Running through the orthodox accounts of the depression was the view that the interventionist trend of government activity in the post-war period was itself a prime reason for the failure of adjustment mechanisms to work as well as they had in the past.'[26] At the time Keynes and Robertson had no alternative economic theory with which to reply. There was, of course, no recommendation from economists that the government should invest in factories and thus compete with private industry. For this would have meant admitting the total failure of the capitalist social order. Conveniently enough, 'objective' economic theory indicated that the blame lay elsewhere!

In the United States, however, despite expert advice to the contrary, President Roosevelt launched his 'New Deal' policy in 1933. Although accompanied by many conservative measures, it did include a large programme of such public works as housing estates, roads, dams, harbours, irrigation works and similar projects. The budget deficit increased (there had been a surplus in 1929–30) but industry failed to respond; unemployment did fall to 14% in 1937 but rose again to 19% in 1938. As the experts had warned Roosevelt, a budget deficit caused by public works would not induce an economic recovery—on the other hand, things had not got worse. But the capitalist economies were not to achieve full employment again until after the onset of the Second World War. As Keynes himself lamented in 1940: 'It is, it seems, politically impossible for a capitalist economy to organise expenditure on the scale necessary to make the grand experiment which would prove my case—except in war conditions.'[27] But perhaps, he informed his American audience, the establishment of a war economy would be 'the stimulus, which neither the victory nor the defeat of the New Deal could give you, to greater consumption and a higher standard of life'. So it proved to be.

In 1883, the year Karl Marx died, John Maynard Keynes was born. Whereas Marx had set out to criticize and destroy capitalism, Keynes set out to be one of capitalism's saviours in its hour of need. And with Keynes' achievements (although applied in a way he never intended) Marx's dream of a just and humane world society appeared to recede into the indefinite future. With what consequences for mankind we shall explore in later chapters. We turn now, however, to a discussion of the revolutionary achievements of a man politically committed to the defence of capitalist society.

4. The Keynesian Paradigm

For many years before he published his revolutionary work, Keynes had advocated unorthodox economic policies as a means of reducing unemployment. He was, however, unable to give his policies the theoretical support that radical departures from the orthodox tradition obviously required if they were going to be taken seriously by his fellow economists and by governments responsible for ensuring public acceptance of the social and political structure. For Keynes, unemployed men and idle factories* indicated a fundamental weakness in the classical paradigm. He could not believe that it was the greediness of those workers who were employed that prevented those on the dole from finding work; neither could he believe that a large programme of public works paid for by the government by borrowing money from those prepared to lend it would bring even greater economic and social disharmony. But he was unable to argue these convictions from a well-articulated economic theory. As, however, he sought to escape from the stranglehold of the classical paradigm, his critical thoughts increasingly focused on the foundation stone itself of classical theory—Say's Law.

In order to focus attention on an essential feature of Keynesian economic theory, let us consider an imaginary situation in which, after the economy has reached a full employment equilibrium position, all households suddenly decide to save an extra fraction of their income so causing the original saving schedule $S(r)$ of Fig. 5.3 to jump to the right as shown in Fig. 5.4.

* Unemployment in advanced capitalist countries is clearly to be distinguished from that in the so-called underdeveloped countries. We shall say much about underdevelopment in chapter 8; we remark here, to prevent misunderstanding, that it is impossible to discuss the plight of the 'third world' without studying the latter's relationship with the advanced capitalist countries.

The immediate effect is a rise in the volume of saved income to S' and since intended investment expenditure remains temporarily at I_o, there is an unintended investment (of inventory) to the value of $(S' - I_o)$. This would be an unstable situation. According to the classical economists, the excess of savings in the capital sphere would force down the interest rate, thus making saving less profitable but investment more so. Ultimately a new equilibrium position O' would be reached at which the new interest rate r'_o, lower than r_o, equilibrates intended investment and savings at $I'_o = S'_o$. Employment

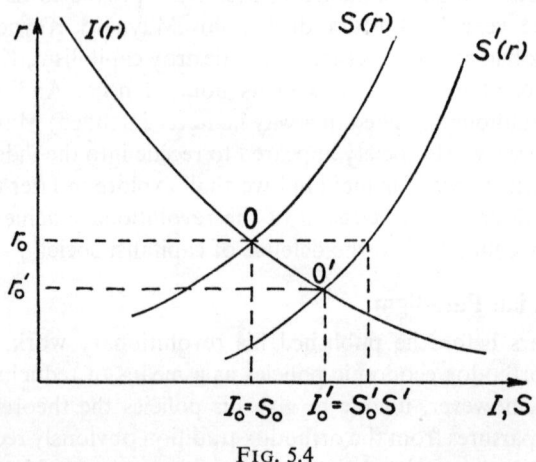

FIG. 5.4

remains unchanged since the temporary redundancy caused by the fall in sales of commodity goods (firms making such goods would dismiss workers) is eventually compensated by increased employment in the capital goods industries following the rise in investment as the interest rate declined. The national product remains temporarily the same but the increased investment expenditure each period leads to faster economic growth than formerly. As always, the classical economists would argue, saving is a virtue and the more people save the faster the economy will grow. This is Say's Law: income saved will always be spent.

It was on this kingpin of the classical theory that Keynes launched his attack. It would indeed be most convenient if as saving increased and the interest rate declined, investment expenditure automatically rose to match the increased saving. 'It may well be,' wrote Keynes, 'that the classical theory represents the way in which we should like our Economy to behave. But to assume that it actually does so is to assume our difficulties away.'[28] How in fact might the economy behave in such a hypothetical situation—would investment increase as the interest rate declined? Not at all likely, thought Keynes. Since investment is ultimately undertaken for the purpose of satisfying a customer, an increase in saving may not only reduce present consumption-demand but investment-demand as well! After all, a decision not to have dinner today 'does *not* necessitate a decision to have dinner or to buy a pair of boots a week hence or a year hence or to consume any specified thing at any specified date'.[29] It simply, remarked Keynes, 'depresses the business of preparing today's dinner without stimulating the business of making ready for some future act of consumption'. Thus why should industry decide to invest *more* just when consumers are making a decision to consume *less*? It would be convenient if industry did so, but perhaps somewhat foolish from the entrepreneurs' viewpoint unless they felt convinced that the public was only consuming less today in order to consume more tomorrow. Failing this long-term expectation, and despite a drop in the interest rate, the entrepreneurs might well decide not to increase investment expenditure. What would happen then? The public has taken a decision to save more, industry has taken a decision to keep investment expenditure the same. For equilibrium, investment expenditure per period must equal the income saved per period. But how can this equilibrium result if investment expenditure does not rise to compensate for the extra income saved? Let us examine the situation carefully.

Let us introduce a radically new hypothesis by supposing that 'households' spend a constant fraction c of the national income Y_o and therefore save a constant fraction $s = 1 - c$ of this income. (The classical economists had assumed S to be a function of r, the rate of interest; however, we are now assuming S to be primarily a function of Y the national income itself.) Then the saved income over the period in question is $S_o = sY_o$. Since the economy is in equilibrium (by hypothesis) we must have that intended investment I_o is equal to S_o and hence

$$S_o = I_o = sY_o, \qquad (5.3)$$

giving
$$Y_o = I_o/s. \qquad (5.4)$$

Now suppose that the 'propensity to save' s suddenly increases to s' but that investment expenditure I_o stays constant. In that case the new equilibrium position will be when $S_o = I_o$ once again. But since s has increased to s', something must have changed—it is the national income Y! The new Y, say Y', is given by

$$S_o = s'Y' \qquad (5.5)$$

and, since $S_o = I_o$, we have

$$Y' = I_o/s'. \qquad (5.6)$$

Furthermore, since s' is greater than s, the national income has shrunk and unemployment has increased! Let us look at what has happened dynamically, analysing step by step the transition to the new equilibrium of less than full employment.

We represent the equilibrium situation by the flow diagram of Fig. 5.2

except that we now amalgamate the capital and consumer goods industries together in one box as shown in Fig. 5.5.

Let us begin by supposing that in period 0, as in former periods, the households save a fraction s of their income (and thus spend a fraction $c = 1 - s$

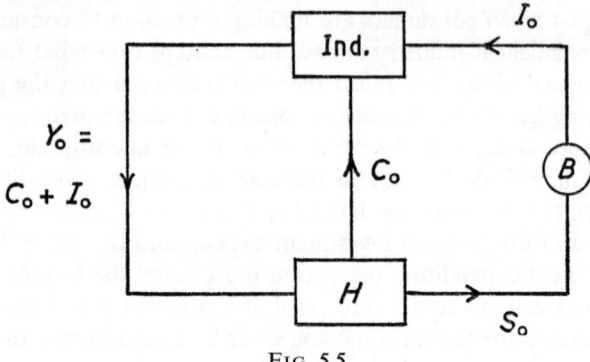

FIG. 5.5

of it) but that in period 1, and for all subsequent periods, the households save a fraction s' of their income where s' is greater than s. In the first period, therefore, the amount spent on consumption will be less than before by the extra amount saved, namely $(s'Y_o - sY_o)$, as shown in Fig. 5.6 (i). (For

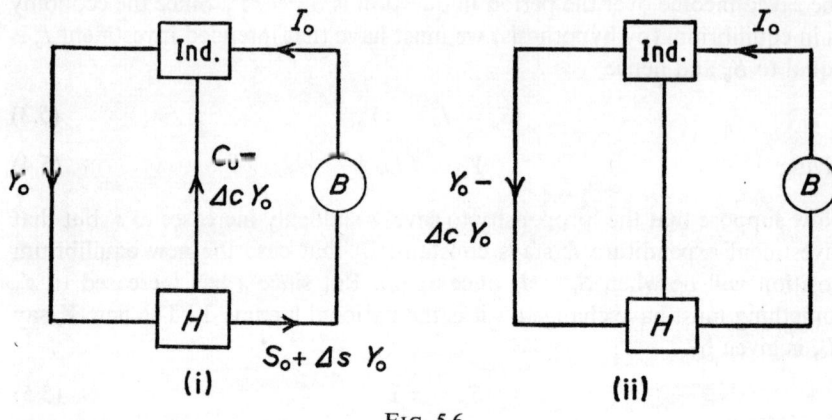

FIG. 5.6

convenience, we write $(s' - s)$ as Δs and $(c - c')$, the decrease in the propensity to consume, as Δc.) Of course, $\Delta s = \Delta c$. Clearly $\Delta s Y_o$ accumulates in the capital sphere (in the banks we have supposed). Industry now receives less revenue than before by the amount $\Delta s Y_o = \Delta c Y_o$ and this loss of revenue is passed on to the households as loss of income, as shown in Fig. 5.6(ii). The national income is now $(Y_o - \Delta c Y_o)$. The households then spend $c'(Y_o - \Delta c Y_o)$ and hence the firms' total revenue decreases by an additional $c'\Delta c Y_o$ (i.e.

CAPITALISM IN CRISIS

in addition to the original decrease $\Delta c Y_0$). National income is now ($Y_0 - \Delta c Y_0 - c' \Delta c Y_0$). Once again households spend c' of this and the firms make yet another loss $c'^2 \Delta c Y_0$. Clearly the net result of this multiplier effect resulting from the original decision on the part of the households to cut back on consumption is a drop in the national income (and therefore a drop in national consumption since investment stays constant) by the amount

$$\Delta c Y_0 (1 + c' + c'^2 + \ldots) = \frac{\Delta c}{1 - c'} Y_0. \tag{5.7}$$

What, however, has been happening to savings while the national income has been decreasing? In the first period there was an *increase* in savings by $\Delta s Y_0$. But in the second period the drop $\Delta c Y_0$ in the households' income (because their additional savings are not returned to them) means that only $s'(Y_0 - \Delta c Y_0)$ is saved and thus although income saved in this period is still up on period 0 it is so by the decreased amount $(\Delta s Y_0 - s' \Delta c Y_0)$. In the next period the households' income decreases yet again by $c' \Delta c Y_0$ so that income saved decreases by $s' c' \Delta c Y_0$. In the next period it will decrease by $s' c'^2 \Delta c Y_0$ and so on. Hence after many periods the increase in income saved per period will be

$$\Delta s Y_0 - s' \Delta c Y_0 (1 + c' + c'^2 + \ldots)$$
$$= \Delta s Y_0 - s' \Delta c Y_0 \frac{1}{1 - c'} = \Delta s Y_0 - \Delta c Y_0 = 0. \tag{5.8}$$

This is the result we expected. Since the increase in saved income was not returned to industry (through increased investment expenditure on the part of industry) and hence to households, the attempt by households to increase permanently the volume of savings each period was destined to be a failure.

There is a further point to consider. In the above blow by blow account we assumed that investment expenditure remained constant as the production and sale of consumer goods steadily decreased. But as the national income fell it would be strange indeed if the more cautious business men did not decide to temporarily postpone intended investment. If this happened, I_0 would fall to, say, I' and the national income would fall even lower than $Y' = I_0/s'$ to $Y'' = I'/s'$. There would now be serious unemployment in both the consumer goods *and* investment goods industries. Furthermore, as we noted in the discussion on the business cycle, a small decision to cut back investment in the consumer goods industries means a disproportionately large cut back in the capital goods industries. Hence the final result of a downswing in the economy may mean drastic unemployment in those sensitive industries making capital equipment. We could then have a situation of idle machines, idle factories and a large fraction of the labour force fruitlessly seeking employment—exactly the situation in the thirties! Clearly in such a situation recovery would start if only industry would begin to invest more. But it would be a somewhat intrepid capitalist who would build an additional factory when those already existing are not being used! The economy would appear to be firmly stuck at the bottom of a depression.

How could industry in such a situation be encouraged to invest more? The classical economist, apart from advocating a decrease in the workers' real wage, would reply that the public should be advised to save more so that industry would have more funds available for investment. Keynes' reply was that people should save less! Indeed, as Lerner wrote in 1936: 'It is not so very long ago that we had Professor Robbins and Keynes on the wireless, respectively advising the world to save more and to spend more!'[30] We are now approaching the heart of the Keynesian diagnosis of a capitalist economy in depression. The increased saving advocated by Robbins would only make matters worse and a cut in the workers' wages would at best do no good at all. For the depressed economy is suffering from a lack of demand for goods and therefore, in such a situation, demand for goods must be stimulated and thrift discouraged. Should industry not increase demand through additional investment expenditure, then the government could do so (and should do so, said Keynes) by giving jobs to the unemployed, thus providing them with an income and transforming their desire for goods into an *effective* demand—a demand backed by money! Let us look more formally at the Keynesian analysis of how a capitalist economy functions—or dysfunctions, as the case may be!

The simple Keynesian model

At one point the classical economists and Keynesians are in agreement: the economy is in equilibrium when income saved is equal to income invested. For the classicists, however, the norm will be a full employment equilibrium, with fluctuations in the interest rate ensuring the equality of I and S. For the Keynesians, on the other hand, the interest rate plays only a minor role, if any at all—equality between I and S is ensured by variations in the national income itself! There is, for Keynesians, no special reason why I should equal S at full employment. On the contrary, unless the government intervenes directly in the economy, there is every reason to suppose that equilibrium will be established with idle machines and considerable unemployment. Indeed, given a constant investment expenditure I' and a constant propensity to consume c of the households there is only one equilibrium value for Y. If this should coincide with the full employment equilibrium income, the only conclusion to be drawn from the simple Keynesian model is that the gods of chance have blessed the economy.

Suppose, however, as is more likely, that chance has not been kind to the economy. How can the economy be stimulated so that Y increases from the less than full employment Y' to the full employment income Y_0? Clearly, if only the entrepreneurs would increase investment expenditure from I' to I_0, then the problem would be solved. From equation (5.4) we see that the necessary increase ΔI is given by

$$\Delta I = s(Y_0 - Y') \qquad (5.9)$$

which is only a fraction s of the desired increase in national income. To see this result dynamically, suppose that in period 1 the capital goods industries increase investment expenditure by ΔI in order to build extra factories or to produce the extra investment goods ordered by the consumer goods industries. This extra income flows into the households who spend $c\Delta I$ of it and save $s\Delta I$. In the second period, an extra ΔI is again received by the households (the entrepreneurs maintain each period the new level of investment expenditure). However, the original $c\Delta I$ spent in the first period is income for shopkeepers and businessmen whose households thus receive this extra income $c\Delta I$ together with the increment ΔI. Clearly a fraction c of this extra income $(\Delta I + c\Delta I)$ will be spent and a fraction s saved. Together with the regular increment ΔI, this fraction $c(\Delta I + c\Delta I)$ is returned to the households and the whole process is repeated. Clearly the total increase in the national income will be

$$\Delta Y = \Delta I(1 + c + c^2 + c^3 + \ldots) = \frac{\Delta I}{1 - c} \quad \frac{\Delta I}{s} \qquad (5.10)$$

as given by equation (5.9). Furthermore, savings will automatically increase by the increment

$$\Delta S = s\Delta I(1 + c + c^2 + \ldots) = s\frac{\Delta I}{1 - c} = s\frac{\Delta I}{s} = \Delta I. \qquad (5.11)$$

Obviously according to our model this must happen. Although the households save the same proportion of their income, the national income rises to such an extent that their savings increase to match the additional investment expenditure ΔI. This is the typically Keynesian result: savings are brought into equality with investment through the movement of the national income, in this case a welcome upward movement.

Clearly the rise to full employment would be quickened and would demand less investment by entrepreneurs if households decided to increase the propensity to consume during the transitional period. As Keynes advised in the radio talk referred to, the way out of a depression is to spend *more*, not *less*.

One additional point. We have assumed that the rise in national income corresponds to a rise in the real national product, i.e. not merely to a rise in both wages and prices leaving no one better off. That this can happen is made possible by the fact that unused resources, men and factories, are idle and unsold inventories exist in warehouses waiting to be sold. Should the economy already be at full employment when entrepreneurs decide autonomously to increase investment expenditure, then the extra income would not initially find extra goods available in shops and warehouses and the result would be a rise in the national income through a rise in prices that would therefore not correspond to an equivalent rise in the *real* national product. In other words, the economy would undergo inflation.

5. The Classical versus the Keynesian Paradigm

For the classical economist all members of the labour force seeking work at the going wage find it; the national product is limited only by the stock of capital equipment K_0. This stock K_0 thus gives rise to the full employment national income Y_0, part of which S_0 is saved and invested, which in turn gives rise to a greater K_0 and in the long run to a greater Y_0. Thus we can envisage the economy working in the following dynamic pattern according to the classical economists:

$$K_0 \longrightarrow Y_0 \longrightarrow S_0 \longrightarrow I_0.$$
$$\uparrow \text{\_ \_ \_ \_ \_ \_ \_ \_ \_} |$$

For the Keynesians the (autonomously decided) investment expenditure and the propensity to save on the part of the public together determine the level of the national income Y' and the amount K' of capital equipment in use. Income saved per period necessarily follows investment expenditure through the movement of Y. Thus, according to the Keynesians, the appropriate dynamic pattern for the economy is:

$$I' \searrow$$
$$\nearrow Y' \leqslant Y_0$$
$$s \longmapsto sY' = S' = I'.$$

Although, of course, the classical and Keynesian models described above are greatly simplified, nevertheless, one can see the revolutionary nature of the Keynesian critique of classical theory by observing that Keynes directly negated two fundamental assertions of the classical economists:

Classical economists	*Keynes*
1. Increased savings necessarily lead to increased investment	Increased savings do NOT necessarily lead to increased investment
2. Prolonged unemployment is impossible with flexible wage rates	Prolonged unemployment is NOT impossible with flexible wage rates.

Let us now attempt to make the Keynesian model slightly more sophisticated and show in greater detail its differences from classical theory. First of all the classical economists assumed that investment expenditure and savings depend very strongly on the interest rate, so much so that the investment and savings schedules certainly intersect at some (positive) interest rate. The Keynesians, however, argued that I and S are not strongly dependent on the interest rate; on the contrary, they argued, S in particular is a function of Y rather than of r, and thus there may well be a level of the national income Y_c above which the investment and saving schedules simply do not intersect. If this is the case, as for the schedules of Fig. 5.7, then equality of S and I can be achieved by changes in the interest rate only if Y does not exceed Y_c. If, for example, the interest rate happens to be r_0 and the national income Y_0 (as shown in Fig. 5.7), then S is considerably greater than I and the national income will certainly drop. Even if r falls to zero, then Y must still fall to Y_c for I to equal S. If r falls not to zero but only to r', then Y must fall even further to Y' before equality of I and S is achieved. Clearly, the further the interest rate can be made to fall, the smaller will be the final drop in the level of the national income.

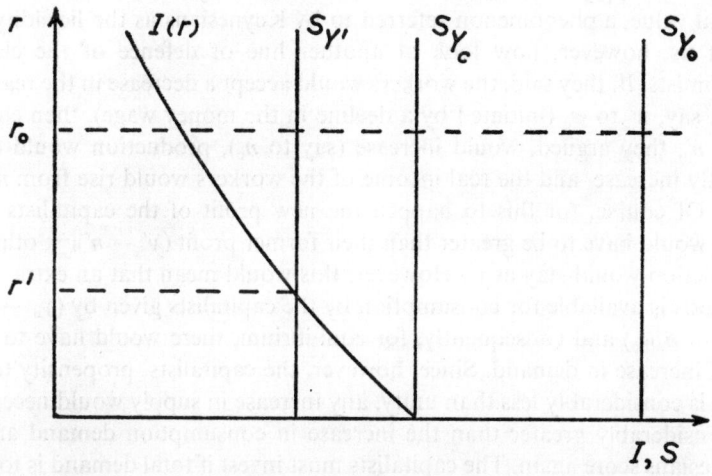

FIG. 5.7

However, to make matters worse for the classical economists, Keynes argued that there are very good reasons for supposing that the interest rate will in general not fall, and indeed cannot ordinarily be made to fall, to very low values. For suppose in a depressed economy a concerned government tries to lower the interest rate by increasing the amount of money in the capital sphere, e.g. by purchasing bonds formerly sold to wealthy members of the public. Normally, Keynes agreed, the greater the supply of money then the lower is the interest rate—provided, of course, (and this is an apparently reasonable assumption to make) the demand for money remains constant. After all, no one hoards money (the classical economists assumed) when any form of investment brings a return; money in the form of cash is needed only for *transactions*, other reasons for needing it being negligible. However, in much depressed times, argued Keynes, there is a *speculative* demand for money and the lower the interest rate the greater this demand. For if and when the interest rate rises, ready money can be put to profitable use. On the other hand, bonds bought when the interest rate is low only decrease in value when the interest rate rises.* Hence when r is low it is financially wise to hold

---

* A bond entitles the holder to a fixed sum of money each year for a stated number of years. Clearly the higher the rate of interest, the smaller the value of a bond.

money and not bonds. Consequently when r is low the envisaged government intention of lowering it even further by buying bonds on the open market is thwarted; holders of bonds are anxious to sell while the government wishes to buy but bond prices do not rise and neither does the interest rate decrease—the supply of money increases, to be sure, but the demand for money is such as to absorb it all. In fact, the Keynesians later argued, there may be a critical value below which the interest rate cannot be forced, no matter by how much the money supply is increased; the demand for money is in effect infinite at this critical value, a phenomenon referred to by Keynesians as the liquidity trap.

Let us, however, now look at another line of defence of the classical economists. If, they said, the workers would accept a decrease in the real wage from, say, w' to w_0 (initiated by a decline in the money wage), then employment n', they argued, would increase (say to n_0), production would consequently increase, and the real income of the workers would rise from $n'w'$ to n_0w_0. Of course, for this to happen the new profit of the capitalists ($y_0 - n_0w_0$) would have to be greater than their former profit ($y' - n'w'$), otherwise production would stay at y'. However, this would mean that an extra supply of goods is available for consumption by the capitalists given by ($y_0 - y'$) $-$ ($n'w' - n_0w_0$) and consequently, for equilibrium, there would have to be an equal increase in demand. Since, however, the capitalists' propensity to consume is considerably less than unity, any increase in supply would necessarily be considerably greater than the increase in consumption demand and the Keynesians score again. The capitalists must invest if total demand is to equal total supply but the only way of guaranteeing this is by invoking Say's Law! According to the Keynesians, as we have seen, there is in a depressed economy a speculative demand for money; when the money wage drops, any increased income accruing to wealth holders is not all spent—far from it. Real demand may fall! And, in addition, capitalists are reluctant to invest when machines and factories are already idle. Hence, the Keynesians were able to conclude, a decrease in the workers' money wage would in no way serve to restore full employment; on the contrary, they reasoned, it would be followed by a decrease in the price level until the real wage was the same as before, while the national product y' remained constant and unemployment continued.[31]

The classical economists, however, attempted a comeback for their paradigm by declaring that as wages and prices continued to fall, then wealth holders would see their bank balances increasing in real value and would begin to increase consumption. Eventually a stage would be reached when the increased propensity to consume would lead to full employment. So ran Pigou's answer to Keynes' accusation that falling wages and prices would not reduce unemployment. The Keynesians were not impressed. They pointed out that no one knew to what extent wages and prices would have to be lowered before the Pigou effect became important, if it ever did; and in the process, the workers might decide that capitalism had had a long enough run for its profit. In any case, Keynes' analysis suggested an altogether different course.

Better by far, the Keynesians therefore recommended, to take this much surer way out of a depression, the way of increased government expenditure financed by the sale of short-term government bonds. For when private investment is insufficient to create full employment aggregate demand, then, said Keynes, it is the duty of the government to remedy the deficiency. Before going on, however, to look at the political implications of Keynes' recommendations, let us show how government expenditure and taxation can be included in our (still much simplified) model of a perfectly competitive capitalist economy.

6. The Inclusion of Government Expenditure and Taxation

Let us consider an equilibrium state of a perfectly competitive capitalist economy in which in each period the government buys goods of value G from industry and receives a tax revenue T consisting of T_1 from industry and T_2 from households. In each period (let us assume) industry makes and sells goods not only of value G to government, but also consumer goods of value C to households, and capital goods of value I to industry itself. Of this income $(C + I + G)$ received by industry, $(C + I + G - T_1)$ is paid to households and T_1 to government in taxes. Out of their income the households pay T_2 in taxes, spend C, and save S. In each period the consumer goods industries borrow I to purchase investment goods of value I from the capital goods industries. These flows of income are shown schematically in Fig. 5.8.

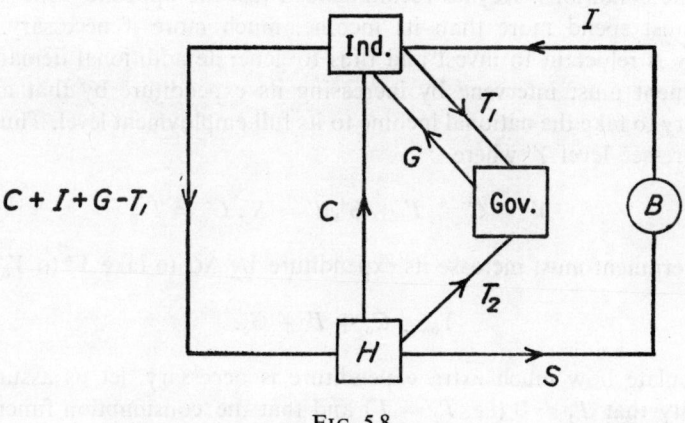

Fig. 5.8

The principal circular flow is between industry and households. During each period income to the value $(S + T_1 + T_2)$ is withdrawn from this flow; on the other hand in each period income to the value $(I + G)$ is injected into this flow. The circular flow remains constant since (by assumption) income withdrawn each period is equal to income injected, i.e.

$$S + T = I + G. \tag{5.12}$$

Provided this condition remains satisfied the value of the national product will remain at

$$Y = C + I + G \qquad (5.13)$$

and national expenditure will likewise remain at this value (all goods produced are sold, there being no *unintended* investment of inventory).

Now the important point to observe about equation (5.12) is that no longer need I equal S for equilibrium. Clearly, investment expenditure per period can be much less than income saved by households per period, *provided* there is an excess of government expenditure over tax revenue by an exactly compensatory amount—in other words, provided

$$S - I = G - T. \qquad (5.14)$$

The government is able to finance such a budget deficit either by printing money or, more usually, by borrowing that income saved by households but not required by industry for investment purposes.

Now we can see how, according to Keynes, a government can act so as to restore a depressed economy to a full employment equilibrium. Consider a situation in which the national income Y' is well below that necessary for full employment but in which the government budget is balanced, i.e. $G' = T'$, an act of policy strongly recommended by classical economists in adverse economic conditions. Keynes recommended just the opposite—the government must spend more than its income, much more if necessary. Since industry is reluctant to invest and thus to generate additional demand, the government must intervene by increasing its expenditure by that amount necessary to take the national income to its full employment level. Thus from the depressed level Y' where

$$Y' = C' + I' + G', I' = S', G' = T' \qquad (5.15)$$

the government must increase its expenditure by ΔG to take Y' to Y_o where

$$Y_o = C_o + I' + G_o. \qquad (5.16)$$

To calculate how much extra expenditure is necessary, let us assume for simplicity that $T_1 = 0$ (i.e. $T_2 = T$) and that the 'consumption function' is given by

$$C = c(Y - T) \qquad (5.17)$$

i.e. households wish to spend a fraction c of their *disposable income* $(Y - T)$. Let us further assume that the government rate of taxation is such that the tax revenue T rises proportionately with Y, i.e.

$$T = tY. \qquad (5.18)$$

CAPITALISM IN CRISIS

Hence we have

$$Y' = c(Y' - tY') + I' + G' \qquad (5.19)$$

and

$$Y_o = c(Y_o - tY_o) + I' + G_o.$$

Therefore, the additional government expenditure necessary is

$$\Delta G = G_o - G' = [1 - c(1 - t)](Y_o - Y'). \qquad (5.20)$$

Hence to achieve the required rise in the national income $\Delta Y = Y_o - Y'$ the additional government expenditure needed is less than ΔY; the multiplier effect occurs as described before, the multiplier in this case being $(1 - c(1 - t))^{-1}$. Furthermore, we note that the government's tax revenue rises (although not enough to cover the additional expenditure) by the amount $\Delta T = t\Delta Y$. In addition we note that the rise in consumption expenditure (presumably gratefully received by a surprised industry which might then be stimulated to increase its own investment expenditure) is given by

$$\Delta C \equiv C_o - C' = c(\Delta Y - \Delta T) = \frac{c(1-t)}{[1 - c(1-t)]}\Delta G. \qquad (5.21)$$

While on the economic effect of the government budget, let us note the effect of a budget in which the government increases expenditure by ΔG but at the same time increases the taxation rate from t to t_o in order to keep the budget balanced, i.e. to ensure $\Delta G = \Delta T$. In this case Y' will rise only by the amount $\Delta Y = \Delta G$ and there will be no multiplier effect and no increased consumption. Nevertheless, the additional government product (homes, schools and hospitals, let us hope) will benefit the entire community. Hence the increased balanced budget is clearly not economically neutral: it increases the national income above what it would be without the additional government expenditure, it provides employment, and therefore results in a redistribution of disposable income from the employed to the unemployed.

There will, of course, be no need for a government to increase the taxation rate in order to obtain a balanced budget if additional budget expenditure stimulates industry to undertake further investment. Suppose, for example, that an increase in government expenditure of ΔG induces a rise in private investment of $\Delta I = i\Delta G$. Then the total rise in investment will be $(1 + i)\Delta G$, the national income will rise by $[1 - c(1 - t)]^{-1}(1 + i)\Delta G$, and the government will receive back an increase in taxation revenue of

$$\Delta T = \frac{t}{[1 - c(1-t)]}(1 + i)\Delta G. \qquad (5.22)$$

For the United States (very approximately indeed) c is of the order of 0·9 and t about 0·3. Hence

$$\Delta T \approx 0·8(1 + i)\Delta G$$

and thus, according to this simple model, if industry undertakes investment expenditure of more than about a quarter of ΔG, the government eventually finds itself with a budget surplus.

On the use of a simple model

To be sure we have in these few pages done no more than sketch some of the broad outlines of the at one time competing 'classical' and Keynesian paradigms. To what extent, however, can *sketches* of complicated paradigms be useful not only for identifying causes of problems but in suggesting further research-action programmes for the solution of these problems?

Of considerable use! Naturally such sketches as presented in this chapter are completely useless from the point of view of giving detailed advice on problems of day-to-day economic policy. But such is not the purpose of this book. Its purpose is to present an overall paradigm centred on the question as to whether problems generated by capitalist society can be resolved within the class structure and competitive ethos of capitalism or whether they can be resolved only by the construction of a radically different society. And, as the reader will, I hope, ultimately agree, the sketches presented within this overall paradigm framework are of considerable relevance with respect to the answering of such a fundamental question.

7. The New Deal Policy and the Second World War

After this long but very necessary theoretical detour, let us return to the United States in the depression years of the thirties, allowing Keynes to remind us of the basic issues as he saw them. In 1933 he wrote:

> If our poverty were due to famine or earthquake or war—if we lacked material things and the resources to produce them, we could not expect to find the Means to Prosperity except in hard work, abstinence and invention. In fact, our predicament is notoriously of another kind. It comes from some failure in the immaterial devices of the mind, in the workings of the motives which should lead to the decisions and acts of will necessary to put in movement the resources and technical means we already have.

Two years earlier he had proposed a remedy to a government investigating committee:

> Government expenditure will break the vicious circle. If you can do that for a couple of years, it will have the effect, if my diagnosis is right, of restoring business profits more nearly to normal, and if that can be achieved, then private enterprise will be revived. I believe you have first of all to do something to restore profits and then rely on private enterprise to carry the thing along.[32]

The cause of the Great Depression was a lack of aggregate demand, maintained by industry's unwillingness to invest. The remedy was government expenditure on socially useful projects—the surest and quickest remedy; all other prescriptions were at best doubtful and at worst unemployment-inducing. The government budget must go into deficit and, if necessary, stay in deficit until full employment was reached. So ran Keynes' diagnosis and prescription. Let us look at what happened.

During the boom years of the twenties productivity rose rapidly; in manufacturing, for example, output per man hour increased between 1920 and 1929 by as much as 60%. Real wages did not increase as fast as productivity; money wages rose very little although prices fell by about 15%. Hence while real wages rose between 15 and 20%, profits rose much more. In 1919 the top 1% of the population enjoyed 12·2% of the total disposable income; by 1929 the top 1% was enjoying 18·9% of the total income. Over the same years the fraction of the total income enjoyed by the top 5% increased from 24·3 to 33·5%.[33] How could the boom be sustained with such an inequitable distribution of income? The majority in need could not consume adequately since they lacked sufficient income; the wealthy few, on the other hand, had no need to consume—their excess income was used in the speculative purchasing of company shares in the hope of turning a small fortune into a large one. The boom, however, could be sustained only if capitalists were prepared to invest the ever-increasing volume of savings in the building of extra plant, new factories and homes. Indeed, between 1925 and 1929, the rate of increase of capital goods production was almost twice that of consumption goods. But since all investment is made in the expectation of future sales, the relevant question was whether an increasing abundance of goods would continue to find purchasers. What would the housing industry build when all Americans who could afford to do so had acquired homes? In 1917 some 1¾ million cars were sold; in 1929 nearly 4½ million cars found buyers, bringing the total number of cars owned by Americans to 26½ million—one car for every six Americans. But how could the sale of cars continue to increase as the distribution of income became more and more inequitable? The total output of electrical goods and supplies tripled in value between 1921 and 1929; while only some 200 000 radios were marketed in 1923, in 1929 the number was 5 million! According to the economic historian, Peter d'Alroy Jones, advertisers attempted to maintain the sale of durable goods 'by encouraging the speedy casting off and replacement of consumer "durables" and by forever introducing real or feigned "latest improvements" '.[34] Indeed, advertising expenditure in 1929 was $3·4 billion, some 4% of the total disposable income. But no amount of advertising would have been able to maintain aggregate demand without a redistribution of the national income. The poor could not spend; the rich saved too much. And industry was not prepared to invest all the savings. Indeed, again quoting Jones, 'business could make enough profit to finance itself by "plow-back" and did not need to get to the banks for loans to extend plant or expand facilities. . . . Even the large outflow of capital funds that had found investment opportunities abroad still left idle bank resources at home.'[35] Clearly the economy was heading for a serious downturn. To make matters worse, 'overproduction' had already occurred in the very important agricultural sector of the economy. Agricultural productivity had increased greatly since the end of the First World War but demand had not increased proportionately. For even the rich cannot be advertised into making eating a way of life and, even if they could, there were not enough of them to eat the quantity of food American farmers were attempting to sell. The excess of supply over demand thus forced down prices and reduced the income of American farmers to scarcely a living wage. As their income declined, so their demand for durable goods declined, and with it the income of the producers of durable goods. Only heavy credit financing sustained the consumption of durable goods until 1929. 'In July of that year,' wrote Arndt, 'the volume of industrial production began to decline. Employment rose slowly till July, remained constant in August and fell in October.'[36] In that month the stock market collapsed. According to Arthur Lewis, 'the news was received with some relief by the authorities, and a mild recession was expected to give place rapidly to continued expansion. In the first half of 1930 a rally did occur, but to everyone's surprise it gave way in the second half of the year.'[37] In 1929 less than one in thirty Americans was unemployed; four years later one in four was without work. In these four years investment goods output fell by nearly 90% including residential construction. Nine million Americans lost their saving accounts as banks closed. Fifteen million unemployed Americans were told at the gates of silent factories that no work was available for them.

As we have already seen, Roosevelt's agricultural policy was to increase the income of farmers by paying them to restrict production and by destroying already growing crops. Since farming was not a monopolistic industry, the farmers by themselves could not restrict supply in order to raise their income. This the government did for them. In a world in which quite literally hundreds of millions of human beings were hungry food was destroyed and farmers were paid not to produce. The problem—the farmers had produced an abundance of food! The solution—the deliberate creation of scarcity! Farm prices rose as a result. From 1932 to 1936 the price of wheat all but tripled, cotton doubled in price and the net income of American farmers rose from $2·5 billion to $5 billion.[38]

In the teeth of nearly all expert advice, Roosevelt went ahead with a programme of public works, slum clearance and massive relief for the destitute. As a result, and against his better judgement, he saw the budget deficit increasing yearly. But unemployment slowly declined. From 25% in 1933 it declined to 22% in 1934, 20% in 1935, 17% in 1936 and 14% in 1937. The gradual improvement convinced Roosevelt that in 1937 he should balance the budget, a feat he almost accomplished. The result was the sharp recession of 1938, described by Winch as 'the most rapid decline in American economic history'[39]: unemployment rose to 19% and private investment dropped 50% from its value of $10 billion in 1937 to $5 billion in 1938. Unemployment was still 17·2% the year the Second World War started in Europe. Although real consumption expenditure in 1939 was 6% above the 1929 level, non-residential fixed investment expenditure was still 42% below the level in 1929 and residential construction 20% below. The capital goods sector, the motor and driving force of capitalist society, remained inactive. From 1929 to 1940 unemployment never fell below 14%. According to Roosevelt's critics, the New Deal Policy had been a failure and had proved conclusively that budget deficits are not the way to get a capitalist economy out of a depression.

The critics could not have been further from the truth, as the Second World War was to show. Despite Roosevelt's New Deal Policy, government spending had simply not been great enough. Assailed by the orthodox economists on the grounds that budget deficits would impede recovery and attacked by private industry for trespassing on its preserve, Roosevelt was never able to implement his humanitarian programme. The end of the housing boom in 1928 had left millions of poorer Americans in sub-standard homes, a condition aggravated by the large scale unemployment of the depression. Accordingly the United States Housing Authority was set up by Roosevelt with a capital of half a billion dollars (and later trebled) in order to assist local communities in slum clearance and provide low-cost homes for the poor. But the Authority's programme was bitterly fought and held up by private real-estate interests so that by 1941 only 120 000 homes had been completed. In this way private interests opposed much of Roosevelt's social welfare programme. Government deficit expenditure, opposed by orthodox economist and industry alike, was simply not enough to eliminate unemployment. From 1929 to 1939 total government spending increased from $10·2 billion to $17·5 billion, an increase of only 70%. In Britain the government did not even attempt a 'New Deal' policy. Indeed, a year after the British Chancellor of the Exchequer, Neville Chamberlain, had triumphantly announced that 'by cuts, by economy and by severe taxation the Budget was balanced',[40] Keynes wrote sarcastically that were 'the Treasury to fill old bottles with bank-notes, bury them . . . and leave it to private enterprise on well-tried principles of *laissez-faire* to dig up the notes again . . . there need be no more unemployment'. Of course, Keynes agreed that 'it would, indeed, be more sensible to build houses and the like; but if,' he pointed out, 'there are political and practical difficulties in the way of this, the above would be better than nothing.'[41] Neither alternative (!) was to be adopted on a sufficiently large scale by either the American or British governments, although Roosevelt's limited efforts were at least Keynesian in direction. The third alternative, forced on both governments by the outbreak of the Second World War (Germany by 1936 had already practically eliminated unemployment through the deficit financing of the Nazi rearmament programme), provided an almost laboratory demonstration of the validity of Keynes' advocacy of deficit government spending. In the United States the budget deficit rose from under $4 billion in 1939 to over $57 billion in 1943. As a result, the real value of the nation's output rose over 70% between 1939 and 1944 as unemployment fell from over 19% to 1·2%. At the same time, private industry more than doubled its share of the total output. In the production of tanks and aircraft, the advanced capitalist nations were able to achieve the objective which for over a decade of peace had eluded them: full employment and a rising standard of living. The political consequences of this seemingly paradoxical situation we shall discuss in a later chapter.

Since we should, however, not completely pass by Germany's success in eliminating unemployment and reviving the flagging fortunes of the capital goods industries, we end this chapter by quoting from some of the conclusions drawn by Arndt in his well-known report *The Economic Lessons of the Nineteen-Thirties*. With respect to the Nazi period, Arndt writes:

> One of the main reasons for the revival of business confidence in 1933 was that German businessmen had no doubt whatever that, as soon as internal and international conditions permitted, the Nazis would embark on a vast rearmament programme. In the eyes of the German capitalists who helped Hitler to power, his rearmament programme, together with his anti-socialist policy, provided the *raison d'être* of his régime. . . .[42]
>
> Rearmament was an ideal 'public works' policy. It stimulated directly the most depressed sectors of the German economy, the heavy industries and the constructional and building trades. Since the demand of the State for armaments represented completely additional demand, it did not in any way diminish existing investment opportunities and thus discourage private investment activity. Again, rearmament was the one method of large-scale public expenditure which was not only immune against the charge, however misguided, of 'extravagance' but which was certain of general approval on patriotic grounds.[43]

Of the major ruling élites in Western countries, German capitalists felt themselves most threatened by the possibility of radical proletarian action and the construction of a socialist society. Their reaction was support of fascism and a heavy rearmament programme, the former in order to eliminate socialist opponents and the latter to eliminate unemployment and underwrite the capital goods industries. The way to full employment and economic prosperity had been found by the German capitalist class. The Second World War was to demonstrate these lessons to the corporate élite of the United States at a time when the socialist alternative to capitalism seemed far more threatening than in the 1930s.

Chapter 6

OBJECTIVITY AND COMMITMENT IN THE SOCIAL SCIENCES

> But men must know that in this theatre of man's life it is reserved only for God and the Angels to be lookers on.
> FRANCIS BACON *De Augmentis Scientiarum*

1. On the Physical and Social Sciences

As we saw in the first three chapters, the nature of the physical sciences is a controversial matter. We should therefore not be surprised to find that the nature of the social sciences is subject to even greater controversy. For if, with all necessary qualifications, we can agree that one of the principal aims of the physical scientist is the discovery of unchanging, immutable laws of nature, to what extent is it meaningful to maintain that his colleague in the social sciences should pursue the same objective? If planets can be said to obey the Newtonian laws of motion (to a certain degree of accuracy) or Einsteinian laws of motion (to an even greater degree of accuracy), what are the corresponding immutable laws of nature which human beings must obey? Or is the search for such laws a wild goose chase? The evidence suggests the latter. For as Barrington Moore has confessed:

> Social science, after some two hundred years, has not discovered any universal propositions comparable in scope or intellectual significance to those in the natural sciences. Whatever the variety among the different disciplines, it is safe to assert that the generalizations of social science nowhere approach the range and cogency of those in physics or chemistry.[1]

The American sociologist Robert Merton mournfully suggests that 'perhaps sociology is not yet ready for its Einstein because it has not yet found its Kepler—to say nothing of its Newton, Laplace, Gibbs, Maxwell or Planck'.[2] But perhaps sociology should not be aiming to produce its Kepler for the simple reason that since human beings can think, feel, create and act purposefully, it is, to say the least, not immediately obvious that such beings obey natural laws *in the same sense* that planets are said to obey them. At the very least it is meaningful to ask, Is a social *science* possible?

Pure and applied physical science: the commanding of nature

To explore the implications of this question let us first refresh our memory with respect to the role of the physical scientist in 'the search for truth'. We first note that apart from usually small-scale laboratory experimentation, the pure physical scientist does not alter the physical environment. Indeed, the attitude underlying his approach to nature is that nature, *if looked at in the right way* (his task), will present no problematic behaviour. And yet, as we have seen, the ever-present mismatch between the physical scientist's paradigm and reality continually confronts him with a multiplicity of problems demanding solution. Now since nature is peculiarly indifferent to appeals for better behaviour, methods of solution are limited to articulation of the existing paradigm or the development of new ones. The use of force would be thought improper: if planetary orbits are not as predicted by the existing paradigm, then it would not be considered a satisfactory solution, techniques permitting, to *force* the planets into the prescribed orbits. For it is the aim of the pure astronomer to predict the 'natural' motions of the planets using a theory which appears aesthetically pleasing to him; if ever the planets are to be 'commanded', the task is that of the (future) *applied* astronomer.

The pure physical scientist searches for that aesthetically pleasing theory which, in a given domain of phenomena and to a certain accuracy, correctly predicts the behaviour of matter; the applied physical scientist can, in principle, use such theories in order to control and dominate nature. Thus, as Francis Bacon appreciated, 'knowledge is power' (provided, of course, the applied scientist also has political control over material resources). And yet it is not immediately obvious, even granting the proviso, how knowledge of theories does enhance men's power over nature. For nature 'obeys' whatever theory it does obey. And since knowledge of that theory does not mean that nature can be made to obey a different theory more convenient to man, it would appear that such knowledge allows men only to predict the course of nature but not to control and dominate her. Yet this is manifestly not the case. From whence comes, then, man's power over nature?

Knowledge of a physical theory T allows the pure physical scientist to predict how an isolated physical system will evolve with time provided he knows the relevant initial conditions I_1 characterizing the system at some time t_i; he can then predict the final state of the system F_1 at some future time t_f. Suppose, however, that the final state F_1 is not to the liking of an applied physical scientist who prefers the result F_2 at t_f. In that case the applied scientist can use the theory T to predict what must be the initial conditions I_2 characterizing the system at t_i so that it will evolve to the desired state F_2 at t_f. The actual initial conditions are, however, I_1 not I_2. But it is precisely here that the applied scientist can attempt to *intervene* in the physical system in order to change the initial conditions from the undesired I_1 to the desired I_2. Should he be able, by his intervention, to establish the initial conditions I_2 then, if the theory T is valid, the desired result F_2 will ensue without further intervention on his part (although of course further intervention is always possible). It is precisely because man can intervene in nature that knowledge of valid theories increases man's command over nature.

OBJECTIVITY AND COMMITMENT

Provided man is able to establish, through his physical interaction with nature, the desired initial conditions, then nature, while obeying her own laws, evolves towards that state desired by man. To quote yet another of Francis Bacon's aphorisms: 'Nature to be commanded must be obeyed.'

Clearly man, although a part of nature, is distinct from inanimate matter. For man's constant interaction with inanimate matter is mediated through his consciousness which, as knowledge of the behaviour of nature increases, allows man's interaction with his environment to become more successfully goal directed. To be sure, theories of limited validity may not always give intended results. But this is another problem. The important point remains: knowledge of theories obeyed by nature gives the knower the possibility of intervening in nature so as to achieve a desired result that could otherwise be achieved only with considerably greater physical effort or could even not be achieved at all.

On the activity of social science: the commanding of men?

With this discussion in mind, we can now ask whether pure and applied *social* science are possible in the same way as are pure and applied *physical* science. It would seem that the answer must be in the negative. For let us consider the dilemma of a social scientist who seeks to predict his own society's future development but who wishes to do so (since he claims to be a 'pure' social scientist) without in any way altering the 'natural' course of events. Let us even suppose our 'pure' social scientist to have devised *and tested* a theory of how his own society 'functions' without having disturbed in any way the 'natural' development of the society (in itself a *most* remarkable achievement), and who has (as a scientist is ethically obliged to do) published his findings. Now it is with respect to this act of publishing that we are obliged to recognize that a crucial distinction must be drawn between the physical and social sciences. For publication of research results by physicists has not as yet been known to alter the behaviour of the moon, the planets or the 'elementary particles'. But men are *qualitatively* different from inanimate matter: men make their own history, the planets do not; men react to newly discovered knowledge about themselves and their historical situation, the planets do not. Hence we note the (not very surprising but nevertheless important) result that the 'pure' social scientist can and does influence social behaviour 'merely' by publishing his research results. Indeed, it is quite possible that the investigations of a 'pure' social scientist into the nature of his own society can lead to significant social changes which might otherwise not have occurred and which, by definition, such a 'pure' social scientist could have had no intention of so causing. Hence we must view the publication of research results as a form of social *intervention* and conclude that, since there is *at least* this component of 'applied' social science in 'pure' social science activity, the distinction between 'pure' and 'applied' social science cannot be made as sharply as in the physical sciences. Moreover, since the gathering of 'social data' and the testing of social theories are themselves social activities of a 'perturbing' kind we may in general conclude that a scientific investigation of one's own society necessarily implies social intervention either with intended or unintended social consequences—but consequences just the same. From the sending of questionnaires, through publication of results, to full blooded political action, social science is an activity in which the concept of the detached, non-interfering scientist is a fiction. Social life and development is a process in which the social scientist is himself a participant. The question therefore arises, how ought the social scientist to participate and with what social objectives?

But, it may first be objected, if there exist deterministic social laws, i.e. laws that human beings have no choice but to 'obey' whether or not they know of the laws' 'existence' (just as people must continue to 'obey' the law of universal gravitation even after the discovery of its 'existence'),* then the discovery of such social laws and their subsequent publication would not necessarily give rise to social changes. Indeed, in the extreme limit of such a philosophical position, one in which *all* human behaviour is held to be as completely determined in the same sense as is the motion of the planets, the very concept of intervention becomes meaningless, together with such concepts as responsibility and freedom. The fully 'determinist' social scientist does not therefore have to ask himself how he *ought* to act in society and with what social objectives; he simply *does* act and *does* have objectives.

The 'free-will' versus 'determinism' debate is age-old. I make only the following remarks. In practice, the fully 'determinist' social scientist interacts *purposefully* with his environment, *processes* information, *evaluates* theories, experiences moral dilemmas and *makes choices*. He therefore behaves *as if* he were a free agent. Nevertheless, by replying that all such actions and choices are 'in reality' determined, including the declaration in which this is asserted and so on into an infinite regress, the 'determinist' social scientist can therefore maintain that he cannot do other than what he does, that he is not a free agent.

This position is as 'unfalsifiable' as it is 'unprovable'. However, for our purposes we note that the 'determinist' social scientist does act—and acts purposefully in society—whatever the 'causes'. The question that interests us is, how is such a scientist likely to act and with what social objectives? For if, on the one hand, he devotes his life to looking for deterministic social laws— an unchanging, immutable social reality behind the changing flux of social appearances—he is unlikely to be successful (for, I believe, no such laws 'exist'). On the other hand, if he 'chooses' to 'intervene' in society with the aim of making his fellow citizens behave in a law-like way, he may be able to

---

* Ben-David, a sociologist of science, has written that 'scientists are engaged in discovering "laws of nature" that cannot be changed by human action . . . [and thus] their systems have to fit the structure of natural events. In principle, this is true also of social scientists and scientific students of culture . . .'[2n]

achieve success—for although men are not robots, it is conceivable that they can be so mutilated as to thereafter always behave like robots. A very real danger, therefore (one to be investigated in chapter 10), is that the 'determinist' social scientist may seek to 'command' other men and, if successful, will be in a very handy position to convince himself that his triumph demonstrates the truth of his philosophical position.

In welcome opposition, however, to such determinism and its possible social implication is the 'actor' or 'pilot' image of man. For in this self-portrait man sees himself as creative agent and *potential* controller of his own life, a self-portrait that holds it possible for men, in making their own history, to become increasingly self-aware and in the process to liberate themselves from patterns of behaviour that at one time appeared 'innate' and, indeed, immutable. According to this view, therefore, since social laws can themselves be made the *object* of man's actions, a social science is thus possible that can have as its aim (in the words of one social scientist) the 'emancipation of all self-conscious agents from the seemingly "natural" forces of nature and history'.[3] Indeed, from this point of view the most important distinction to be made between the physical and social sciences is that although it makes no sense to talk in terms of liberating electrons, it makes excellent sense to talk in terms of liberating people or, possibly, of liberating *some* people and oppressing others. Again we confront the question, What ought, or are likely, to be the goals of social scientists?

Social scientists, like their colleagues in the physical sciences, are born and brought up in societies. Their education and social position will ensure that they have definite social interests and objectives, which they may at some time reject for different ones. As students of social science, they will learn about those social problems considered important by their teachers, about their causes and possible solutions. As practising social scientists enjoying (usually) a relatively high social status, their attention will be drawn to certain problems in, let us suppose, their own societies. Their analyses as to the causes of these social problems will suggest possible solutions, the acceptability of which will depend on their own interests and aims. Clearly an acceptable solution of a problem for one social scientist, if implemented, could well pose a problem for another social scientist with radically different values and goals. For different solutions imply different futures. This is not so obviously the case in the physical sciences. For here the choice between competing paradigms that confronts the pure physical scientist appears to entail 'merely' a choice between different ways of looking at the same physical reality (although see chapter 10 for a discussion of possible social and economic factors underlying such changes of world view). But, for the social scientist, the choice between competing social paradigms is not 'merely' one between different *images* of the existing social reality but, very explicitly, a choice between mutually exclusive future *societies*. Furthermore, as we later argue, the choice in practice is one between widely different degrees of liberation: between, at one extreme, continuing or increased oppression and, at the other extreme, possible liberation for all. Indeed, since existing societies are far from having attained a maximum possible degree of human emancipation, the fundamental choice confronting social scientists is essentially that between commitment to programmes of 'social engineering' *within the established structures of power and control* (thereby perpetuating existing class or élite control) or commitment to programmes of revolutionary political action with the intention of helping to build societies significantly less exploitative and manipulative than existing ones.

The reader will observe that having first emphasized qualitative differences between the physical and social sciences, we are now making use of Kuhn's analogy between paradigm changes in the physical sciences and social revolutions. For certainly in one respect the pure physical scientist always faces the same dilemma as the social scientist—which is whether to attempt to continue to solve problems within the established paradigm (or social framework) or to attempt to articulate a radically different paradigm (or to help construct a radically different society). In this chapter we develop Kuhn's analogy. For physics has experienced two major crises and has undergone two major revolutionary changes since the triumph of the Newtonian paradigm. But what of the social order that nourished and has continued to nourish the growth of science? Is it still the case, as John Stuart Mill declared in the 19th century, that 'the principle of private property has never yet had a fair trial in any country'? Is his plea still reasonable that 'political economists [should seek] not the subversion of the system of individual property but the improvement of it'?

To begin to answer these questions and to develop ideas concerning the nature of the social sciences we take up the themes of chapters 4 and 5 to remind ourselves of the *commitments*, theories and actions of three social scientists: of the political radicalism of Marx on the one hand and the political conservatism of Robbins and Keynes on the other. The two questions posed concerning the validity of Mill's beliefs we return to in the last chapter.

2. Three Social Scientists and Their Research-Action Programmes

On Marx

Marx loathed capitalist society. Although he believed that the capitalist stage in the development of the productive forces was necessary in order to provide an industrial base sufficient to eliminate scarcity, nevertheless he did not see how capitalist relations of production could ensure a truly human life for all members of society, indeed for any member. On the contrary, they enshrined a way of life making money the measure of all human achievement. Marx therefore looked for possible agents of change in capitalist society which would be able both to overthrow the existing class structure and to construct a classless and therefore radically different social order, one in which each man would be free to develop to the full his own individual potentialities while at the same time cooperating with all other men in the realization of common goals. Marx's theory of English capitalist society and of the effect of his own actions in that society led him to believe that the bourgeoisie was producing its own grave-diggers; that during an inevitable economic crisis the proletariat would seize power in order to construct the new social order he so greatly desired. At Marx's funeral in 1883 Engels summed up very accurately the life-struggle of his best friend:

> For Marx was before all else a revolutionist. His real mission in life was to contribute, in one way or another, to the overthrow of capitalist society and of the state institutions which it brought into being, to contribute to the liberation of the modern proletariat, which *he* was the first to make conscious of its own position and its needs, conscious of the conditions of its emancipation.[4]

On Keynes

Keynes, born the year in which Marx died, committed himself to the preservation of that social order which Marx had striven to destroy. He was very explicit about his interests, declaring in 1925:

> If I am going to pursue sectional interests at all I shall pursue my own. When it comes to the class struggle as such, my local and personal patriotisms, like those of everyone else, except certain unpleasant zealot ones, are attached to my own surroundings. I can be influenced by what seems to me to be Justice and good sense; but the *class* war will find me on the side of the educated *bourgeoisie*.[5]

In the same year he singled out Marx's *Capital* for direct attack:

> How can I accept the doctrine which sets up as its bible, above and beyond criticism, an obsolete textbook which I know not only to be scientifically erroneous but without interest or application to the modern world? It is hard for an educated, decent, intelligent son of Western Europe to find his ideals here, unless he has first suffered some strange and horrid conversion which has changed all his values.[6]

In 1936, writing at the height of the Great Depression, Keynes expressed his fear of the possibility of that radical social transformation which Marx had devoted his life to achieving: 'At the present moment people are unusually expectant of a more fundamental diagnosis; more particularly ready to receive it; eager to try it out, if it should be even plausible.'[7] And hence the orientation of his research commitment: 'But it may be possible by a right analysis of the problem to cure the disease whilst preserving efficiency and freedom.'[8]

On Robbins

Robbins was similarly preoccupied with, as he saw it, the threat of socialism; his intention was to avert this threat and thus save capitalist society. 'If the direction of policy in Great Britain, and the modern world generally, today is overwhelmingly socialist', he wrote in the early 1930s, 'this is not because it is dictated by the objective facts of the situation, or because the masses with one accord have willed a socialist reorganisation of industry. It is because men of intellect, with powers of reason and persuasion, have conceived the socialistic idea and gradually persuaded their fellows.' Robbins was of the opinion:

> It may be that the forces which have been released by the ideas of forty years ago, have become so powerful ... that it is now too late to arrest them. It would be unwise to ignore the very strong probability that this is so. But until the case, which experience and more recent developments of knowledge have shown can be made against them, has been argued with as much patient and disinterested intelligence as went to the establishment of their ascendancy, we are not justified in concluding that reason and persuasion have reached the limit of their effectiveness. At all events it is worth trying.[9]

His work *The Great Depression* was a major effort to convince his colleagues that the causes of the Great Depression were as he, Robbins, understood them and to show his colleagues how full employment could be restored. His book was written as a contribution to the efforts to ensure a future capitalist England.

Crisis, paradigms and commitments

The economic crisis of the 1930s was a traumatic experience for all economists. Both Robbins and Keynes committed themselves to finding solutions of the pressing social problems within the existing class-structured society. However, whereas Robbins committed himself to articulation of, and action based on, the neo-classical economic paradigm, Keynes wrote mockingly in 1933 that 'many people are trying to solve unemployment with a theory which is based on the assumption that there is no unemployment'.[10] But this was precisely the point the neo-classical economists wished to make. As Schumpeter put it in 1934, 'What we face is not merely the working of capitalism, but of a capitalism which nations are determined not to allow to function.'[11] Robbins in particular was telling workers that if they wished their unemployed colleagues to find work they would have to be prepared to let wages move freely downwards to reach their full employment equilibrium level. Robbins' general conclusion was that 'if recovery is to be maintained and future progress assured, there must be a more or less complete reversal of contemporary tendencies of governmental regulation of enterprise. The aim of governmental policy in regard to industry must be to create a field in which the forces of enterprise and the disposal of resources are once more allowed to be governed by the market.'[12] For Robbins could see little wrong with neo-classical economic theory; it was simply that people were not behaving in a way that would allow capitalism to function adequately—the remedy was therefore to persuade people to behave in the manner required by that theory in which, he, Robbins so wholeheartedly believed. Indeed in 1932 Robbins confidently declared that, 'At the present day, as a result of the theoretical developments of the last sixty years, there is no longer any ground for serious differences of opinion on these matters, once the issues are clearly stated.'[13] In the second edition of his *Essay* (1935), he did not see fit to revise this judgement: 'The efforts of economists during the last hundred and fifty years have resulted in the establishment of a body of generalisations whose substantial accuracy and importance are open to question only by the ignorant or the perverse.'[14]

But Keynes was 'perverse'. For him the situation was critical. Not only was it futile to attempt to persuade people to behave in a way they were obviously unwilling to, but, much more important (as discussed in chapter 5), he believed that prolonged unemployment could occur even in a perfectly competitive capitalist society. The fundamental principles on which the neo-classical theory had been built were, he believed, simply wrong. Thus for Keynes, both the orthodox economists and the Marxist revolutionaries presented a threat to the survival of capitalist society. For the actions of the orthodox economists such as Robbins would, Keynes deduced from his theory, only make matters worse and this would only be to the advantage of those revolutionaries who wished to construct a socialist society. In his open letter to Roosevelt in December 1933 Keynes warned the newly-elected President of the awesome responsibilities that he, Roosevelt, had assumed (and which Keynes felt he similarly shared):

> You have made yourself the trustee for those in every country who seek to mend the evils of our condition by reasoned experiment, *within the framework of the existing system*. If you fail, rational change will be gravely prejudiced throughout the world, leaving orthodoxy and revolution to fight it out. But if you succeed, new and bolder methods will be tried everywhere, and we may date the first chapter of a new economic era from your accession to office (emphasis added).[15]

As has been argued, the Second World War was to prove Keynes right. After only two years of war, the American national income had risen by more than in the previous ten years while unemployment had all but disappeared. Robbins' 1935 statement that 'the waging of war necessarily involves the withdrawal of scarce goods and services from other uses, if it is to be satisfactorily achieved' could scarcely have been further from the truth.[16] During the war, however, the Keynesian theory won an important convert: 'It has afforded an interval,' wrote Robbins, 'in which, our entanglement in the controversies of the past being suspended, we could consider old positions without that acute attachment to interested intellectual capital, which in normal times, makes it so difficult to change one's position.' 'I owe much,' he admitted, 'to Cambridge economists, particularly to Lord Keynes and Professor Robertson, for having awakened me from dogmatic slumbers in this very important respect.'[17] Indeed, after the war, as we shall see in succeeding chapters, Keynesian economic theory became the new orthodoxy, bringing with it, however, the 'permanent war economy' on which, it will be argued, the stability of the modified capitalist social order depends.

After so describing the activities of Robbins, Keynes and Marx, we can now briefly summarize why we believe these three men were practising social science, with different abilities perhaps, but nonetheless social science. For they did not look for *universal* social laws which men must obey whether they know of their existence or not; on the contrary, they focused their attention on the problems of the society in which they lived, hoping thereby to be able to solve those problems which most worried them in a manner compatible with their own values and goals. Their detailed theories were therefore not applicable to all societies that have existed throughout history; on the contrary, they were applicable only to capitalist society, implicitly or explicitly suggesting ways in which capitalism could be either preserved and improved, or overthrown and replaced.

All three men had in their minds images of futures they considered desirable. According to the theory espoused by Robbins, a perfectly competitive capitalist society would not be prone to prolonged and severe unemployment of labour: it was, however, necessary to test his theory by persuading people not only to behave differently but in particular to behave in such a way as to achieve and allow to function that society which he, Robbins, would personally welcome. Indeed, according to Robbins,

> It is not really to be believed that the majority of men in democratic communities are so in love with poverty and instability that if they were convinced that certain policies led in that direction they would continue to support them. On the contrary, it is clear that they at present support these policies because they believe—wrongly it has here been argued—that they lead to greater stability and progress. If they were convinced otherwise, can it be doubted that they would abandon them?[18]

Robbins sought to convince men otherwise.

According to his theory Robbins was able to make two conditional predictions: (1) if men continue to behave as at present, then the future will be as bad as, if not worse than, the present (and will be followed by calamitous mass movements); (2) if men behave in the way advocated by Robbins, then the future will be that of a properly functioning capitalist society, i.e. no prolonged unemployment and a steadily rising standard of living for all. Thus, seen through the eyes of his theory, Robbins had two images of the future, one which he deplored and one which he welcomed. He was convinced his theory was right; he therefore sought to alter men's behaviour and thereby to ensure the future he welcomed. He was equally convinced that Keynes' theory was wrong and that action based on this theory would bring only further chaos.

According to Keynes, however, Robbins' theory was wrong and action based on the policy recommended by Robbins would bring disaster. A change in policy was necessary. Indeed as early as 1930 Keynes wrote that

> unless we bend our wills and our intelligences, *energised by a conviction that this diagnosis is right*, to find a solution along these lines, then, if the diagnosis *is* right, the slump may pass over into a depression, accompanied by a sagging price level, which might last for years, with untold damage to the material wealth and to the social stability of every country alike [first emphasis added].[19]

Believing in the validity of his own theory, Keynes therefore sought to convince others and to persuade governments to act so that, as predicted by his theory, full employment would be restored and a modified capitalist society preserved, a future which he personally welcomed.

The aim of Marxists, however, was not to preserve capitalism in any form but to transcend its class structure in the realization of communist society. To be sure, many post-Keynesian Marxists agreed that the application of Keynesian theory could well restore full employment to a depressed capitalist economy and so preserve the capitalist social order in the short run. But in the long run the effects of such a policy would be catastrophic. The political pressures in capitalist society would mean that government expenditure could not be devoted solely, or even mainly, to public welfare; the class structure would remain and so would capitalist exploitation of the underdeveloped countries; further conflict with the Soviet Union would be inevitable. From the Marxist point of view, the total collapse and replacement of capitalist society was immeasurably preferable to its preservation in modified form; it would not be possible to solve the economic problems of humanity, to say nothing of man's non-material needs, if capitalism in advanced industrial societies continued to exist. To be sure, there could be no absolute guarantee that a just and humane world could be constructed but, according to Marxist theory, the future would be black indeed if capitalist society could not be transcended. Of course, radically changing a social order is no light matter; but then neither is clinging on at all costs to the existing one. For Marxists, attempts to preserve capitalist society come what may would have disastrous consequences for mankind; but, believed Marxists, the proletariats of the advanced capitalist countries would not allow such attempts to succeed.

Thus writing on Keynes in 1946 the American Marxist Paul Sweezy declared that 'while it is right to recognize the importance of Keynes, it is no less essential to recognize the shortcomings'. These were, according to Sweezy,

> the unwillingness to view the economy as an integral part of a social whole; the inability to see the present as history, to understand that the

disasters and catastrophes amidst which we live are not simply a 'frightful muddle' [Keynes' expression] but are the direct and inevitable product of a social system which has exhausted its creative powers, but whose beneficiaries are determined to hang on regardless of the cost.

It was Sweezy's opinion, however, that Keynes 'could never have recognized, let alone transcended, the limitations of the society and the class of which he was so thoroughly a part'. But some of Keynes' younger followers will recognize, Sweezy predicted, 'that patching up the present system is not enough, that only a profound change in the structure of social relations can set the stage for a new advance in the material and cultural conditions of the human race'.[20]

The Copernican and Keynesian revolutions

One word more on some of the similarities and differences between the physical and social sciences—a brief look at the Copernican and Keynesian revolutions.

For Copernicus the Ptolemaic system of the world was aesthetically objectionable; therefore it could not correspond to the true physical reality. For Marx capitalist society was morally and ethically objectionable; therefore it could not be, and could not be allowed to be, the permanent social reality. Copernicus imaginatively created a theory, radically different from the Ptolemaic, describing a more harmonious world system; therefore, despite difficulties which had to be overcome, he believed his theory corresponded much more closely to physical reality than Ptolemy's. Marx imaginatively created a theory describing the possibility of *constructing* a non-exploitative, more harmonious social system, radically different from the one in which he lived; therefore, despite difficulties which had to be overcome, he would act in such a way as to make more probable the construction of the radically different social order he envisaged. Marx's own comments are illuminating: 'A spider conducts operations that resemble those of a weaver and a bee puts to shame many an architect in the construction of her cells. But what distinguishes the worst architect from the best of bees is this, that the architect raises his structure in imagination before he erects it in reality.'[21]

While Copernicus, however, did not have to transform his mental creation into reality (it already existed!) Marx, on the other hand, had to help erect in reality the structure he had first created in his imagination. To be sure, the Copernicans had to act so as to win men over to their way of seeing the universe but for the Marxists it was not enough to win the proletariat over to their way of viewing the existing society, however much this was clearly a prerequisite for successful revolution. The revolution had still to be undertaken and successfully carried through.

Problems certainly existed in the Ptolemaic theory but throughout the 16th century Ptolemaic astronomers tried to resolve these problems within the conceptual framework of their theory. Indeed, the empirical evidence seemed to point against the reality of the Copernican system. Likewise throughout the 19th century the majority of economists attempted to resolve problems within the conceptual framework of the existing theory, hoping thereby to establish a theory that would provide theoretical justification for the existing social order. However, just as the discovery of the phases of Venus proved difficult to explain within the framework of the Ptolemaic system, so the prolonged unemployment of the 1930s proved similarly troublesome for the classical economists. Nevertheless, they attempted to resolve the problem within the framework of their established theory by declaring that it was not their theory which needed correction but rather the actions of those people preventing capitalist society from functioning properly. The Ptolemaic astronomers were, however, scarcely able to declare that the phases of Venus needed correcting. But radical change is always unwelcome to some. Hence the compromise solution of Tycho Brahe which, lacking geometrical harmony, was rejected as beneath contempt by the scientific revolutionaries, Kepler and Galileo. Likewise Keynes who, unable to accept the classical diagnosis and yet wishing to prevent the construction of a radically different society, committed himself to the articulation of a compromise solution which would justify government intervention within the framework of a basically capitalist society. Unlike Tycho Brahe, however, Keynes was successful, winning the support of those economists who believed reform of the capitalist social system possible. For non-Marxist socialists, writes Winch, 'here was an effective weapon for use against the Marxists on the one side and the defenders of old-style capitalism on the other; a real third alternative, the absence of which before the General Theory had driven many into the Communist camp'.[22] Keynes had succeeded in his objective; he had helped preserve a society *which could have been radically changed*. While the future of the solar system was not at the disposal of the scientific revolutionaries, the future of capitalist society was, at least to some extent, in the hands of the professional economists. They committed themselves to different research programmes and to different futures. History has, however, yet to write its final verdict on the wisdom of the Keynesian research programme. But one thing is certain: Keynes was a most influential social scientist.

Before, however, looking at how Keynesian theory was applied in the United States (the subject of the next chapter) we first give an indication of the way that both Marxist and 'value-free' social scientists would surely object not only to many of the comments I have made concerning Marx on the one hand, and Robbins and Keynes on the other, but also to my overall interpretation of the nature of social science activity.

3. On Marxism

In his *The ABC of Socialism* the American Marxist Leo Huberman contrasts 'utopian' socialism against the 'scientific' socialism of Marx and Engels. He writes:

> 1. The socialism of the Utopians [Owen, Fourier and Saint-Simon] was based on a humanitarian sense of injustice. The socialism of Marx and Engels was based on a study of the historical, economic and social development of man.
> 2. Unlike the Utopians, Marx spent no time on the economic institutions of Tomorrow. He spent almost all of his time on a study of the economic institutions of Today.
> 3. With the Utopians, socialism was a product of the imagination, an invention of this or that brilliant mind. Marx brought socialism down from the clouds; he showed that it was not merely a vague aspiration, but the next step in the historical development of the human race—the necessary and inevitable outcome of the evolution of capitalist society.
> 4. Marx transformed socialism from a utopia to a science. Instead of a visionary fantastic blueprint of a perfect social order, he substituted a down-to-earth theory of social progress; instead of appealing to the sympathy, goodwill and intelligence of the upper class to change society, he relied on the working class to emancipate itself and become the architect of the new order.

Some further comments by Huberman are interesting:

> 5. History is not chaotic—it conforms to a definite pattern of laws which can be discovered. Karl Marx discovered those laws of development of human society. That was his great contribution to mankind.

And finally:

> 6. Now what is it that brings about social and political revolution? Is it simply a change in men's ideas? No. For these ideas depend on a change that occurs first in economics—in the mode of production and exchange.[23]

The implications of these and similar comments are, for me, the following (although I am not suggesting that Huberman necessarily held these views, I do believe that the following views represent a typical attitude of many Marxists to Marx's 'scientific socialism'):

(*a*) That Marx did not base his socialism on ethical principles
(*b*) That scientific socialism is concerned with the institutions of today, not tomorrow
(*c*) That Marx showed socialism to be the *inevitable* outcome of capitalism
(*d*) That Marx showed that the proletariat is necessarily the historical agent of change

(e) That Marx discovered the laws of development of human society
(f) That changes in the 'economic base' are primary, changes in men's ideas secondary.

From such statements it is not difficult to see how the following 'vulgar' Marxist position could have arisen, namely that men's social behaviour is determined by natural law just as is the behaviour of planets, and that inevitably the proletariat will construct the new socialist order, just as it can be predicted that Mars will be at a certain position at a given time in the future. Of course, it may at first sight appear paradoxical for a party to exist in order to ensure the inevitable; after all, no party is deemed necessary in order to assist the daily rising of the sun since this daily event is guaranteed by the law of conservation of angular momentum (provided, of course, man does not intervene in nature so as to stop the earth's daily rotation!). But the paradox is only apparent. For, it can be replied, the party's existence is itself determined by natural law and likewise the actions of all party members are themselves links in an immense causal chain. (The infinite regress is apparent.) While moral and ethical judgements will continue to be made, since they are utterances as fully determined as all other human utterances, human responsibility remains an illusion.* Man's consciousness and goals are determined for him as are his actions. According to the iron laws of history, capitalism is destined to be replaced by socialism. History's agents are the proletariat and men like Marx.

Was this Marx's view?

The above is antithetical to my conception of Marx's 'Marxism'. Nevertheless, can it be argued from Marx's own writings that he believed in historical determinism and that his own actions were fully determined by forces over which he could exert no control?

(i) THE INEVITABLITY OF SOCIALISM

'Capitalist production begets, with the *inexorability* of a law of Nature, its own negation,' wrote Marx.[24] 'What the bourgeoisie, therefore, produces, above all, is its own grave-diggers. Its fall and the victory of the proletariat are equally inevitable'—thus the Communist Manifesto. In the Preface to the first edition of *Capital*, Marx expressed himself at greater length:

> Intrinsically it is not a question of the higher or lower degree of development of the social antagonisms that result from the natural laws of capitalist production. It is a question of these laws themselves, *of these tendencies working with iron necessity towards inevitable results*. The country that is more developed industrially only shows, to the less

---

* An attitude I regard as dangerous. It condones atrocities committed in the past and makes future atrocities that much more likely.

developed, the image of its own future. . . . And even when a society has got upon the right track for the discovery of the natural laws of its movement—and it is the ultimate aim of this work, *to lay bare the economic law of motion of modern society*—it can neither clear by bold leaps, nor remove by legal enactments, the obstacles offered by the successive phases of its normal development. But it can shorten and lessen the birth pangs (all emphases added).

(ii) THAT CONSCIOUSNESS IS NOT A DETERMINING FACTOR IN SOCIAL DEVELOPMENT

'It is not the consciousness of men that determines their existence,' wrote Marx, 'but, on the contrary, their social existence determines their consciousness.'[25] And in the preface to the second edition of *Capital* he not unapprovingly quoted a critic: 'Marx treats the social movement as a process of natural history, governed by laws not only independent of human will, consciousness and intelligence, but rather, on the contrary, determining that will, consciousness and intelligence . . .'

(iii) THAT MARXISM IS NOT BASED ON AN ETHICAL COMMITMENT

'Communists preach no morality at all', wrote Marx and Engels in *The German Ideology*. 'They do not put to people the moral demand: Love one another, be not egoists, etc.; on the contrary, they know very well that egoism, like sacrifice, is under specific conditions a necessary form of the individual's struggle for survival.'[26] In the Preface to the first edition of *Capital*, Marx asked for

> one word more to avert misunderstandings. The persons of capitalists and landowners are not, in my book, depicted in rose-tinted colours; but if I speak of individuals, it is only in so far as they are personifications of economic categories, representatives of special class relations and class interests. Inasmuch as I conceive the development of the economic structure of society to be a natural process, I should be the last to hold the individual responsible for conditions whose creature he himself is, socially considered, however much he may raise himself above them subjectively.

This was not Marx's view

Despite the apparent force of these citations from Marx's writings it is, I believe, relatively easy to demonstrate that Marx's view of man was not as crudely determinist as it has been so often interpreted. To begin with, Marx himself explicitly denied that men are products of their environment:

> The materialist doctrine that men are products of circumstances and upbringing, and that therefore changed men are products of other circumstances and changed upbringing, forgets that it is men that

OBJECTIVITY AND COMMITMENT

change circumstances and that the educator himself needs educating . . . The coincidence of the changing of circumstances and of human activity can be conceived and rationally understood only as *revolutionary practice*.[27]

And as for the inevitability of socialism, Marx wrote that it was not possible to metamorphose his 'historical sketch of the genesis of capitalism in Western Europe into a historico-philosophical theory of the general path every people is fated to tread', maintaining that he had not provided a 'general historico-philosophical theory, the supreme virtue of which consists in being suprahistorical.'[28] For Marx had concentrated his intellectual powers on a critique of capitalist society; he had not looked for laws of history which of necessity human beings had to obey whether or not they knew of their 'existence'. To be sure, Marx and Engels agreed that 'the history of all hitherto existing society is the history of class struggles' but the struggles between 'oppressor and oppressed' had no predetermined outcomes; each struggle ended 'either in a revolutionary reconstitution of society at large, or in the common ruin of the contending classes.'[29] But if this is the case why, it must be asked, did Marx write so often that the victory of socialism was *inevitable*? We attempt to answer this question after the following comment on Marxist ethics.

Marx, the 'scientific' socialist, and Owen, Fourier and Saint-Simon, the 'utopian' socialists, subscribed to very different theories as to how socialism could be achieved. Marx did not believe that setting 'a good example' to capitalists would cause them to relinquish their privileges; for this reason, although he based his socialism on ethical humanism, as is manifestly evident throughout his writings, he was emphatically opposed to moralizing. For capitalism, which lived by the sword, could only be abolished by use of the sword. Only in a classless society would it be possible 'to love one's neighbour as oneself'; to preach this commandment in capitalist society would at best be futile and at worst would militate against the decisive action that alone could bring a non-exploitative society into being. Marx was very explicit about this: 'The social principles of Christianity preach cowardice, self-contempt, debasement, subjugation, humility, in short, all the properties of the canaille, and the proletariat, which does not want to be treated as canaille, needs its courage, its consciousness of self, its pride and its independence, far more than its bread.'[30] Only revolutionary action by the proletariat could end the degradation suffered by all members of capitalist society and in particular by the proletariat. Marx again is very explicit:

> In the conditions of existence of the proletariat are condensed, in their most inhuman form, all the conditions of existence of present-day society. Man has lost himself; however at the same time he has acquired not only a theoretical consciousness of his loss, but he has also been forced by an ineluctable, irremediable and imperious *distress* to revolt against this inhumanity. It is for these reasons that the proletariat can and must

emancipate itself. But it can only emancipate itself by destroying its own conditions of existence. It can only destroy its own conditions of existence by destroying all the inhuman conditions of existence of present-day society, conditions which are epitomized in its situation.[31]

Like the utopian socialists, Marx conceived the possibility of a non-exploitative society. But since, unlike the utopian socialists, he did not believe that appeals to the capitalists' sense of justice would be in any way effective, it would be necessary to make a very careful study of the existing society in order to determine how best to achieve the desired goal. There was, Marx concluded, only one way of overthrowing the bourgeoisie and achieving a classless society, the way of proletarian revolution.

Marx viewed the future through the lenses of his own theory and through those lenses the future appeared promising indeed. For the capitalists were caught in a trap. Competition between capitalists would inevitably result in constant improvement of the productive-forces, combined with growth of monopolies and increasingly serious crises of over-production. Such a social system could not continue indefinitely, Marx thought. His theory suggested to him no way by which the bourgeois class could prevent a universal crisis from occurring and *certainly Marx had no intention of attempting to find such ways*. He was no Keynes. Certainly it was *possible* that the crisis would end in the ruin of the contending classes but for Marx this was scarcely conceivable. For the proletariat was becoming aware of itself as a class and would, when the inevitable crisis came, be able to use its superior numerical strength in overcoming the bourgeoisie's by then demoralized police force and army. He, Marx, and others like him would do their utmost to help ensure that when the universal crisis happened, the proletariat would emerge victorious. The triumph of socialism was therefore *inevitable*! For how could man not take the last remaining step to freedom.

In *The Eighteenth Brumaire of Louis Bonaparte*, as Horowitz has pointed out,[32] Marx made quite clear his approach to an understanding of historical change. Criticizing Hugo's subjectivism (which made Napoleon appear a really great man, a bolt from the blue) and Proudhon's objectivism (which made the *coup d'état* appear as historically inevitable) Marx wrote that 'I, on the contrary, demonstrate how the class struggle in France created the circumstances and relationships that made it *possible* for a grotesque mediocrity to play a hero's part' (emphasis added).

Likewise Marx thought that he had demonstrated beyond any doubt that capitalist society would inevitably undergo a severe and universal crisis which would thus make it *possible* for the proletariat to seize power from the bourgeoisie and abolish itself as a class. But would the proletariat thus seize the opportunity when it presented itself? Revolutionary action would be necessary. But would it occur? It was clear that only disaster would result in perpetuation of the capitalist social order. The choice confronting the proletariat would be between socialism and barbarism. And yet, even so, there could be no absolute guarantee that the proletariat would develop a revolutionary consciousness and would be able to act decisively when the universal crisis occurred. If Marx truly believed that the success of socialism is inevitable, guaranteed by natural law, then he was surely wrong. But, much more likely, Marx felt himself able to predict with certainty only the universal crisis of capitalism; on the other hand, the triumph of socialism was inevitable, not because of natural law, but because Marx could not conceive of the possibility that man's long and arduous road to freedom would end after all in disaster and barbarism. To be sure, action would be necessary to help the proletariat develop a revolutionary consciousness. But the proletariat would be receptive and in the end would liberate itself—would have to liberate itself. The alternative was unthinkable. And yet when in the 1930s the universal crisis occurred, the proletariats of the advanced capitalist countries either had no revolutionary consciousness or, if they had, were unable to achieve political power.

The consequences of this momentous failure will be examined in the remaining chapters. For the remainder of this chapter, however, *without claiming in any way to have more than touched the surface of problems in the interpretation of Marx's writings*, we turn our attention to a doctrine totally opposed to the spirit of this book—to the widely-held belief that the social sciences should be 'value-free'.

4. On 'Value-Free' Social Science

The case for value-free social science

There are many social scientists who subscribe to the view that the aim of social science is to obtain objective knowledge about the social world and that therefore of logical necessity social science must be 'value-free'. For objective knowledge, they claim, has only truth value; it is either true or false and logically can have no ethical or political implications. To be sure, social scientists must understand whatever values are involved in the actions and institutions they are discussing but, it is claimed, it is none of their business as objective scientists to approve or disapprove either of those values or of those actions and institutions. The proponents of value-free social science believe that in this respect there is complete identity between the natural and social sciences. Just as the natural scientist aims at understanding the physical world without passing judgement on whether or not, for example, the law of gravity should or should not be what it is, so the social scientist must aim at understanding the social world, without passing judgement on the various societies, institutions and social practices he is studying. The only legitimate aim of social science is to understand what *is*, not what *could be* or *should be*. The only value-judgement allowed, therefore, is that value-judgements should be excluded. As one sociologist puts it: 'Once a person accepts a value he becomes somewhat biased or prejudiced thereby.'[33] Another sociologist, M. Albrow, has expressed the opinion that a social scientist, far from being committed to any form of social and political action, is on the contrary 'committed to the belief that to be objective is to be free from moralizing and to avoid recommending courses of action'. Indeed, since 'the "sociological standpoint" insists that human beings must be studied in respect of how they actually act rather than how they might ideally act . . . the perspective of the sociologist is "impractical". His search is for understanding, untrammelled by the needs of men of affairs.' The value-free standpoint could not be more clearly expressed. Furthermore, 'That sociology should be value-free', says Albrow, 'is taken by most sociologists to be axiomatic.'[34] According to another sociologist, R. Bierstedt, 'The scientist, as such, has no ethical, religious, political, literary, philosophical, moral or marital preferences. . . . As a scientist he is interested not in what is right or wrong or good or evil, but only in what is true or false.'[35] Bierstedt's colleague in sociology, A. Green, is even more categorical: 'At no time does the scientific investigator state his own opinion of what ought to be, or of what is right and what is wrong. His own biases and preferences do not enter the picture.' What, then, is the task of the social scientist? It is, says Green, 'to analyse, to explain, and to increase the store of knowledge. His job should not be confused with the wielding of political or economic power, power which he does not possess.'[36] With respect to economics, we see that Pigou has stated that economics 'is a positive science of what is and tends to be, not a normative science of what ought to be.'[37] Similarly Lipsey, in his much-used textbook *An Introduction to Positive Economics*, takes pains to distinguish between what he calls positive questions (those that can in principle be settled by an appeal to empirical observation) and normative questions (those that imply making a value-judgement—what should or should not be the case—and that therefore, claims Lipsey, cannot be decided by reference to experiment or observation). He then informs his reader that this 'separation of the positive from the normative is one of the foundation stones of science and that scientific enquiry, as we normally understand it, is confined to positive questions'.[38] We return later to Lipsey's formulation of the nature of scientific enquiry. To sum up, I quote Myrdal's succinct appraisal of the attitude of 'value-free' social scientists, as he understands it (and to which he is opposed):

> The student should have no ulterior motives. He should confine himself to the search for truth and be as free as possible from both the pressures of tradition and of society around him and his own desires. More particularly he should in his research have no intention of influencing the political attitudes of his readers, either inside or outside the countries whose conditions he is studying. His task is to provide factual information that will help them all reach greater rationality in following out their own interests and ideals, whatever those are. In his scientific work

he should have no loyalties to any particular country or group of countries or any particular political ideology, whatever his own preferences. Indeed, he should have no loyalties at all except to the professional standards of truth-seeking.[39]

Let us give the American sociologist A. W. Gouldner an appropriate (?) last word:

> The myth of a value-free sociology has been a conquering one. Today, all the powers of sociology, from Parsons to Lundberg, have entered into a tacit alliance to bind us to the dogma 'Thou shalt not commit a value-judgement', especially as sociologists. Where is the introductory textbook, where the lecture course on principles, that does not affirm or imply this rule?[40]

It goes without saying, of course, that the above comments on 'value-free' social science serve only to give the flavour of this school of thought. Many of its proponents would no doubt take exception to some of the remarks made and clearly there is not complete unanimity within the school. Rather, however, than making more precise the ideas of committed 'value-free' social scientists we shall attempt to explore systematically the nature of research in the social sciences, asking continually how our conclusions square with various interpretations of what constitutes the methodology and practice of 'value-free' social science. To this end, let us first of all assume that value-free social scientists mean that value-judgements must play no role in scientific enquiry.

Fact finding and viewpoints

The first point that needs to be made is that fact finding is not an activity carried out at random. Indeed, 'research without an actively selective point of view', declared the American sociologist Robert Lynd, 'becomes the ditty bag of an idiot, filled with bits of straw, pebbles, feathers and other random hoardings'.[41] Facts have to be selected and they are selected because they are considered relevant for whatever phenomenon the social scientist is investigating. This much is obvious and presumably few 'value-free' social scientists would deny that value-judgements are made in the selection of data. It cannot be therefore in this sense that value-judgements must be excluded from scientific enquiry.

In addition, we assume that social scientists wish to do more than simply state facts relevant to a particular phenomenon, e.g. the statement that the unemployment rate in the United States in 1933 was 25%. If the making of such statements is all the value-free social scientist aspires to then we already have an explanation as to why social science has not achieved successes comparable with those of the physical sciences. Science is a problem-solving activity and if social scientists have no problems or, if they have, are not interested in solving them, then clearly there is no social *science*. Since, however, this is not the case, or not completely the case, we must examine quite carefully the attitude of the value-free social scientist to problem solving.

Positive economics and problem-solving

As to the nature of problems in the social sciences, it would seem at first sight that for the social scientist a problem is any social phenomenon which is inexplicable and/or undesirable given his paradigm—his ideal of social order. Clearly, however, a value-free social scientist *qua* scientist can never declare that a social phenomenon is undesirable, for to do so would mean assuming a definite value commitment—a commitment which he has forbidden himself. How then does the value-free social scientist identify problems? Since Lipsey's textbook is an influential one and is recommended by the economist M. Blaug as one 'written from a consistent positive standpoint',[42] perhaps we can gain insight by examining how Lipsey identifies and deals with problems in economics.

'In most societies,' he writes, 'goods and services are not regarded as desirable in themselves; no great virtue is attached to piling them up endlessly in warehouses, never to be consumed.' And then in a footnote he reassures the reader that 'this is intended as a statement of fact, a statement about what is, it is not intended to imply any value-judgement about what ought to be'.[43] In other words, for Lipsey *qua* scientist, it is irrelevant whether societies wish to consume or wish not to consume the goods they produce; he is concerned only with *what is*, not with *what ought to be*. However, with one eye on the prolonged unemployment of the 1930s, Lipsey goes on to declare that 'it is one of the most important problems of economics to discover why it is that free-market societies produce such periods of involuntary unemployment which are *unwanted by virtually everyone in the society*' (Lipsey's emphasis).[44] Clearly if economists cannot explain why prolonged unemployment sometimes occurs in capitalist societies, i.e. if, on the contrary, their theories predict only temporary unemployment at most, then obviously such a phenomenon is a problem for them. Of course it may well be that what attracts the attention of the economist to the phenomenon of unemployment is because, as Lipsey chooses to emphasize, virtually no one wants it. Well and good, but the value-free social scientist must not himself desire or not desire the phenomenon. His job as a scientist is simply to explain it. His sole concern must be, as a value-free scientist, to discover why it happens. But in the very next sentence Lipsey informs his reader that 'having discovered why this is so, the next problem is to investigate how such unemployment can be prevented from occurring in the future'! Now how is it possible for a value-free social scientist to wish to influence the future course of events? Is this not taking a value-stand in favour of full-employment rather than of unemployment? Lipsey leaves us in no doubt of his humanitarian feelings: in the previous paragraph he speaks of the 'terrible effects on human beings who are unable to find work for prolonged periods of time' and then, in case his readers may not be aware of these 'terrible effects', he advises in a footnote that 'the student with no personal experience of unemployment or depression should [sic] attempt to gain some idea of this experience by reading one or two of the many books on the Great Depression of the 1930s'; and Lipsey recommends in particular Orwell's *The Road to Wigan Pier* and Steinbeck's *The Grapes of Wrath*. Thus if Lipsey's claim that 'scientific enquiry . . . is confined to positive questions' is accepted, then Lipsey's text contains more than scientific enquiry! Unemployment is a problem for Lipsey because he takes a value-stand against unnecessary suffering and for this reason tries to convince his readers that economists should 'investigate how such unemployment can be prevented from occurring in the future'. It is, as he says, 'the next problem'. Once again, let me repeat—I am not quarrelling with Lipsey's value-judgements; what I wish to quarrel with is his assertion that scientific enquiry excludes normative questions. Lipsey himself is guided by value-judgements in his identification of problems and in prescriptions for the future prevention of unwanted phenomena. Clearly, however, Lipsey's values are easily identifiable and should the reader have different values, then he is at liberty to direct his scientific enquiries towards the achievement of different goals, towards the construction of a different society from the one desired by Lipsey. Faced, then, with the *necessity* of research orientated towards the achievement of certain goals, how is it possible for the social scientist to maintain that he is value-free?

Social change and social theory

Before discussing this question further, let us examine why it is that the social scientist, unlike his colleague in the physical sciences, finds that problem-solving in his own society inexorably involves him in social and political action. As we have seen, problems confronting the pure physical scientist are in a certain sense not real: nature behaves in whatever way it does behave and, if looked at in the right way, will not present problematic behaviour to the observer. The only aim of the pure physical scientist is to look at nature in this 'right' way. For the social scientist, however, problematic behaviour can be such not merely because the established theory does not predict it but because it can be morally, ethically or politically unacceptable to the social scientist as a social being. Even if the established theory predicts the unwelcome phenomenon, a moral problem still exists for the social scientist who believes that the phenomenon *is preventable*. For human beings can be reasoned with, an appeal can be made for different behaviour, force can be used. The classical economists believed, for example, that their paradigm would remain a valid description of reality if, among other things, workers would behave differently and accept a decrease in wages. Workers could be appealed to, the government could use force against trade unions—the paradigm could be preserved by changing the social reality by the use of either persuasion or force. Physical scientists cannot save their paradigm in this way! But for the social scientist who dislikes large-scale unemployment a successful solution will necessarily entail a change of the social reality—for unemployment must be eliminated. Clearly, however, there are *different* political frameworks in which this can be achieved. But political commitment of one kind or another is unavoidable. Even in a crisis situation physical scientists, despite different commitments to competing paradigms, have no intention of altering the way Nature behaves—for they cannot do so! But the social scientist *can* attempt to alter the way people behave and there is, for him, no escape from this dilemma. One way or the other, he must decide what his aim is. Should the social scientist aim only at predicting what will happen (say continuing unemployment), then he takes a value-stand against the help which he believes it is in his power to give. On the other hand, should he explain how according to his theory unemployment can be eliminated, then he takes a value-stand in favour of full-employment. There is no way in which the social scientist can be ethically and politically neutral. The refusal to give advice is as value-motivated as is the giving of it.

On 'value-free' advice

Nevertheless, the claim persists that value-free advice can be given! The argument is the following, expressed admirably by Klappholz in a recent review article: 'Value-free advice requires that we be given ethical premises which, in conjunction with empirical hypotheses and initial conditions, enable us to deduce a prescription.'[45] The problem of value-free advice, declares Klappholz, 'is a matter of deducibility; how the requisite value premises come to be formulated—if they can be formulated at all—is altogether irrelevant to it.' Now in a sense I am in agreement with Klappholz. *Given* ethical premises, then prescriptions can be deduced using only empirical hypotheses. There seems little possibility of disagreement here. Yet it is only in the abstract that such advice could possibly be value-free; indeed, note how Klappholz assures us that it is irrelevant how the requisite value premises come to be formulated in the first place. But in being prepared to carry out the necessary scientific investigation and give the advice so deduced, the social scientist has clearly accepted the given value premises as being sufficiently compatible with his own. Why should such advice be called value-free? By all means, let us call it scientific advice. But because the values implied are acceptable to the social scientist, it is misleading, to say the least, to call such advice value-free. *Logically*, to be sure, 'the statements about means-ends relations are value-free',[46] as Nagel writes, but the social scientist in making such statements is not. This is therefore the fundamental point as issue between those who believe social science can be value-free and those who believe it cannot be. *Logically* it *can* be value-free but social scientists do not live in a world of logic only. They, as human beings, are not value-free, and therefore cannot be indifferent to the value premises 'given' them. They will, on the contrary, choose those research programmes which at the very least do not contain ethical premises contrary to their own. Of course, if a social scientist has transformed himself through mutilation into a computer, then clearly he no longer has values — or, rather, his 'values' are, in each run, those of his programmer. However, such a being would be regarded by all his fellows as a psychopath — hardly a state to be desired. Yet it would appear that some social scientists *do* desire this state. For the insistence that scientific advice be called 'value-free' suggests that social scientists should aim at achieving an attitude of moral indifference to the human situation (learning, in other words, to regard the rest of the human race as biologists learn to regard the greenfly) with the ultimate aim of being prepared to place their expertise unreservedly at the service of whoever commands or pays. Indeed, it is exactly this apparent equating of scientific objectivity with intellectual prostitution that has provoked Gouldner into making the impassioned outburst that 'if today we concern ourselves exclusively with the technical proficiency of our students and reject all responsibility for their moral sense, or lack of it, then we may some day be compelled to accept responsibility for having trained a generation willing to serve in a future Auschwitz.'[47] Perhaps human beings can be turned into thing-beings so that they accept whatever ethical premises are fed into them. Nevertheless, I think it reasonable to query whether or not this should be the legitimate goal of social science.

Lundberg and 'value-free' social science

Lundberg, however, hopes to maintain the moral and political neutrality of science by claiming that it is not as scientists but as citizens that men and women (who otherwise call themselves scientists) advocate a particular policy. In their activity as scientists such people, Lundberg declares, must remain morally and politically neutral; only as citizens can they favour one point of view rather than another. He writes that social scientists should 'agree that the sole function of scientific work is *to grind out and publish systematically related and significant "if . . . then" propositions which are demonstrably probable to a certain degree under given circumstances*'. After which, 'scientists may then *in their capacity as citizens* join with others in advocating one alternative rather than another, as they prefer' (Lundberg's emphases).[48] Clearly one major trouble with this approach is that it is unworkable (apart from its being amoral). The social scientist cannot simply 'grind out' significant 'if . . . then' propositions and then as citizen advocate one of them. The articulation of a theory is the work of several years, if not of a lifetime. The theory the social scientist commits himself to and chooses to articulate is the one which appears to promise acceptable solutions to those social problems of most concern to the social scientist.

Unfortunately, the practice of 'value-free' social science tends to focus the attention of social scientists on to those solutions of problems which do not radically change the established social order and thereby the value system

which the 'value-free' social scientist unconsciously holds. Clearly such commitment to the preservation of the *status quo* serves, at the same time, to prevent awareness from developing that radical solutions presupposing a qualitatively different social order are both possible and perhaps desirable. Given, however, a spectrum of values and therefore a spectrum of commitments on the part of social scientists, many possible solutions will be proposed and discussed, some solutions presupposing the existing social order, others presupposing a radically different social order. Such controversy among social scientists is much to be desired. Changing a social order is no light matter but neither, of course, is the policy of preserving a social order, come what may.*
In such weighty matters the availability of many different perspectives makes a reasoned choice of action so much the more likely. When, however, the vast majority of social scientists either refuse to do more than diagnose problems or only attempt to articulate those solutions to problems that are compatible with the existing social framework, claiming that thereby they are being value-free and therefore 'scientific', we see the enormity of their self-deception; since they act in such a way as to prevent awareness of the possibility of radical change they thus make more likely the continuation of the present into the future. Their refusal to speculate on the feasibility of new social orders is not, as they claim, a commitment to objectivity and truth; it is a commitment to the perpetuation of the existing social order. In other words they clearly perform, despite all protestations to the contrary, an ideological service in favour of the *status quo*.

Science versus speculation?

If the 'value-free' social scientist declares that it is not scientific to speculate, this immediately betrays, I believe, a basic misunderstanding of the nature of scientific thinking. Dahl, for example, in his *Modern Political Analysis* contrasts speculative against scientific thinking in the following interesting way. 'Two of the most important ways of exploring alternatives,' he tells us, 'are by speculation and by science.' Nevertheless, it is admitted that 'speculation is an important prior stage in the growth of scientific knowledge'. We are then informed that 'philosophers had speculated about the Universe centuries before Copernicus; indeed, even with Copernicus the theory of a heliocentric solar system rested on highly speculative reasoning'. Thus it is that 'profiting

---

* Popper, for example, argues that it is impossible to possess the experimental knowledge needed for what he calls Utopian engineering. In any case, it is his opinion that 'every attempt at planning on a very large scale is an undertaking which must cause considerable inconvenience to many people, to put it mildly, and over a considerable span of time'.[48a] Opposition, he argues, must be suppressed and with it reasonable criticism. Popper dogmatically commits himself to the belief that piecemeal social engineering is *always* preferable to radical social change. But what Popper overlooks is that the former kind of social change may 'inconvenience' far more people over a longer period of time than the latter. There can, however, be no general rule. Sometimes piecemeal change will be sufficient; at other times only radical social change will suffice. The social scientist must decide according to his values, goals and his theoretical analysis of the situation he finds himself in.

by speculation and chance discoveries, science opens up new alternatives and often closes out old ones'. '[Although] frequently,' we are told, 'the alternatives suggested by speculation are irrelevant until they have been more fully developed by science.'[49] The emphasis is revealing. Speculation is to be distinguished from science. It is true, says Dahl, that speculative reasoning first gave rise to the idea of a heliocentric planetary system but from then on, the impression is created, science took over in first demonstrating the reality of this planetary system and then in discovering more truths about the Universe. Speculative reasoning may or may not be useful. But it is not scientific.

Dahl, of course, is free to define 'science' as he thinks fit, but his definition of science is one conducive to the misunderstanding of the nature of science we are concerned with in this chapter. Indeed, even Popper and Kuhn find common ground in agreeing that speculative reasoning is an essential ingredient of at least the physical sciences. For Kuhn, speculative thinking automatically emerges as the existing paradigm is articulated into crisis. As he writes, 'Since no experiment can be conceived without some sort of theory, the scientist in crisis will constantly try to generate speculative theories that, if successful, may disclose the road to a new paradigm and, if unsuccessful, can be surrendered with relative ease.'[50] For Popper, speculative thinking is the very life-blood of science. 'Out of uninterpreted sense-experiences science cannot be distilled,' he writes, 'no matter how industriously we gather and sort them.' On the contrary, 'bold ideas, unjustified anticipations and speculative thought, are our only means for interpreting nature: our only organon, or only instrument, for grasping her'.[51] If, then, 'value-free' social scientists refuse to speculate on the possibility of radically new social orders because they feel that thereby they are being truly scientific, truly within the tradition of the physical sciences, they are simply wrong. Speculative reasoning is an essential part of scientific reasoning and indeed those episodes in the history of physics, sketched in chapters 2 and 3, amply suffice to show that it is a *sine qua non* of scientific progress.

Perhaps, however, the 'value-free' social scientist declares that scientific hypotheses must be in principle capable of empirical testing and that speculation concerning new social orders is thereby ruled out as unscientific since such speculation cannot be tested by experience. The argument is fallacious, but it is revealing to examine its logic in greater detail. The diagnosis of a problem within a given social order can be tested as to its validity through the efficacy of remedial actions based on the diagnosis itself. Either the 'value-free' social scientist himself is inconsistent and recommends action (as does Lipsey) or the problem is diagnosed in the form 'the phenomenon A occurs because of B'. Either way the established authority receives advice as to appropriate action. A can be prevented by suitable action with respect to B. If the power of the established authority is such that action can be taken with respect to B, then the diagnosis can be tested. All this seems very scientific and proper—and it is, except for the claim of the social scientist that he has been value-free. Consider now, however, the advice of an explicitly value-committed social scientist who declares that, for the reasons and values he states, the phenomenon A can and should be prevented by the construction of a society S' qualitatively different from the existing one S. The new society S' is, let us suppose, forbidden by the constitution of S. In that case the speculations of the value-committed social scientist will not easily be tested in practice since the established authority will use its superior force to see that S' is not constructed. This, however, can scarcely mean that the predictions concerning the feasibility of S' were unscientific in the first place. For since when have predictions been declared unscientific simply because an established authority prevents the appropriate experiments from being carried out? In such a dilemma the social scientist, if his values permit, must show how the authority's use of force can be overcome and S' constructed regardless. There is after all nothing in science which states that ideas can be tested only by the established authorities in ways approved by them. That social scientists are so often defenders of the *status quo* in no way implies the conclusion that only violence by the established authority is 'scientific'. To be sure, if that is the way it so often appears to the 'value-free' social scientist it is simply because he has so internalized a set of values that the existing social order appears to him as the universal standard against which all non-conformist behaviour appears as problematic, deviant and, above all, corrigible.

Social science and physical science

One suspects that 'value-free' social scientists believe that they are being true to the scientific method practised by physical scientists, a method that has obviously led to a deep understanding of the physical world and one which, it is popularly believed, 'ignores all judgements of value; for instance, all aesthetic and moral judgements'. Now even if this were true it would seem extremely arbitrary, not to say timid, to restrict the scope of the social sciences in the name of the physical sciences. But such a characterization of the 'scientific method' is extremely misleading and our brief look at the history of physics has sufficed not only to show this but also to demonstrate how totally unrealistic is the attempt to justify 'value-free' social science by analogy with the physical sciences. Indeed the nature of progress in the physical sciences indicates exactly the opposite—that social scientists cannot be expected to make the same degree of progress that physical scientists have achieved until they acknowledge and welcome the fact that value commitments are an essential feature of the scientific quest.

Revolutions in the physical and social sciences

Revolutions have occurred in the physical sciences and will most probably continue to occur provided physical scientists are not determined to maintain the community's paradigm, *come what may*. To be sure, the paradigm is never lightly surrendered. For commitment to the paradigm is, after all, a defining characteristic of 'normal science'. But when puzzles remain insoluble over an appreciable period of time, then a crisis period ensues in which physicists commit themselves to different and competing paradigms, guided by aesthetic criteria and by intuition as to the 'true' nature of reality. The point is that paradigm commitment eventually leads to subversion of its own principles and thus prepares the way for revolutionary progress. Chomsky goes as far as asserting that 'honest inquiry is inherently "subversive" in any field.' He believes that, 'the physicist working at the borders of current knowledge will attempt to challenge assumptions that retard understanding' and it is Chomsky's provocative opinion that 'the same would be true of serious social inquiry, if it existed on any significant scale in the universities'.[52] At this point, however, we should note a major difference between the physical and social sciences. The physical scientist is free to be as revolutionary as he wishes—funds are readily available; the more revolutionary he is, the greater his chances of winning a Nobel Prize and world acclaim. For revolutions in the physical sciences are no longer thought to undermine the political structure of any society.* This is hardly the case in the social sciences. For radical research in these disciplines would tend to be politically subversive, highly likely not only to suggest that the existing institutional framework is in need of qualitative change but also to indicate the general framework of a society in which present problems would be soluble or non-existent. Should the social scientist actually suggest an action programme for such a change, then in addition to finding himself rather quickly without research funds he would also find himself running the risk of severe political repression, particularly if his programme is judged by a ruling élite to represent a serious threat to its political and economic power. As Lynd has clearly pointed out: 'The dangers inherent in making one's intelligence explicit in terms of its full implications for going institutions is immediate and real. It is much safer either to avoid dangerous hypotheses or, when one does touch them, to leave the implications of one's data to be read between the lines, if and as the reader so elects.'[53] Quite so. But the 'value-free' investigation of what is and how things have changed, chosen in preference to a value-oriented investigation of how things could be, and indeed should be, must not be confused with 'scientific objectivity'. If the majority of social scientists do not choose to investigate the possibility of qualitatively different societies, and to suggest ways of realizing such societies, then this is a political and moral decision which

* It was, of course, not always so, as Galileo found out to his cost, and during two unhappy decades in the Soviet Union certain aspects of relativity and quantum theory were regarded by the Soviet government as 'decadent, bourgeois' science. The biological sciences fared even worse in the Soviet Union. Vavilov, a leading Soviet geneticist, was condemned to death for his support of 'bourgeois' genetic theory, a sentence later commuted but too late to save Vavilov's health and life. Many of his colleagues met a similar fate. In general, however, provided natural scientists stick to their subject matter, their revolutionary activities now meet with governmental approval, independently of the government's political colour.

the practice of science in no way obligates. However it is a decision that explains why the social sciences have as yet not achieved the spectacular successes of the physical sciences.

To be sure, physical scientists are in several ways more fortunate than social scientists. To be sure, no matter how chaotic and complex physical reality appears to be, the motivating faith of physical scientists is that harmony can be found, that the multiplicity of phenomena making up that reality obscures a simple, beautiful harmony discoverable if only reality is looked at in the right way. It is therefore not unreasonable for Nobel laureate Feynman, in answer to the question why science is possible at all, to state that in his opinion 'it is because nature has a simplicity and therefore a great beauty'.[54] But this can scarcely be the motivating faith of social scientists; if harmony is to be found in society, then it must be a *constructed* harmony—the existing social structure will have to be changed. Only if the social scientist is indifferent to the sufferings of his fellow men and indifferent even to their (and his) possible disappearance, could he conceivably obtain aesthetic satisfaction from a theory correctly predicting the state of today's world. Of course, indifference to the human situation is not impossible; and manifestly some human beings exist who like to watch other human beings suffer. Furthermore, it is clearly possible to see in poverty, disease and illiteracy a certain desirable order, if not even harmony, if their abolition would threaten one's own privileged position. But those social scientists who do see the phenomena of hunger, deprivation and oppression as ugly, and, furthermore, as ugliness that could and should be eliminated, will become committed to a research programme that at the very least challenges the basic institutional frameworks of several countries, and perhaps those in which they live. They will then expose themselves to physical danger. The position of the social scientist is not enviable. Nevertheless, there is no way in which the responsibility of making moral and political choices can be avoided. The choice of moral and political 'neutrality', usually taken from the vantage point of material affluence, is a commitment all the same and it is a commitment in favour of the *status quo* and of service to the *de facto* political power. The attempt to establish that science itself dictates a choice of moral and political 'neutrality' is at best a self-deception and at worst a deliberate attempt to use the prestige of 'science' as a means of underwriting the *status quo*.

Chapter 7

PROBLEMS OF POST-WAR CAPITALISM IN THE UNITED STATES

> We are in a double crisis—the crisis of our internal character as a nation and the crisis of the relationship between America and the world. After so many years of overweening confidence in our ability to fix up all the troubles of mankind, we are now suffering increasing doubt that we can even heal the ills of our own national community. The time has surely come for a reassessment of our institutions and values.
>
> ARTHUR M. SCHLESINGER JR.
> *The Crisis of Confidence*, 1969

THE first half of this essay has in part served to provide a historical and conceptual framework in which it is possible to take up the questions with which we began this essay. Why is the present world characterized by so much hunger, disease and violence? Above all, what needs to be done in order to construct a world in which, *at the very least*, there is no longer an arms race, no longer widespread malnutrition, disease and oppression.

Fortified by the argument of chapter 6, we emphasize at this point that we are not looking for universal laws that all societies obey, for universal social laws that human beings must obey whether they know of their existence or not. We are not engaged in what we consider to be a wild-goose chase. Instead we are looking at a specific historical period faced by specific problems in an attempt both to identify the causes of these problems and to find possible solutions based on the values implicit in the above goal.

It is with this end in view that we return to the theme of chapters 4 and 5.

1. The Nature of Advanced Capitalist Society

An essential aspect of the key to an understanding of the nature of post-war capitalist societies lies in an understanding of the causes and consequences of the Great Depression—that traumatic decade for capitalist countries which came to an end with rearmament and the Second World War. Not surprisingly, a post-war objective of all capitalist countries has been to attempt to ensure that the disaster of the 1930s will not be repeated.

Let us first recall a result of the analysis in chapter 5, namely that an important driving force in capitalist society is the capital goods industry. It was this industry which was most badly hit during the 1930s, net investment being actually negative for five consecutive years; even in 1939 non-residential fixed investment expenditure was still more than 40% below its 1929 level. The fact that rearmament and war revived the fortunes of these sensitive industries reinforced the lesson for capitalist ruling circles that, unless in time of peace they underwrite the capital goods industry against cut-backs in investment expenditure, prolonged and extensive unemployment will inevitably occur.

The necessity of growth in advanced capitalist societies

A cut-back in investment expenditure tends to produce consequences destabilizing for a capitalist society. For if lack of investment expenditure and unemployment resulting from redundancy in the capital goods industries lead to decreased consumption expenditure by unemployed workers and lower profits earned by the consumption goods industries, then a further cut-back in investment expenditure becomes likely taking the economy on to the second round of a downward spiral leading, if not arrested by government intervention, to prolonged, extensive unemployment and unused productive capacity. Clearly the existence of the latter only serves to inhibit increased investment expenditure. As Keynes argued, one way out of such a depression is for the government concerned to run budget deficits in the financing of public works. Although we shall later look at the development of the United States in its historical and global context (an advanced capitalist society which through government expenditure did partially attempt to resolve its problems in the way Keynes was suggesting) we first of all abstract the United States out of this context to ask what must be the policy of its ruling members in order to prevent the occurrence of widespread unemployment and excess capacity once full employment has been achieved with little government intervention in the economy. Clearly a necessary feature of an adequate policy must be the underwriting of a high level of investment expenditure sufficient to maintain full employment. Furthermore, if such investment also results in a steady growth of national income *per capita*, it will be doubly desired. For in so far as growth provides a means of raising the living standards of both 'poor and rich' together, not only is growth welcomed by all income brackets but politically destabilizing demands for redistribution of the national income are thereby blunted. Thus, in their well-known textbook McDougall and Dernburg write that 'economic growth is, of course, a great blessing. It is what enables the individuals in society to look forward to rising living standards. It is, moreover, a great social solvent. When all can look forward to rising levels of well-being, the pressure for redistribution of the existing pie is reduced.'[1] Since, therefore, a high rate of economic growth requires a high level of investment expenditure, then provided growth *can* be maintained at a sufficiently high rate, the probability of the occurrence of widespread unemployment and excess productive capacity is made that much smaller while at the same time all income groups experience a rising material standard of living.

Barriers to continued growth

Can, however, growth be maintained indefinitely?* It would appear improbable. For while rising investment expenditure creates employment and maintains a high level of effective demand it also, by bringing into existence an enlarged stock of capital goods, tends to lower the rate of profit, thus impeding employment and prosperity. For example, as noted by Klein in his analysis of Keynes' *General Theory*, '... the housing boom of the 1920s in the United States led to such an accumulation of residential capital that rents began to fall, and new housing investment remained low for a long time'.[2] Clearly there is always the danger that once the wealthier members of society have bought the houses, cars and other material possessions they need, their expenditure on commodity goods will tend to decline as, at the same time, industry is experiencing difficulty in finding profitable investment outlets for the increasing volume of savings.

As Joan Robinson writes:

> An abrupt increase in thrift leading to a decline in outlay on consumption goods is acutely distressing to entrepreneurs; some find themselves unable to make the sales in expectation of which their productive capacity was built up, and if, in consequence, they dismiss workers or curtail investment plans, others also suffer a shrinkage of markets. A gradual increase in thrift, leading to a gradual fall in the ratio of outlay on consumption to outlay on investment would do no harm if the amount of outlay on investment gradually increased to the corresponding extent. But there is no reason to expect this ever to happen ...[3]

Her conclusion is that 'thrift, in short, makes possible a high rate of accumulation and yet sets obstacles in the way of achieving it'.

There are further problems. For even if we suppose that consumption expenditure remains a constant proportion of the national income, continued growth will still tend to generate barriers against its own continuation. For although 'today's' investment guarantees both today's employment and profits for the industries which invested 'yesterday', today's investment also increases, as it was intended to, the productive capacity of tomorrow's economy. This means that a greater supply of goods becomes potentially available (say of real value Δy). However, since the propensity to consume c merely stays constant (and does not increase), then not all these extra goods will be bought by households (but only a fraction of value $c\Delta y$). Demand will therefore not be equal to supply *unless there is an extra investment demand* (of

---

* We ignore for the moment the very relevant considerations that the earth is finite, that supplies of raw materials are finite, that they come in the main from underdeveloped countries and that industrial growth produces undesirable ecological changes. We consider these and other factors in due course. The reader is, however, asked to bear in mind ecological considerations throughout the reading of this chapter.

value $(1 - c)\Delta y$. In other words, if tomorrow's economy is to be used to full capacity, tomorrow's volume of investment *must* exceed today's. Otherwise unused capacity will exist, so causing investment to decrease, unemployment to rise, and a spiral into depression to begin. 'The economy finds itself in a serious dilemma: if sufficient investment is not forthcoming today, unemployment will be here today. But if enough is invested today, still more will be needed tomorrow.' Thus reasoned Domar in 1947. His conclusion was that 'as far as unemployment is concerned, investment is at the same time a cure for the disease and the cause of even greater ills in the future'.[4] Not only must the economy grow, but each year it must grow *by a larger amount* than in the previous year. However, each year that larger amount of investment expenditure will become increasingly difficult to achieve. And hence serious depressions will become increasingly difficult to avoid. In 1960 the Executive Vice-Chairman of the Cabinet Committee on Price Stability and Economic Growth explained to his audience of business and financial journalists that '*the real growth imperatives* arise from the fact that a strong economy is a growing economy. An economy with a high *per capita* income such as ours generates a large volume of private saving which must flow into capital accumulation if the economy is to sustain itself. In other words, the continued vitality of the system requires growth'.[5]

Prescriptions for ensuring continued growth

Let us look at some proposals to maintain growth by (i) keeping the propensity to consume high, and (ii) by ensuring expanding and profitable investment opportunities. We notice that according to Keynes: 'The remedy would lie in various measures designed to increase the propensity to consume by the re-distribution of incomes or otherwise; so that a given level of employment would require a smaller volume of current investment to support it.'[6] The measures adopted to maintain a high propensity to consume have been chiefly 'otherwise'.

(i) ON THE PROPENSITY TO CONSUME

(a) *Advertising*

In August 1960 Advertising Business bought a third of a page in the New York Times to explain its role to the public:

WHAT WOULD HAPPEN IF ALL ADVERTISING STOPPED?

... Without advertising our national economy, our national *life*, would be bleak indeed. In many ways, advertising is the *power plant* of our society ...

Advertising not only gives people news about new products, but provides the *urge* for people to own and enjoy these products. The wider and deeper the penetration of our products into the life of America, the greater the need for more production. This means more jobs. More jobs mean more people able to enjoy what we make.[7]

Schon makes, I believe, an accurate assessment: 'There has been an ironic reversal in the relationship between producers and consumers, so that it is no longer possible to say whether producing industries exist in order to satisfy consumer needs or whether consumers, goaded by ever more persuasive advertising, exist as appendages to the system of industrial production.'[8]

One thing, however, is clear. Whether or not it is the 'nature' of man to wish to consume an increasing volume of goods each year, advanced capitalist society cannot survive unless he does so.* As Heilbroner explains in scholarly language: 'The endless and relentless exacerbation of economic appetites in advanced capitalist society is not merely a surface aberration but a deeply rooted functional necessity to provide the motivations on which the market system depends.'[9]

We note here that a somewhat inequitable distribution of income enables advertisers to attempt to maintain a high average propensity to consume by the technique of associating the possession of commodity goods with social status. Thus, if such techniques are successful, all income groups will attempt to maintain or even to increase their propensities to consume as their incomes rise, the lower income groups in order to achieve not only a higher material standard of living but also the superior social status of the higher income groups who possess the status-imparting commodity goods, the higher income groups in order to retain their lead in the possession of commodity goods and therefore their superior social status. The necessity for growth thus becomes self-perpetuating. (Of course, a *too* inequitable distribution of income could lead to excessive saving by the rich which would be clearly destabilizing. On the other hand, an *insufficiently* inequitable distribution of income, in so far as it would lead to material equality, could at the same time lead to material contentment and therefore to economic instability and finally to poverty and political discontent in the midst of potential affluence. It would seem that there are problems connected with the distribution of income in a capitalist society!)

Thus it is that the advertising industry employs a large number of intelligent people whose principal aim is to annul each other's efforts in all directions but one; for all-pervasive advertising makes the possession of goods in general seem important and therefore, in helping to maintain the propensity to consume at a high level, helps at the same time to preserve full employment and an unchanging social structure.[10]

* In speaking of underdeveloped countries a former head of the Economic Planning Unit of Nigeria has declared: 'In many countries it *is* difficult to get going. There *is* a problem of insufficient ambitions, not dissimilar to the problems that advertising tries to cure in highly advanced economies. The creation of wants may present problems.'[8n]

(b) *Built-in and psychological obsolescence*

Marketing consultant Victor Lebow is similarly clear on the needs of advanced capitalist society: 'Our enormously productive economy demands that we make consumption our way of life . . . that we seek our spiritual satisfactions, our ego satisfactions, in consumption. . . . We need things consumed, burned up, worn out, replaced and discarded at an ever-increasing rate.'*[11] 'Indeed, Robert Theobald has argued that: "So long as the present socio-economic system is not changed, abundance is a cancer and the various parts of the system *must* continue to do their best to inhibit its growth' (Theobald's emphasis).[12]

Thus in order to help prevent possible abundance—the 'calamity' of market saturation which, despite advertising pressure to buy, is always an unwelcome possibility—goods can be built not to last for too long and models can be frequently changed. One example is sufficient to illustrate this functional wastefulness of advanced capitalist society. Ignoring the question as to whether an 'automobile civilization' is desirable or not, we note that according to a study made by three economists the cost of model changes in cars over the period 1956–60 came to about $700 per car (more than 25% of purchase price) or about $3·9 billion per year.[13] It would appear that it is not sufficient for each family to have a steadily increasing number of cars; it is also desirable that each year present models be sold or discarded and new models bought.

(c) *Product innovation*

As emphasized in the early 1960s by Herbert Holloman, the then Assistant Secretary for Science and Technology, the industrial role of scientists is an important one. For not only are model changes important, new products are even more so. 'We must encourage,' declared Holloman, 'the establishment of the institutions and environment that most effectively . . . encourages even higher levels of innovative activity.'[14] Now nearly every home in the United States has, for example, a black and white television, status-imparting colour television is being marketed, to be bought at first by pace-setters in the higher income brackets. Once nearly every home has a colour television, three-dimensional colour television can be introduced. As the market for this product nears saturation, industrially employed scientists will be expected to have developed new products of even greater attractiveness for civilized living.

(ii) On the propensity to invest

(a) *Population growth*

According to Kurihara 'a capital-rich society need not fear the depressing effect of capital accumulation on the marginal efficiency of capital if this

---

* A policy, one feels, with certain ecological implications.

effect is offset by the stimulating effect of population growth on the demand for final output.'*[15] Perhaps this is true, provided corporate managers are convinced that the growing population will find jobs and hence enjoy an income which will make consumption possible. With increasing automation, however, this is by no means likely. (On the contrary, as the rate of technological innovation increases, and as automation and cybernation continue—for the purpose of cutting costs and so increasing profits—so paradoxically an ever-decreasing number of blue and white collared workers will be needed in the automated factories and offices. Hence unemployment will tend to increase and progressively fewer people will be able to buy the always potentially increasing quantity of goods available. Automation therefore poses in the long run a serious threat to the stability of capitalist society.)

(b) *Investment outlets*

Keynes himself believed that ever-expanding investment outlets could not be found indefinitely. Indeed he believed that an advanced capitalist society could within a generation or so produce sufficient capital equipment to satisfy all its material needs. Hence the problem.

Wrote Keynes ironically: 'Ancient Egypt was doubly fortunate, and doubtless owed to this its fabled wealth, in that it possessed two activities, namely, pyramid-building as well as the search for the precious metals, the fruits of which, since they could not serve the needs of man by being consumed, did not stale with abundance. The Middle Ages,' he continued, 'built cathedrals and sang dirges.'[16] Noting that 'two pyramids, two masses for the dead, are twice as good as one; but not so two railways from London to York', Keynes pointed out that 'each time we secure today's equilibrium by increased investment we are aggravating the difficulty of securing equilibrium tomorrow.'[17] What is needed therefore is the production of useless goods. Keynes' ironical conclusion (and it is one scientists might well note) was that 'in so far as millionaires find their satisfaction in building mighty mansions to contain their bodies when alive and pyramids to shelter them after death, or repenting of their sins, erect cathedrals and endow monasteries or foreign missions, the day when abundance of capital will interfere with abundance of output may be postponed.'[18]

We shall take up again the problem of investment outlets for advanced capitalist societies. We simply note here that, according to Klein (1966), Keynes' 'pessimistic outlook for investment does not appear to be entirely justified by recent events. . . . New inventions inspired by the development of World War II, such as radar, automated devices, and atomic power were not foreseen.' He writes that the 'prospects are that new discoveries will continue to flow: new investment opportunities will continue to arise . . . Who is to say', he concludes triumphantly, 'what the economic prospects of the moon are?'[19] What indeed! In a more sane and humanitarian mood,

* Once again we have occasion to note how ecologically unsound is capitalist society.

however, Klein also writes: 'Decay of U.S. cities and problems of urbanization have become worse in the past two decades. More than ever before, we need imaginative public investment on a large scale. This will simultaneously attack the urban problems and the unemployment problem.'[20]

The quality of life in advanced capitalist society

Although we have not completed our sketch of advanced capitalist society, it is already apparent that among its essential features are the institutionalization of competitive behaviour, the consumption of commodity goods as a way of life, and the functional production of waste. Moreover, for as long as the basic foundation stone of capitalist society is preserved—production for private profit rather than for the general good—these somewhat disagreeable features of capitalist society must necessarily characterize Western societies. Paul Mazur, a prominent New York investment banker, has described his fear (as reported by Baran and Sweezy) of what would happen, for example, were Americans not goaded into making consumption a way of life: 'Clothing would be purchased for its utility value; food would be bought on the basis of economy and nutritional value; automobiles would be stripped to essentials and held by the same owner for the full 10 to 15 years of their useful lives; homes would be built and maintained for their characteristics of shelter without regard for style or neighbourhood.'[21] Baran and Sweezy observe that Mazur does not seem to realize that a way of life based on the production of commodity goods for use rather than profit would give people the opportunity to improve the quality of their lives, to build a society based on principles other than commercial ones. But quite rightly Mazur asks: what would then 'happen to a market dependent upon new models, new styles, new ideas?' The answer is that capitalist society could no longer exist.

Some global considerations

Thus far we have been looking at an advanced capitalist society in the abstract, as if it were self-contained and non-interacting with the outside world. This major defect in our analysis must now be removed. In the first place we should note that the inequitable distribution of the United States' national income is paralleled by the inequitable distribution of international income. For the United States with some 6% of the world's population and covering approximately 6% of the world's land surface produces some 40% of the world's annual industrial output and consumes likewise some 40% of the world's raw materials used each year. Such inequitable use obviously generates and will continue to generate 'international tensions'. Furthermore, supplies of raw materials are finite; unless therefore there is spectacular scientific and technological advance industrial growth cannot continue indefinitely. As an additional complication, the ever-increasing production and consumption of industrial goods, combined with the need to increase agricultural productivity to feed an ever-increasing world population, are together changing with potentially calamitous consequences the ecological balance between man and his physical and biological environment. The recent dangerous levels of air pollution in New York and Tokyo (August, 1970), are considered to be but one aspect of the visible tip of an ecological iceberg. Indeed some ecologists argue that industrial man is in the process, rapidly reaching completion, of converting the earth into an enormous Titanic.[22] Prudence would therefore suggest, on ecological grounds alone, that industrial growth in the advanced industrial societies be stopped for at least one generation (preceded by redistribution of income) while further research is carried out. Humanistic concern would suggest that the advanced industrial societies also devote their energies to helping to improve the standard of living in underdeveloped countries, while at the same time both developed and underdeveloped countries cooperate together in the attempt to achieve population stabilization. Needless to say such a programme would demand as a prerequisite radical social and political changes in all Western societies and especially in the United States (we shall consider the position of the Soviet Union in chapter 9).

On the desirability of a stationary material state

Surely not a fate worse than death. Perhaps even an objective to be desired. Keynes himself had looked forward to the time when the 'Economic Problem' would be solved and the art of living would take first priority, as had John Stuart Mill nearly a century earlier. Keynes' contemporary Lewis Mumford expressed himself firmly on the matter:

> Humanly speaking, it [capitalist society] has worn out its welcome. We need a system more safe, more flexible, more adaptable, and finally more life-sustaining than that constructed by our narrow and one-sided financial economy.... Our goal is not *increased* consumption but a vital standard: less in the preparatory means, more in the ends, less in the mechanical apparatus, more in the organic fulfilment. When we have such a norm, our success in life will not be judged by the size of the rubbish heaps we have produced: it will be judged by the immaterial and non-consumable goods we have learned to enjoy, and by our biological fulfilment as lovers, mates, parents and by our personal fulfilment as thinking, feeling men and women.[23]

To choose such a way of life in place of possible, if not probable, world suicide would scarcely seem to me to be the worst of all possible bargains.

But the problem is political. For capitalist societies, as we have seen, must either change and change radically, or they must continue to promote economic growth in order to prevent change and will therefore inherently court global disaster.

Growing concern by economists

It would be unfair to suggest that all economists are in favour of promoting growth rather than redistribution of resources and a qualitatively different way of life. There have recently been at least cries of increasing unease. Gabor, for example, although not an economist but an applied scientist, expressed the opinion in 1961 that

> Fifty years ago an engineer could have a perfectly good conscience when he bent over his job, fully satisfied that he was a benefactor of humanity when he produced more or cheaper machines and all sorts of consumer goods. He can still have a good conscience if he happens to be a Russian, an Indian or a Chinese. But in Western industrial countries at the present time such myopic self-satisfaction amounts to blindness.[24]

In 1962 the economist Schumacher wrote that 'Present-day economics, while claiming to be ethically neutral, in fact propagates a philosophy of unlimited expansionism, without any regard to the true and genuine needs of man, which are limited.'[25] A year later Theobald declared that it is 'ludicrous to continue to define economics as the art of distributing scarce resources'.[26] In his well-known 1967 book entitled *The Costs of Economic Growth*, Mishan expresses a similar opinion that 'the classical description of an economic system makes sense in today's advanced economy only when stood on its head. Certainly the American economy presents us with a spectacle of growing resources pressing against limited wants'.[27] Likewise the American economist Galbraith believes that 'economics, as a discipline, has extensively and rather subtly accommodated itself to the goals of the industrial system'.[28] He tells us that

> It would, *prima facie*, be plausible to set a limit on the national production that a nation requires. The test of economic achievement would then be how rapidly it could reduce the number of hours that one needed to meet this requirement. Were economists to advocate this goal, with the revolutionary effects that it would have on the industrial system, there would be grounds for complaint. None has been so uncooperative.[29]

Galbraith insensitively reminds his fellow economists that 'the questions that are beyond the reach of economics—the beauty, dignity, pleasure and durability of life—may be inconvenient but they are important'.[30]

Government expenditure and welfare

The world that such concerned economists portray is one (incompletely but correctly) characterized by an obsessive preoccupation with commodity goods, by uncontrolled technological innovation, and by a combination of private affluence and public squalor. However, although the first two features appear to be intrinsic to capitalist society, it would not at first sight appear that widespread poverty and public squalor are necessary features of an advanced capitalist society. And indeed economists often argue that when the level of private investment expenditure is insufficient to maintain full employment, then government expenditure on public welfare could and should fill the breach. In this way, they argue, not only could a full employment level of effective demand be permanently maintained but adequate schools, homes, hospitals, parks and other necessities of life could be provided for all.

Unfortunately, such economists tend to overlook that there are political constraints imposed on the direction of government spending. Indeed, according to Heilbroner (1968), 'The nature of class interests in a capitalist system is not mentioned in any textbook [on modern economic theory], so that nothing in the nature of political or social constraints confines the free movement of the economic model.'[31] This feature of 'objective' economic theory is so important that it is worthwhile to quote Heilbroner at further length. He continues that there is

> the systematic exclusion of matters that might connect the functional model with the pressures and resistances of the political world. This exclusion, which accounts for so much of the irrelevance of economics, is by no means accidental. Rather, it results from a fundamental failure of vision on the part of modern model-builders, *who do not see that the social universe that they are attempting to reproduce in a set of equations is not and cannot be adequately described by functional relationships alone, but must also and simultaneously be described as a system of privilege* [Heilbroner's emphasis].

It is this restriction of vision which allows economists to believe that the capitalist social order can be reformed by minor adjustments into a just and humane society, that no radical changes are necessary. There are, they admit, problems that remain to be solved, but solutions are possible, they claim, within the institutional framework and class structure of capitalist society.

A reformist government intent on welfare expenditure in a capitalist society will, however, be met with severe resistance. In the first place, pressure groups will join forces to prevent such expenditure. Although the medical profession, for example, will have no objection to more homes for people, and the real estate agents no objection to free medical service for all, nevertheless it may quickly become apparent to the two groups that it is in their mutual interest to present a united front against both welfare programmes. Moreover, massive welfare expenditure runs contrary to the desire of members of the middle class 'to get ahead and keep ahead'. There is therefore considerable reluctance from those who have achieved social success to see a reformist government borrowing or taxing money from wealth owners in order to raise the standard of living of those whose very poverty confirms the material success and status of the middle and upper classes. Furthermore, the abolition of poverty—a finite task—would be destabilizing for capitalist society. For once every family had achieved an adequate home and high material standard of living, how would the ever-growing volume of savings find profitable investment outlets? A materially contented, non-consuming public would clearly mean either the end of the capitalist system or, once again, depression on a vast scale and the deliberate destruction of food and goods. Most important of all, of course, and the large corporations are well aware of this, government construction of homes, schools, hospitals, parks, etc., could lead people to ask if it would not be better were the government to nationalize all large industrial firms and produce only for use, not for profit. Not even Keynes was so heretical as to suggest this.

On corporate power

Over the past century the dynamic of capitalist society has given rise to the development of giant corporations. To get an idea of the concentration of economic power it is sufficient to realize that in 1963, for example, 0·1% of the 300 000 manufacturing companies in the United States performed 80% of industrially sponsored research and development, accounted for 60% of the sales of all manufacturing companies and just over 60% of total manufacturing employment.

The leading corporations are immensely large and powerful, as the following information with respect to General Motors will indicate.

Chairman Donner of General Motors owned in 1964 only 0·017% of the total stock, but its worth was $4 million. His salary in that year was $200,000 plus a bonus of $453,750. (We see that corporate managers have considerable financial interest in the success of their companies.) According to Heilbroner, General Motors makes

> almost as much profit per car as it pays out for wages on the car, and for every dollar of increased labour costs since 1947 it has raised automobile prices by about $3·75. Its profit 'target' is 20% on capital after taxes—on the assumption that it will operate its plants only 36 weeks out of 52. This goes a long way to explaining its after tax profits for 1965 of 2 125 606 440 dollars.[32]

In that year the revenue of General Motors was eight times that of the state of New York.

We should note that such giant corporations are essentially autonomous organizations within American society, answerable to no one. For each year, according to the interests of their respective corporations, a small number of corporate managers decide how much capital is to be invested, what products are to be made, where they will be made, who will make them, and at what prices they will be sold. The American people are required to adjust themselves to these autonomously made decisions as is also the government, one of whose principal functions is to maintain the services necessary for the smooth running of corporate life together with that level of effective demand at which corporations can make their desired profits.

We are now touching the nerve centre of American capitalism, the nerve centre that is responsible for the most wasteful and dangerous features of the world's most scientifically-advanced capitalist country.

Government expenditure and arms

Because as yet we have not attempted to trace in a global context the historical development of the United States since the outbreak of the Second World War, our discussion, although at times referring to actual features of the United States economy, has nevertheless been carried through at an excessively abstract level. As soon, however, as we situate the United States in a historical and global context then we immediately perceive the overwhelming probability of a feature of its post-war economy that, out of an a-historical and 'non-global' discussion, emerges only as an unpleasant possibility—namely its militarization.

Let us recapitulate. Severe unemployment in an advanced capitalist society can always be reduced, according to Keynesian theory, by government deficit purchasing of industrial goods, a process that leads eventually not only to increased tax revenue but also to increased private investment. However, while industry welcomes government underwriting of the sensitive capital goods industries, it clearly requires that such underwriting should in no way reduce industry's non-government market. There are other considerations. If social stability is to be achieved, the end product of government expenditure must win the approval not only of industry but also of the general public. Clearly, expenditure in the purchase of any form of waste would be adequate were it not for the necessity of gaining public approval and forcing international competitors to follow suit. Pyramid building, for example, would fail to meet both conditions. Similarly the pyramid-building equivalent of massive expenditure on space exploration, although it would no doubt win the support of a technocratic 'priesthood', would also not necessarily win popular assent in a society needing considerable urban reconstruction. And other countries might be reluctant to follow suit. There is, however, one form of 'waste' which, with skilful propaganda, readily gains public approval and satisfies all other conditions—and that is the manufacture of armaments. For once the public is convinced that the threat of armed attack is real, 'defence' expenditure will meet with its approval, the more so as new jobs are created and employment increases. The other 'advantages' of appreciable arms expenditure are obvious. The manufacture of armaments, while contributing to the full utilization of industrial capacity, at the same time plays its part in ensuring the relative scarcity of consumer goods. Thus abundance is prevented and profitable market prices can be maintained for consumer goods, the armaments themselves being sold at a profit by the larger corporations to the main purchaser—the government. Finally, arms expenditure, in causing other countries to arm, is self-justifying and self-perpetuating. Abundance is not to be feared. If 'two railways from London to York' are not twice as good as one, two tanks are. In any case technological and scientific 'progress' will ensure a satisfactory obsolescence rate so that ever-expanding investment outlets for the capital goods industries will always exist to be financed by the government.

To be sure, this is in the abstract. However, once we realize that it was the Second World War that ended the decade-long depression in the United States, that allied the giant corporations and military together, that brought a country aspiring to socialism into prominence as a world power, and that demonstrated to the corporations their increasing dependence on supplies of raw materials from the (politically unstable) underdeveloped countries, then we have the necessary framework for an understanding of post-war developments. For continuation of the arms expenditure forced on the United States by the Second World War was then recognized to be not only politically necessary (if only in order to guarantee sources of raw materials) but also an economic guarantee against a repeat of the 1930s, a guarantee, moreover, that could be made ideologically acceptable to all—or to nearly all.

What I am therefore suggesting is that Western societies are not only inherently wasteful in themselves but are *primarily* responsible both for the misery in the underdeveloped countries and for the escalating arms race with the Soviet Union. (Obviously this is not to say that there are not endogenous factors in underdeveloped countries that make for economic and human backwardness—there are—that the Soviet government has not been and is not repressive—it has been and it is. But, I believe, Western societies remain primarily responsible for the increasingly serious world situation.) Let me attempt to substantiate this assertion.

2. The Militarization of the United States Economy

Capitalism in peace

In chapter 5 we had occasion to remark many times on Keynes' pessimism concerning capitalism's ability to solve its economic problems. Since as late as 1939 some 17% of the American labour force was unemployed, it was clear that Roosevelt's New Deal Policy, with its limited public expenditure, had been far from successful. A year earlier Keynes had written to the President urging an ambitious programme of public works but had received only a perfunctory reply. Keynes was therefore led to believe that only a war could prove his theory correct. He was to be proved tragically right. In his 1936 work he had recognized that there were political difficulties in the way of, for example, building houses for people, since 'every house which is built serves to diminish the prospective rents obtainable from further house building'.[33] Furthermore, as Keynes recognized, inequality of income distribution in an advanced capitalist society tends to lead to an excess of savings that does not readily find investment outlets. Since during the 1930s industry would not invest, it was the government's responsibility, reasoned Keynes, to undertake investment and so restore full employment. But in what could the government invest? Roosevelt's housing programme was bitterly fought by private interests, as indeed was his entire New Deal policy. And yet as Schlesinger has stressed, 'the New Dealers . . . believed in capitalism. They wanted to reform the system, not to destroy it. Their social faith was in private ownership tempered by government control'.[34] The exasperation of the New Dealers was understandable. As expressed by Jerome Frank of the Department of Agriculture: 'We socialists are trying to save capitalism, and the damned capitalists won't let us.'[35] In the 1936 election campaign Roosevelt became convinced that 'the business classes were out to destroy anyone who threatened their wealth or prerogatives'.[36] Indeed, according to businessman Wier, writing against Roosevelt in 1936, 'Today he is opposed, not by a "small minority" as he says, but almost unanimously by the business and professional men of the country.' Roosevelt's anger against the determination of such men to deprive people of, for example, social security led him in his election address to declare that 'only desperate men with their backs to the wall would descend so far below the level of decent citizenship' and led him also to an overwhelming victory over his Republican opponent.[37] In this critical election, let it be noted, the Communist Party received only 80 000 votes and the Socialist Party 200 000 out of a total vote of more than 44 million (in the 1932 elections the two Parties had gained respectively 120 000 and 887 000 votes). Truly Roosevelt could observe in 1938 that 'As a nation we have rejected any radical revolutionary program. For a permanent correction of grave weaknesses in our economic system we have relied on *new applications of old democratic processes.*'[38] Unfortunately, however, the 'permanent correction' was not to come through the New Deal policy but through massive government expenditure on arms. And, just as Keynes had predicted, the unemployment problem disappeared and the American standard of living rose! The New Deal simply faded away; 'the relief problem and the recovery problem both disappeared with the coming of war prosperity'.[39] Capitalism had found the answer to the threat of abundance. For war and the preparation for war, financed by the government, would provide investment outlets *that would meet in every way the approval of industry.*

Capitalism in war

It is not surprising therefore that the response of American industry to the challenge of war was, to say the least, impressive. Industrial production nearly doubled between 1940 and 1944. In the period 1929–39 government expenditure had increased only some 70% (from $10·2 billion to $17·5 billion, measured in current dollars) but in the period 1939–44 government expenditure increased by over 580% to a maximum of $103·1 billion in 1944). Although small businesses did not profit greatly from this massive increase in government spending, big industry did. Unemployment fell; the way out of a depression had been found. According to one commentator in his essay *The New Age of Roosevelt*: 'It had been easy during the thirties to underrate the abilities of the American businessman. It was of prime importance now, when the nation faced a great problem, that he rose to the occasion.'[40] But the nation faced 'a great problem' all through the 1930s and yet the American businessman did not rise to the occasion. *Capitalism apparently could not, and certainly did not, respond successfully to the challenge of more homes, schools and hospitals; it could, and did, respond magnificently to the challenge of more tanks, aircraft and machine-guns.* According to the same historian, 'the great positive change on the economic side was in the increased confidence felt by the business classes, as a result of their immense achievements and high profits during the war years'.[41] To be sure, the war certainly taught the business man that provided an enemy existed there could be no depression. And as Jones has appreciated, 'the economy of the 1950s and 1960s [has been] sustained by the unexpected renewal of population growth and boosted by "local" hot wars, cold war and immense military expenditures in the public sectors'. It is his opinion, however, that 'the international situation dictated that the economy remain on a quasi-war footing'.[42] While this is not incorrect as far as it goes, it is only half the picture. It is also the case that the capitalist economy dictated that the international situation remain on a quasi-war footing.

The marriage between industry and the military

Not only did the business man emerge from the war with great confidence but so did the military. During the war, for example, it was the Services of Supply, headed by General Somervell, which granted or withheld contracts to industry. However, there were also important personnel from the large industrial corporations on Somervell's advisory staff. Cook relates that acting jointly the large corporations and the military were able to prevent implementation of the reconversion plan prepared in 1944 by Nelson, Chairman of the War Production Board.[43] The Vice-Chairman of the Board, Charles E. Wilson, on leave from the presidency of General Electric, argued that any attempt at reconversion before the war ended would disrupt 'established commercial relationships' since small manufacturers would be able to enter the civilian market before the giant corporations. The military, for their part, expressed their fear that reconversion would result in 'people throwing up their war jobs' and declared there to be a shortage of war supplies. The reconversion plan was dropped. Later Somervell himself conceded that shortages had not interfered with the conduct of the war while a Senate investigating committee concluded that 'insufficient production ... has not been the cause of the shortage of weapons and ammunition at the front'.[44] Nevertheless, a major battle had been won by the giant corporations; having greatly profited from the years of war, they were now able to retain their control of the civilian market.

The power of the military did not go unnoticed. A 1946 official government document *The United States at War* declared that during the war the Army had sought 'total control of the nation, its manpower, its facilities, its economy'; the military leaders 'never abandoned the sincere conviction that they could run things better and more expeditiously than civilians'. The former Chairman of the War Production Board (WPB), having complained in his book *Arsenal of Democracy* that 'from 1942 onward the Army people, in order to get control of our national economy, did their best to make an errand boy of the WPB', warned an uncomprehending public that 'the question of military control will confront us not only in war but in peace. The lesson taught by these war years is clear: our whole economic and social system will be in peril if it is controlled by the military men.'[45] But what, of course, Nelson and the critics of the military had apparently not recognized is that the capitalist economic and social system will also be in peril if it is *not* controlled, directly or indirectly, by a partnership between the large corporations and the military. For a 'permanent war economy' is, given the political constraints imposed by the capitalist social system, the only seemingly permanent way for an advanced capitalist society to avoid depression and achieve 'prosperity'. There is considerable evidence that this lesson taught by the World War had been quickly understood by at least some of the corporate élite.

In a remarkably frank speech to the Army Ordnance Association (as reported by Dwight Macdonald[46]) Charles E. Wilson outlined a programme to construct a 'permanent war economy'. Given in January 1944 at a time when it was clear that Germany and Japan were all but defeated and that the United States would emerge from the war as the strongest world power, apparently without an enemy in sight, the speech was nevertheless a plea for the United States to remain on a permanent war footing. Thus Wilson argued that 'instead of looking to disarmament and unpreparedness as a safeguard against war—a thoroughly discredited doctrine—let us try the opposite: full preparedness according to a continuing plan. The thought may be unpleasant, but through the centuries war has been inevitable in our human affairs, as a basic element in evolutionary force. We have yet to learn that hard truth, apparently.' Thus there was a danger, declared Wilson, that 'the revulsion against war not too long hence will be an almost insuperable obstacle for us to overcome in establishing a preparedness program and for that reason I am convinced that we must begin now to set the machinery in motion'. The advice was given that 'such a program must be the responsibility of the federal government. It must be initiated and administered by the executive branch—by the President as Commander-in-Chief and by the War and Navy Departments.' 'Of equal importance,' stressed General Electric's on-leave President, 'is the fact that this must be, once and for all, a continuing program and not the creature of an emergency. In fact', he continued, 'one of its objects will be to eliminate emergencies so far as possible.' Wilson was quite explicit concerning industry's role in maintaining the permanent war economy intact: 'The leaders of industry are as much the leaders of their country as are the generals, the admirals, the legislators, and the chiefs of state. Their responsibility for post-war preparedness is certainly no less. The burden is on all of us to integrate our respective activities—political, military and industrial—because we are in world politics to stay, whether we like it or not.' In particular, stressed Wilson, 'industry must not be hampered by political witch-hunts, or thrown to the fanatical isolationist fringe tagged with a "merchants of death" label'. And one final point: 'Those manufacturers called upon to develop and build the new weapons of war in peacetime should carry personnel devoted entirely to this purpose. War equipment should be the business of this group, and since it is their business, it seems obvious to me that it should be operated the same as any other business—at a reasonable profit.' There were other proponents of a permanent war economy. Navy Secretary Forrestal, for example, a former president of the Wall Street Investment Bankers, Dillon, Read and Co., supported Wilson's plea, forming in 1944 the National Security Industrial Association. Its aim was to see that 'the military-industrial team' created during the war should 'remain active and effective in peace-time' so that America should be 'adequately prepared and defended'.[47] The military-industrial objectives were being very clearly stated.

The ascendancy of the military-industrial complex
Despite the intentions of the protagonists of the 'permanent war economy', the end of the war saw the rapid demobilization of all but a million of America's 16 million men in the Armed Services. Accordingly the Pentagon initiated a nationwide campaign for the passing of a Universal Military Training Act. Its propaganda campaign, however, did not go unopposed by those congressmen who had either not understood or had not accepted the fact that the American economy was to be militarized. Thus in June 1947 a Congressional committee reported that 'the War Department is using government funds in an improper manner for propaganda activities supporting compulsory military training', and recommended the Attorney General to start proceedings immediately 'to stop this unauthorized and illegal expenditure of public moneys'.[48] Although no action was taken, nevertheless in late 1947 Congress rejected the Pentagon's plan for 'Universal Military Training', a rebuff which caused the Pentagon to intensify even further its propaganda campaign.[49] Subsequently in June 1948 Congress passed a bill bringing peacetime conscription to the United States for the first time in its history.*
A few congressmen, however, still protested against some of the methods used. 'The considerable degree of hysteria prevalent today,' they wrote, 'is in a large measure due to the propaganda efforts of the Armed Services them-

* The development of the Cold War will be discussed in chapter 9. It suffices here to remark that even Schlesinger now admits that 'if it is impossible to see the Cold War as a case of American aggression and Russian response, it is also hard to see it as a pure case of Russian aggression and American response'.[49n]

PROBLEMS OF POST-WAR CAPITALISM 197

selves. . . . The Army is spending large amounts of the taxpayer's money to obtain a permanent system of conscription. . . . The Army has acted as if it is the policy-making body of the nation.'[50]

And so the Army, led by the large corporations, continued to act. Cook writes that Secretary of Defense Forrestal was quite explicit in the advice he gave in August 1948 to the first graduating class of the Armed Forces Information School, informing the new publicity experts that it was their duty to overcome the traditional American suspicion of militarism. 'It is difficult,' Forrestal declared, 'because our democracy and our country are founded upon an underlying suspicion of armies and of the force that they reflect and represent. . . . Part of your task is to make people realize that the Army, Navy, and Air Force are not external creations but come from and are part of the people. It is your responsibility to make citizens aware of their responsibilities to the services.'[51] No longer, Cook notes, are the services to be responsible to citizens—but exactly the reverse!

Three years later in 1951 Charles E. Wilson, who in 1944 had proposed the creation of a 'permanent war economy' and who had subsequently been appointed by Truman to the post of Defense Mobilization Director, gave a revealing speech at the annual dinner of the Bureau of Advertising of the American Newspaper Publishers Association. In it he praised the publishers for having written

> millions of words laying down the premise . . . that the free world is in mortal danger. . . . If the people were not convinced of that it would be impossible for Congress to vote the vast sums now being spent to avert that danger. . . . With the support of public opinion as marshalled by the press, we are off to a good start. But the mobilization job cannot be completed unless such support is continuous. . . . It is our job—yours and mine—to keep our people convinced that the only way to keep disaster away from our shore is to build America's might.[52]

If the majority of Americans were successfully propagandized, a few were not. Wrote Albert Einstein in 1950: 'I have . . . the impression that our nation has gone mad and is no longer receptive to reasonable suggestions. Its whole development reminds me of the events in Germany since the time of Emperor William II: through many victories to final disaster.' In April 1951 he added: 'I do not consider it reasonable to compare a disease in the medical sense with the hatred and fear toward Russia which have been instilled in the American people since the death of Roosevelt.' The following June he wrote: 'Those who have propagandized against the alleged external enemy have won the support of the masses. Political stupidity has become so widespread that even reasonable people may find it difficult to discover their way back to a saner foreign policy.'[53]

Einstein was right. The giant corporations and military had (temporarily) achieved their objectives. Let us look at the economic consequences.

The American economy 1945–61

In the years after the war the American economy did not plunge into depression and widespread unemployment as had been feared. On the contrary, effective demand actually increased as the American public, who had been starved of durable goods throughout the war, drew on their savings to produce for American industry a 'market in which sales were limited only by the capacity to produce'. By about 1955, however, sales had once again caught up with demand, so much so that, in the words of Schon, 'moral pressure was brought to bear on the American public to fulfill its obligations as consumers to keep American industry going'.[54]

Although such a state of affairs—market saturation, inequitable distribution of income and heavy advertising—was reminiscent of the state of the American economy in the late 1920s, there was this time one important difference. Whereas in 1929 the government expenditure in armaments had been an insignificant fraction of the national income, by the late 1950s it was typically of the order of 10%. The industrial-military complex had achieved its objectives. Profitable investment outlets had been found for the sensitive capital goods industries and the danger of depression had been averted.

Although, of course, the economically unsophisticated would almost certainly suppose such colossal arms expenditure to be a burden on the American economy (expenditure soon to be increased by the Kennedy administration), the needs of humanity are not those of the American 'industrial system'. Frank Pace, a former chairman of General Dynamics, Secretary of the Army under Truman and later a member of President Kennedy's Foreign Intelligence Advisory Board, phrased the matter very tactfully in an address in 1957 to the American Bankers Association: 'If there is, as I suspect there is, a direct relation between the stimulus of large defence spending and a substantially increased rate of growth of gross national product, it quite simply follows that defence spending *per se* might be countenanced on economic grounds alone as a stimulus of the national metabolism.'[55] As we have seen, the idea of changing the 'national metabolism' so that it could be stimulated by the construction of homes, schools, hospitals, parks and other necessities for a humane life is not greatly appreciated by big business. But the idea of a permanent war economy (although not of a permanently escalating war) is received without protest. According to Henry Luce, publisher of the magazines *Time*, *Life* and *Fortune*, the economy 'can stand the load of any defence effort required to hold the power of Soviet Russia in check. It cannot, however, indefinitely stand the erosion of creeping socialism and the ceaseless extension of government activities into additional economic fields.' He urged Congress (in 1957) to 'begin a massive pruning of non-defence outlays all along the line'.[56]

In 1961 a somewhat bewildered President Eisenhower warned the American public that it must be on its 'guard against the acquisition of unwarranted influence, whether sought or unsought, by the military-industrial complex'. He declared that 'the potential for the disastrous rise of misplaced power exists and will persist'.[57] As an ex-soldier President Eisenhower had not understood what was happening in his administration. The next President was more sophisticated.

The permanent war economy and the Kennedy administration

Compared with unemployment levels between 15 and 25% during the years 1929–39 the semi-militarized American economy had been successful in the first fifteen post-war years. Nevertheless, at no time had unemployment been reduced to its wartime level. Winch writes: 'Between 1947 and 1953 the average rate of unemployment was 4%; between 1954 and 1963 this figure had risen to 5·5% with a peak level approaching 7%.' The recession of 1957–9, Winch notes, had no sooner ended than another occurred. Kennedy's response was to set up, under Samuelson's leadership, a special task force of economists 'to make an appraisal of the current recession and to recommend steps to deal with it'. Winch writes that as a 'first line of defence' the task force recommended, in Keynesian spirit, an acceleration of existing expenditure programmes, including increases in defence expenditure wherever appropriate, and an expansion of welfare programmes and public works.[58] Kennedy's principal measures were an increase in arms expenditure and the initiation of the American space programme.

This whole episode is an especially revealing commentary on the application of Keynesian theory to the prevention of recession in the American economy. Kennedy had campaigned in 1960 for 'crash programmes to provide ourselves with the ultimate weapons [to] . . . close the missile gap'[59] and, although in February, 1961 his Defense Secretary McNamara admitted that there was no missile gap (or rather that there was indeed a missile gap but that it was in America's favour!), on March 28, scarcely two months after his inauguration, Kennedy proposed a nearly $3 billion increase in defence expenditure in order to provide for increased production and development of Polaris submarines and underground Minuteman missiles. The effect on the Soviet rulers, who knew that Kennedy was aware of America's missile superiority at the time he became President, was predictable. As McNamara admitted in 1967: 'Soviet strategic planners undoubtedly reasoned that if our build-up were to continue at its accelerated pace, we might conceivably reach, in time, a credible first-strike capability against the Soviet Union.'[60] As he explained: ' . . . in the course of hedging against what was then only a theoretically possible Soviet build-up, we took decisions which have resulted in our current superiority in numbers of warheads and deliverable megatons . . . [a superiority which] is both greater than we had originally planned, and is in fact more than we require'.[61] Indeed Kennedy's build-up, which continued over three years, was the largest and swiftest in the country's peacetime history: the appropriations for fiscal year 1962 involved a total force of 600 intercontinental ballistic missiles, subsequently raised to 800 in the following year, while the number of Polaris submarines increased from 29 in 1962 to 35 in the next year and then to 41 in 1964, bringing the total number of Polaris missiles to 656. Complaining in 1963 about Kennedy's large military expenditures, former President Eisenhower declared that 'the defence budget I left behind provided amply for our security'.[62] But during Eisenhower's Presidency the United States' economy, while not plunging into depression, had, nevertheless, not enjoyed a boom and was suffering from a recession when he left office. However, in the four years following Kennedy's inauguration the output of goods and services increased by a greater amount than in the previous eight; the economy enjoyed, wrote one of Kennedy's biographers, 'the longest and strongest economic expansion in this nation's history'.[63]

The extent of the military-industrial complex

In order to summarize the vast power and influence of the military industrial complex it is sufficient, I think, to quote two authors who have studied American capitalism in some detail. According to Seymour Melman (writing in the early 1960s):

> Across the face of the United States since the Second World War a network of communities has grown up serving as satellite towns for military bases. . . . At least 10% of the annual value of all goods and services in this country is devoted to military activity. Around 20% of the income produced in manufacturing industry nationally is on military order; 11% of the total labour force, or from 6 to 7 million people is now directly engaged in military employment or in supplying occupations. To make a conservative estimate of persons involved both directly or indirectly, this figure is doubled. Thus, 12 to 14 million jobs depend upon military activity at this writing.[64]

Robert Heilbroner adds (1966) that:

> In the 1960s military expenditure has regularly exceeded the sum total of all personal income taxes, has accounted for one-fourth of all federal public works; has provided 30% of all manufacturing jobs in Kansas, 28% in the state of Washington, over 20% in another five states, and between 8 and 10% in seven more; it has subsidized about one-third of all research in the United States; and not least, has come to be accepted as a normal and permanent fixture of American economic, social and political life by all groups including not least the academic.[65]

The military power of this war machine cannot easily be comprehended. Suffice it to say that by 1963 the United States possessed 3 400 'strategic delivery vehicles' equipped with nuclear armaments. Even if, it was estimated, 90% of all aircraft and 75% of major missiles were destroyed during an attack on the Soviet Union, the surviving 'delivery vehicles' would be able to destroy Soviet society by an overkill factor of more than 200.[66] Even in armaments, it would appear, there is the threat of abundance!

The nature of the military-industrial-scientific-political complex

President Eisenhower, in his warning to the American people, also stated that 'in holding scientific research in respect, as we should, we must also be alert to the equal and opposite danger that public policy could itself become the captive of a scientific-technological élite' geared to 'a military-industrial complex ... the total influence of which is felt in every city, every state house, every office of the Federal Government'.

Let us look briefly at this harmony of interests between industrialists, scientists, military and politicians, noting first of all the contribution of scientists.

According to Melman, the military-related research activity of American universities in the period 1960–2 amounted to more than 50% of the total research done.[67] While it is true that, for example, the United States Armed Forces do support much pure research (or that research considered 'pure' by scientists) nevertheless, the Armed Forces are not philanthropic organizations and such financial support stems from the obvious fact that 'pure' research often leads to military applications while at the same time it serves to project a favourable image of the military in academic circles. The result, for the military-industrial complex, is very rewarding: between two-thirds to three-quarters of American scientists and engineers engaged in research and development, some 280 000, were in the early 1960s engaged in military occupations.[68] The National Science Foundation reported, for example, that aircraft and missile companies alone employed more scientists and engineers on research and development than the combined total of the chemical, drug, petroleum, motor vehicle, rubber and machinery industries.[69]

As to the 'cooperation' between the military, politicians and industry, Galbraith has this to say:

> Were it not so celebrated in ideology, it would long since have been agreed that the line that now divides public from so-called private organization in military procurement, space exploration and atomic energy is so indistinct as to be nearly imperceptible. Men move easily across the line. On retirement, admirals and generals, as well as high civil servants, go more or less automatically to the more closely associated industries.[70]

Indeed in 1960 in the United States there were 691 retired generals, admirals, naval captains and colonels on the pay rolls of the 10 largest military suppliers. Cook writes that the company employing the greatest number, General Dynamics, was also the company that received the largest contracts. General Dynamics 'had on its payroll 187 retired officers, including 27 generals and admirals, and it was headed at the time by Frank Pace', whose opinions and qualifications (a former Secretary of the Army) we have already noted.[71] As for the business men, Galbraith tells us (1967):

> that not for many years has any important business executive condemned the prodigality of expenditure on defence. From all pleas for public economy, defence expenditures are meticulously excluded. . . . No more than any other social institution does the industrial system disapprove of what is important for its success. Those who have thought it suspicious of Keynesian fiscal policy have failed to see how precisely it has identified and supported what is essential for that policy.[72]

Keynes and the 'permanent war economy'

In 1936, in the concluding section of his *General Theory*, Keynes repeated his belief that 'a somewhat comprehensive socialization of investment will prove the only means of securing an approximation to full employment; although', as he pointed out, 'this need not exclude all manner of compromises and devices by which public authority will cooperate with private initiative'.[73] Keynes did not altogether foresee the nature of the 'compromises and devices' that the world's most advanced capitalist society would adopt in order to remain capitalist. Indeed, he wrote: 'I see no reason to suppose that the existing system seriously misemploys the factors of production which are in use. . . . It is determining the volume, not the direction, of actual employment that the existing system has broken down.' Were Keynes alive today, I think he would wish to rewrite these two sentences! However, in the introduction to a 1964 collection of reviews and essays concerning Keynes' General Theory, the editor Lekachman refers to the 'compromises and devices' by which 'socialization of investment' has been achieved: 'the defence-oriented industries—electronics, missiles, and airframes—are government enterprises,' he tells us, 'in everything but name and profit distribution'.[74] The latter exception tells all.

The defence of the free world against communist aggression?

Throughout this chapter we have, perhaps unwisely, postponed discussion of the rationale given by owners of industry, the military, politicians, university academics and the mass media for the creation of a permanent war economy, viz. the necessity to defend the 'free world' against 'communist subversion and expansion'. Prominent among defenders of the free world has been, for example, Nelson Rockefeller.* In a press interview in 1961 (as reported by the *Herald Tribune*) Rockefeller emphasized that 'the salvation of the free

---

* We might in passing note that the free world has certainly not been ungrateful to Rockefeller. The personal fortune of the Rockefeller family was estimated in 1956 to amount to some $3 500 000 000; corporate assets controlled by the family or by allied Standard Oil families exceeded $61 000 000 000.[74n]

world lay squarely on its nuclear military might'. It was Rockefeller's opinion that 'the bomb is not a monster to be regarded with abhorrence and fear, but rather constitutes the free world's only hope against communist domination that otherwise might be inevitable'.[75] This point of view is far from atypical. In the next two chapters we therefore turn to look in some detail at the nature of the free world, whose freedom the advanced capitalist countries declare they are defending, and at the nature of the conflict between the advanced capitalist countries and its 'communist' enemies.

Chapter 8

PROBLEMS OF UNDERDEVELOPED COUNTRIES

> We must demonstrate that the underdeveloped nations—now the main focus of communist hopes—can move successfully through the preconditions [for economic growth] into a well-established take-off within the orbit of the democratic world, resisting the blandishments and temptations of communism. This is, I believe, the most important single item on the Western agenda.
>
> W. W. ROSTOW *The Stages of Economic Growth*

IN chapter 7 the argument was made that government expenditure on arms has, in an important way, helped maintain the relatively high level of effective demand on which the overall post-war prosperity of the United States has been based. Clearly the justification usually given for such vast expenditure on arms is not that it stimulates the economy but that it is necessary for the defence of the western hemisphere against communist subversion and aggression. While the possession of missiles and atomic weapons (it is argued) deters the Soviet Union from attacking the United States and Western Europe, more conventional weapons are important for resisting, and preventing, communist guerrilla warfare in the economically underdeveloped countries of Latin America, Africa and Asia. That the side-effects of defence spending are beneficial to the economy does not pass unappreciated, of course. But, it is claimed, the principal reason for arms expenditure is political, not economic—in short, it is to defend freedom against tyranny. In this chapter we investigate this argument by examining the nature of the relationship between the advanced capitalist countries, in particular the United States, and the so-called underdeveloped countries, paying particular attention to the reasons for the appalling poverty in the latter countries. And in the next chapter we shall examine the nature of the conflict between the Western world and the Soviet Union, between capitalism and, it is often claimed, its rival social system.

1. On Imperialism

Past imperialism—European

That the Western nations, and particularly Britain, have been imperialist is presumably disputed by none. A few remarks must suffice. Clive's conquest and plunder of Bengal in 1757 meant, in the words of Barratt Brown, 'the impoverishment and devastation of a once flourishing Indian State'.[1] As India was 'opened up' to British manufacturers, Indian spinners, weavers and metal workers were driven out of business: 'the population of Dacca, Surat, Nurshidabad and other centres for manufacture in India was decimated in a generation and India's balefully increasing dependence on agriculture had begun'. By 1842 China was also opened up to British 'free trade'. After the opium wars, initiated by the British after the Emperor had refused to allow the East India Company to sell opium to the Chinese people (a highly profitable trade for the Company), the once proud empire which 'had possessed all things' was forced to open its main ports to foreign trade and concede Hong Kong as a British imperial base. So, Barratt Brown writes, British capitalism achieved domination and 'divided the nations into rich and poor—advancing industrial manufacturers and declining primary producers'. Africa suffered the same fate as Asia, except that in addition several million Africans were taken as slaves to the Americas; the remainder were available for exploitation—Leopold's coercion of the 'Congolese' into his rubber plantations resulted both in the decimation of the population and a large personal fortune for him.[2]

Past imperialism—American

The United States was not to be outdone. The extermination of the Indians, for long celebrated as part of the country's folk heritage, was but the first step. As Senator Henry Cabot Lodge declared in 1895: 'We have a record of conquest, colonization and expansion unequalled by any people in the nineteenth century. We are not to be curbed now.'[3] Three years later Senator Beveridge of Indiana gave arguments that have a familiar ring: 'American factories are making more than they can consume. Fate has written our policy for us; the trade of the world shall and must be ours. . . . The Philippines,' he declared, 'are logically our first target.'[4] That year the United States attacked the Spaniards in Manila Bay. The aim, as Beveridge made explicit, was to achieve American domination of the immense market that China presented. Mr. E. H. Conger, who in 1898 held the diplomatic post in Peking, was similarly explicit about America's objectives:

> Next to controlling a desirable port and commodious harbour in China, the permanent ownership or possession of Manila and vicinity would be the most invaluable to us in securing and holding our share of influence and trade in the new era just beginning in this country. . . . It would give a convenient and essential base of supplies, where American trade, capital and brain could and would be massed ready for the commercial conquests which Americans ought to accomplish in China.[5]

Such commercial objectives were wholeheartedly endorsed by the American government. In his *History of the American People* (1902) Woodrow Wilson was certain that the China market was the 'market for which statesmen as well as merchants must plan and play their game of competition, the market which diplomacy, and if need be power, must make an open way'.[6] The

President-to-be was later quite explicit about the methods that, if necessary, were to be used in ensuring 'free trade': 'Since trade ignores national boundaries and the manufacturer insists on having the world as a market, the flag of his nation must follow him and the doors of the nations which are closed must be battered down.... Concessions obtained by financiers must be safeguarded by ministers of state, even if the sovereignty of unwilling nations be outraged in the process.'[7] Such methods continued to be necessary.

Now this Western expansion occurred before 1917 (the year of the Russian revolution); there was no question at that time of defending the freedom of the countries of Latin America, Africa and Asia from communist subversion and aggression. On the contrary, the capitalist countries were intent on imposing a social structure on these countries that would allow maximum exploitation of their natural resources and provide capitalist manufacturers with a market for their surplus produce. The entire world was to be re-made so as to provide the most favourable environment for the development of the advanced capitalist nations.

All this, however, was in the past. Today most of the countries of Latin America, Africa and Asia have achieved political independence and the accusation of continued Western imperialism grates on sensitive academic ears. Nevertheless, it is often admitted that Western domination of the non-white peoples of the world did little, if anything, to improve the general standard of living of these peoples. Indeed, in those continents whose populations experienced the impact of Western civilization, two-thirds of humanity now live in brutalizing conditions that contrast dramatically with the high standard of living enjoyed on the average by their former masters and tutors. The peoples of the underdeveloped countries exist for the most part diseased, infested with parasites, suffering from chronic malnutrition, illiterate, inadequately clothed and housed, and with an average life expectancy as low in some areas as thirty years.

Reasons for poverty

Many authors agree that archaic social systems must be held responsible for the appallingly low standard of living. According to Heilbroner: 'Development is much more than a matter of encouraging economic growth *within a given social structure* (emphasis added). It is rather the *modernization* of that structure, a process of ideational, social, economic and political change that requires the re-making of society in its most intimate as well as its most public attributes.'[8] In Latin America, for example, Heilbroner stresses that 'the principal handicap to development is not an absence of national identity or the presence of suffocating cultures (although the latter certainly plays its part), but the cramping and crippling inhibitions of obsolete social institutions and reactionary social classes'. Mikesell, too, warns that 'human resource development is not something which can be dealt with by massive injections of capital—it is largely a matter of teaching and inducing people to help

PROBLEMS OF UNDERDEVELOPED COUNTRIES 207

themselves. What is involved is the whole social and political fabric of the country, particularly at the community level'.[9] This is the crux of the matter: unless there is social and political change in the underdeveloped countries, widespread poverty will remain—despite the possibility of a growing national product. Accordingly Mikesell has thought it important to emphasize that 'the goal of the development plan should not be simply the production of marketable goods and services, but social welfare broadly conceived'. For as he points out in an example: 'The achievement of a relatively high level of *per capita* income in Venezuela has not eliminated the slums surrounding Caracas or the miserable living conditions in a large part of the rural area. Nor would a doubling or tripling of Venezuela's *per capita* income over the next generation guarantee a substantial improvement in these conditions.' Michael Harrington takes the argument a stage further:

> The Market mechanism cannot be the mainspring of Latin-American development in terms either of attracting foreign or allocating domestic capital. There must be ... conscious economic and social planning, not private enterprise. For the rationality of the profit motive directs the foreign giant corporation to distort the economic structure of the developing nation at best, and to keep it backward at worst. The basic infra-structural needs of the poor nations—roads, education, cheap mass housing, etc.—are simply not profitable investments. No one is really interested in building decent homes for the poverty-stricken *within* the United States, and smart money would shun even more such an undertaking overseas. ... To do incalculable harm to the masses of the Third World, the Western politican or businessman need not be evil, but only reasonable and realistic.[10]

Clearly the high rate of population increase in the underdeveloped countries is a major factor contributing to the low average standard of living. Furthermore, the poorest areas of the world have some of the highest birth rates, e.g. India and the north-east of Brazil. Yet, although the mobilization of the peasantry combined with mass education could help to substantially reduce the birth rate, it is exactly such undertakings that most governments of underdeveloped countries either cannot or will not attempt. They cannot or will not see that the solution of the 'population explosion', like the solution of mass poverty, is best sought in radical social and political change.

Western attitude towards social and political change in the underdeveloped countries

However, despite the (in my opinion) excellent advice of Heilbroner, Mikesell and Harrington, those attempts that have been made by the peoples of many underdeveloped countries to change their social and political systems have usually been thwarted by the United States with the approval of the minor

Western nations on the grounds of 'defence of the free world against communist subversion'. For example:[11,12,13,14]

(1) In Persia in 1953 Mossadegh was overthrown by the United States Central Intelligence Agency (CIA). Some relevant (?) information: Mossadegh had nationalized the country's oil resources. Subsequent arrangements gave 25-year leases on 40% of the oil to three United States companies.

(2) In Guatemala in 1954 the constitutionally elected Arbenz was overthrown by an invasion backed by the CIA and the United Fruit Company. Some relevant (?) information: Arbenz had initiated a land reform programme involving the nationalization of 200 000 acres of land owned by the United Fruit Company.

(3) In the Dominican Republic in 1963 the constitutionally elected president, Bosch, was overthrown by a military coup. Some relevant (?) information: Bosch had been opposed to United States economic control of the Dominican Republic and had threatened to cancel molasses and sugar contracts negotiated with the United States. Two years after the coup, when some Bosch-supporting military leaders attempted a counter-coup, the United States government sent 20 000 Marines to the island. The attempted rebellion was crushed. United States investment in the island's economy flourished.

Of course, that United States companies invest in underdeveloped countries and that the United States government is actively engaged in defending American economic interests can come as no surprise. The question, however, we wish to take up here is whether United States investment and intervention in underdeveloped countries serve *only* the interests of the United States and in no way assist the peoples of the underdeveloped countries in developing and exploiting their natural resources also for their *own* benefit. We begin this investigation by first considering the importance of underdeveloped countries as (1) suppliers of raw materials to the United States and other capitalist countries; and (2) as areas of the world not only serving as markets for the surplus produce of the advanced capitalist countries but also providing profitable investment outlets for the large Western corporations.

2. The Importance of Raw Materials

To appreciate the importance of natural resources for the United States alone, we may refer to a few relevant statements. The President's Commission on Foreign Economic Policy reported in 1954 that: 'Both from the viewpoint of our long-term economic growth and the viewpoint of our national defence, the shift of the US from the position of a net exporter of metals and minerals to that of a net importer is of overshadowing significance in shaping our foreign economic policies.'[15] With regard to the stockpiling of 'strategic materials' in the United States, the International Development Advisory

Board, set up by the President in the 1950s, reported that: '... it is to these countries [from the underdeveloped areas] that we must look for the bulk of any possible increase in these supplies. The loss of any of these raw materials, through aggression, would be the equivalent of a grave military set-back'.[16] Rostow, one of President Johnson's closest advisers, put the matter bluntly:

> The location, natural resources, and population of the underdeveloped areas are such that, should they become effectively attached to the Communist bloc, the United States would become the second power in the world. . . . Indirectly, the evolution of the underdeveloped areas is likely to determine the fate of Western Europe and Japan, and therefore the effectiveness of those industrialized regions in the free world alliance we are committed to lead.[17]

Patton, in his 1968 study entitled *The United States and World Resources*, points out that 'for all of the major metals except iron, more than half of American requirements now come from foreign sources'.[18] Not surprisingly, his opinion is that 'many of the international relations of the United States in years to come will be rooted ultimately in these resource questions'. This opinion is shared by Lovering (1969) who writes:

> The dependence of the United States on foreign sources will almost certainly increase greatly during the next generation. Increasing dependence of foreign sources brings increased vulnerability to military, political, or economic action. This is emphasized by the fact that some of the metals most vital to the economic well-being of free enterprise industrial nations are in areas of political instability or in Communist countries. Most of the known resources of tungsten and antimony lie in such areas, as well as a large part of the world's manganese, nickel, chromium, and platinum.[19]

Fisher (1970) gives explicit advice to the United States government: 'The world's trading and investing system will have to be maintained so that this country can import increasing amounts of a large number of minerals and some other products.'[20] His article is entitled 'Natural Resources—Wise Use of the World's Inheritance'.

On the industrialization of the underdeveloped countries

While these quotations serve to confirm that the United States needs access to, and control of, the sources of the raw materials of the underdeveloped countries (and for this reason alone has been prepared, where possible, to intervene militarily in the underdeveloped countries should any of these countries threaten to develop an independent foreign and economic policy) they do not, however, answer the query as to whether the underdeveloped countries will be increasingly able to consume their own minerals and fuels as they industrialize under the 'protection' of the United States. To investigate further, let us note the following three opinions: According to Meier, 'the scarcest metals in prospect are lead, nickel, tin, copper and cobalt. If the resources, as estimated in the most generous sense possible, are to last even that long [from 25 to 100 years], then a very low rate of consumption must be postulated [!] for the rest of the world. The United States with 6 to 7% of the world population is expecting to consume more than 50% of the total supply of these scarce industrial raw materials.'[21] Bailey also makes the point explicitly:

> On the optimistic view it should be possible for the developing countries to double or treble their consumption of minerals without any great difficulty. [In 1960 mineral consumption in the United States came to the value of $40 *per capita*; consumption in underdeveloped countries amounted to less than $1 *per capita*.] At the same time, the pessimistic view would be that if these countries ... even raise their consumption of minerals to anything approaching that of the developed countries, the strain on the world's resources would be enormous.[22]

Similarly, Morris in his *Problems of American Economic Growth* makes the same argument quite frankly: 'America is interested in these countries because of the raw materials they possess which are important to us in peace and especially in war. It is not at all certain that our interest in obtaining these raw materials will be served by world-wide development, which usually means industrialization.'[23]

We are therefore led to the tentative conclusion that the relationship between the capitalist and underdeveloped countries is exploitative in our favour and that, however regrettable, it is 'necessary' since we live in a world characterized by the scarcity, or possible future scarcity, of vital raw materials with regard to which the capitalist countries enjoy, and intend to continue to enjoy, monopolistic control. Furthermore, it would appear that Western interests are best served by maintaining privileged élites in power in underdeveloped countries who have little genuine interest in the kind of industrialization that would benefit the poor of these countries but who are prepared to enrich themselves by exporting the natural resources of these countries to the advanced capitalist nations, thereby sharing in the high profits of the investing Western companies, while squandering this income mainly in conspicuous consumption and in the purchase of that military equipment from the West without which the continuance of their privileged status in the midst of so much suffering, hunger and disease could not be ensured.

On technological progress and substitutes

The matter, however, can surely not be as clear-cut as this. Firstly, severe social conflict in underdeveloped countries is not in the interests of the Western powers; they may not be able to control the situation and may end up losing their investments and their control of the countries' natural resources. Hence it is probably desirable from the Western viewpoint for 'reasonably' progressive governments in underdeveloped countries to undertake a modicum of social reform and industrialization if only to prevent violent social strife. We shall return to this point later. Secondly, the projections discussed by Meier and Bailey neither take into account possible discoveries of further mineral and fuel resources nor future technological breakthroughs that would result in the commercial exploitability of very low quality ores and the availability of substitutes for currently indispensable natural resources. While such discoveries and breakthroughs have occurred, and will undoubtedly continue to occur, it is, however, a brutally simple fact that no industry would be prepared to lose control of those natural resources which are at present vital to its functioning simply because its directors are assured that in the long run scientific and technological change will produce substitutes. Furthermore, as Landsberg points out, 'In spite of recent increasing use of plastics and other synthetics . . . it is improbable that the relative importance of metals will diminish greatly between now and the year 2000. Certainly absolute growth will continue to be large', he concludes, '. . . metal use *per capita* in the United States will continue to rise.'[24] Hence the policy of Western corporations and governments towards underdeveloped countries is, and will continue to be, based on the assumption of possible future scarcity, until such time as available and cheaper substitutes are an established fact of industrial life. (Furthermore, it is obviously advantageous to the Western companies to first exploit the higher-grade, more profitable ores and resources of other countries thereby keeping their own leaner ores and resources in reserve.)

On why the underdeveloped countries cannot be allowed meaningful political independence

But, it may be asked, if there is little likelihood of long-term scarcity of resources (although probably the very opposite is the case), why is direct aggressive control of the natural resources of the underdeveloped countries so necessary to the advanced capitalist nations, since even if these underdeveloped countries were to achieve both political and economic independence it would still be in their interest to export their mineral and other natural resources to the West in exchange for the heavy capital equipment they would undoubtedly need for the construction of an industrial base? This may be so, but since underdeveloped countries are in no position to pay 'prompt and effective' compensation to Western corporations whose capital investment is expropriated, such nationalization might result, and has resulted, in financial loss for those Western corporations involved. In any case their profit rates would undoubtedly decrease, they would have less control over the prices of their finished products and would, in losing their direct control of the sources of raw materials, be in a much weaker position to limit new competition. And, certainly not least, *if underdeveloped countries managed to achieve political and economic agreement among themselves* they would then be in a powerful position to exert economic pressure on the West—for example, to force the West to agree to considerably increased aid in return for the continued export of those raw materials without which Western economies would grind to a halt. Perhaps this is the most important reason why the West is so careful to try to ensure that the underdeveloped countries will not, in the foreseeable future, enjoy meaningful independence. Thus Landsberg, in his 1964 survey entitled *Natural Resources for United States Growth*, concludes that 'the rest of the twentieth century affords the prospect of sustained economic growth in the US supported by an adequacy of resource materials ... *provided* foreign sources of raw materials remain open *through maintenance of a viable world trading system and investing system*' (latter emphasis added).[25]

Natural resources as vitamins

Let us explore this argument a little more fully. The view that the West is essentially self-sufficient is held, for example, by Aron who, speaking in Brazil in 1962, offered the following advice: 'For advanced countries, the cost of raw materials represents less than 15% of the gross national product. ... The Western countries are not condemned to extinction through the loss of their empires and the revolt of the underdeveloped countries. The latter, to ensure their own progress, do not need to declare war on the West, and it is not in their interests to do so.'[26] Of course, as Aron is well aware, the only declaration of war an underdeveloped country can make on the West is one of economic independence. And this is precisely the point. For Aron's argument that the West has no need of its economic empire since the cost of raw materials is only a fraction of its gross national product (GNP) is scarcely sound. Coppock in *Europe's Needs and Resources* expresses the reality of Europe's situation:

> Foreign trade is thus of comparatively minor importance to the United States and to the other three leading countries in international trade—the UK, West Germany and France ... [but] the overall ratio of trade to GNP is, of course, a very crude measure. It does not bring to light that certain goods imported from overseas are, like vitamins or hormones to the human body, indispensable to the smooth functioning of the European economy, though their money value may be trifling in comparison with the GNP. In 1955 imports of raw cotton, for instance, accounted for only 0·5% of Western Europe's GNP, imports of all non-ferrous ores and metals combined for 0·8%; even the value of crude oil and refined petroleum imports barely exceeded 1% of aggregate European GNP. But if the supply of only one of these or of certain other goods were cut off, the European economy would be threatened with a serious breakdown.[27]

Thus it is in the interests of the 'vitamin-importing' West to ensure the economic dependence of the 'vitamin-exporting' underdeveloped countries. Coppock once again:

> [Although] consumption of the major ferro-alloys in 1970 can reasonably be expected to be double the 1955 level ... [and although] the bulk of the requirements will probably have to be imported ... producing regions (including Ghana and the Congo) have strong trading ties with the countries of Western Europe and it can be assumed that no serious problems are in prospect for obtaining the substantially increased volume of ferro-alloys which Western Europe must import.

Nkrumah in his *Neo-Colonialism, The Last Stage of Imperialism* explains what these 'strong trading ties' amount to:

> The prime objective is to monopolize Africa's sources of raw materials not, as it is claimed, to assist the African countries to develop their economies. For the materials are carried off largely in their raw state or as concentrates to enhance the productive output of the imperialist countries and to be returned to them in the form of heavy equipment for extractive industry and the infrastructure for carrying the resources away.

And yet, as Nkrumah points out, 'it is out of the revenue from the trade in these materials that the African countries look to amass part of the capital that will make it possible for them to utilize these same commodities in the service of their own development.'[28] Nkrumah's remarks can be brought up to date. In 1968 the United States Secretary of State for African Affairs explained that the 'respect for our interests' shown by the African nations 'is illustrated by their special facilities and rights made available to us, by our development of Africa's important mineral and other resources ...' As he further explains: 'US investment in Africa has doubled in the last decade and has been involved in Africa's major output of such strategic materials as copper, bauxite, iron ore, uranium, petroleum, manganese and other scarce minerals.'[29]

The example of tin

Let us consider, as an example, the case of tin. Despite the fact that the United States strategic stockpile* had reached in 1962 the total of 349 000 tons W. and P. Paddock, nevertheless, advised in their book *Famine 1975* (their thesis being that the United States would have a food surplus sufficient to save only selected countries from the famine they predict): 'I am told by officials that the uninterrupted export of [Bolivia's] tin is a strategic necessity. Accordingly, I would place this case among the exceptions to receive American food.'[30] More generally we note that, according to Landsberg,

---

* When the stockpile programme was declassified by President Kennedy in 1963, some $7 billion worth of material and ores had been accumulated.

Fischman and Fischer, 'US resources [of tin] are negligible and national interest is therefore focused on the situation in the producing countries.' According to these authors, 70% of the tin reserves in non-communist countries are in S.E. Asia (Malaya, Indonesia, Thailand, Burma), 10% in the Congo, 10% in Bolivia.[31] The ecologist John McHale tells the same story: 'With such key uses [for tin], the conflicts around the control of tin-ore producing areas have furnished the latent background for considerable political and economic manoeuvring, for example, the countries initially occupied by Japan in World War II were those producing over 60% of the world's tin, and we may note that these areas still furnish a central focus for intensified international conflict.'[32] *

Summary with respect to raw materials

The underdeveloped countries are in a difficult position. Acting individually (or in respect to only one or two raw materials) they can achieve very little. Increasing the price of, for example, tin would only ultimately result in the improved efficiency of electroplating and the further development of substitute metals (both of which happened during the Second World War when tin was in short supply). And yet for as long as various raw materials (such as tin) are important for the Western economy, the West will attempt to maintain control of their sources. But when and if supplies of certain raw materials become exhausted or are made redundant, the underdeveloped countries affected will be left without even the meagre earnings from the export of their one-time natural resources. It is a depressing and frightening prospect.

3. Markets and Investments in the Underdeveloped Countries

In chapter 7 we examined some of the means by which profitable investment outlets for the large corporations are guaranteed by the United States government in order to maintain a high level of effective demand and therefore a low level of unemployment. But capitalism is a world-wide system. Provided there is security against loss, large corporations will seek the most profitable outlets for their capital in all countries, and industry in general will seek profitable export markets for its produce. It was, as we have earlier seen, explicit British and American policy to enforce the acceptance of 'free trade' on underdeveloped countries (in particular, China) and to ensure the security of American overseas investments. American policy has not changed in this respect. Indeed, the then Under-secretary of State, Dean Acheson, told a Congressional audience in November 1944 (as the war was seen to be

---

* To avoid misunderstanding let me stress that I am not suggesting that the United States has intervened in Vietnam solely in order to ensure supplies of minerals from South-East Asia. Far from it. What was and is at stake is the Western right to control the economic resources of *all* underdeveloped countries and Vietnam was considered by successive American governments to be a test-case of this policy. Of course, the determined resistance of the Vietnamese people, so unexpected by American corporate and military planners, has been a major factor contributing to the breakdown of Western ideological consensus and to the proliferation of new 'paradigms'.

approaching its end and therefore, perhaps, the war-induced prosperity as well): 'We cannot go through another ten years like the ten years at the end the twenties and the beginning of the thirties, without having the most far-reaching consequences upon our economic and social systems. . . . We have got to see that what the country produces is used and sold under financial arrangements which make its production possible.'[33] Acheson explained that while America remained capitalist, the domestic market would be insufficient to absorb production on a profitable enough basis to maintain a high level of output and full employment. He was, of course, aware that 'under a different system you could use the entire production of the country in the US' but, unfortunately, the introduction of such a system, i.e. socialism, 'would completely change our Constitution, our relation to property, human liberty, our very conception of law. And nobody contemplates that. Therefore, you must look to other markets and those markets are abroad.'

Towards the end of the war there was much debate among American economists and businessmen as to how the United States could avoid depression once the war had ended. It was generally appreciated that an expanding foreign trade would be necessary, together with other measures to prevent heavy unemployment. In 1943 the Assistant Secretary of Commerce Will Clayton declared at a meeting of the Committee for Economic Development that 'it is terribly important to keep in mind that measures adopted for the conducting of international trade are likely to have a very important bearing on the whole future of free enterprise domestically'.[34] The 1944 proposals by Higgins, the former head economist of the New Deal's Federal Works Agency, to increase federal expenditure on domestic public works and thereby to maintain full employment was not sympathetically received. In November of the same year the National Planning Association published a report *America's New Opportunities in World Trade* advocating a large expansion of government-supported foreign investment in order to resolve America's domestic problems. At the end of the war, the Report asserted, the United States

> will have an enormous backlog of idle and underemployed capital at home, and a capital formation capacity which will greatly exceed the amount of new capital which can currently be absorbed by domestic business. . . . We will have a capital equipment industry nearly twice the size which would be needed domestically under the most fortuitous conditions of full employment and nearly equal to the task of supplying world needs.[35]

And, the report stated, 'this idle capital must be used if full employment is to be maintained'. Three years later in 1947 Clayton would be explaining that 'the Marshal Plan is not a relief program; it is a recovery program for Western Europe. Hence our interests rather than our humanitarian instincts should be mainly considered.'[36] For if the United States failed to provide this aid, 'the Iron Curtain would then move westward at least to the English Channel. Consider what this would mean to us in economic terms alone. A blackout of the European market could compel radical readjustments in our entire economic structure . . . changes which could hardly be made under our democratic free enterprise system.'

Thus we see that American government officials like Acheson and Clayton were well aware that unless the world could be kept 'open' to American exports and investments then social and political changes of a revolutionary nature would inevitably take place within the United States.

Foreign markets and war

Keynes himself pointed out in *The General Theory* that one of the economic causes of war was 'the competitive struggle for markets'. However, he hoped that his theory would show governments how to maintain full employment at home without recourse to armed conflict for the control of markets. As he explained:

> But if nations can learn to provide themselves with full employment by their domestic policy . . . international trade would cease to be what it is, namely, a desperate expedient to maintain employment at home by forcing sales on foreign markets and restricting purchases, which, if successful will merely shift the problem of unemployment to the neighbour which is worsted in the struggle, but a willing and unimpeded exchange of goods and services in conditions of mutual advantage.[37]

Importance of foreign markets and investments for the United States economy

The recent growth of United States assets and investments abroad has been remarkable, as shown in Table 8.1.[38] However, although it is often admitted that export markets and foreign investments are undoubtedly important for certain firms and particularly for the larger corporations, it is quite often

Table 8.1

	1950	1955	1960	1965	1967
$ billion	32	43	68	106	122

claimed that because total exports are less than 5% of GNP, and foreign investment per year much less than 10% of domestic capital investment, then export markets and foreign investments cannot be important *for the American economy as a whole*. This is not so however. Such markets and particularly foreign investments are important both for large corporations and for the American economy as a whole. To see this we note that in 1964 (as estimated by Magdoff) the total value of movable goods (as opposed to services, etc.) was equal to $280 billion and that the size of the foreign market in that year was approximately two-fifths of the domestic output of farms, factories and mines—and two-fifths is a very significant fraction.[39] Of that fraction, however, exports play a minor and decreasing role. For example, the production abroad arising out of United States investment was 4·3 times larger than exports in 1950; by 1964 it was 5·5 times larger. Indeed, American owned industry abroad now has a 'GNP' greater than that of any other nation, except for that of the United States itself and the Soviet Union.

The pattern of American investment abroad

Now even if we make the (highly doubtful) assumption that the best path to social and economic progress for the underdeveloped countries does not lie in social and political changes but in their openness to private investment from American manufacturing companies, the actual pattern of American investment shows that extractive rather than manufacturing investment predominates in the underdeveloped countries. In Table 8.2 the percentage

Table 8.2

Industry	Europe	Latin America	Africa	Asia
Mining	0·4%	12·6%	21·9%	1·1%
Petroleum	25·6	35·9	51·0	65·8
Manufacturing	54·3	24·3	13·8	17·5
Other	19·7	27·2	13·3	15·6

distribution of American investment in Europe, Latin America, Africa and Asia during 1964 (as given by Magdoff) is shown for purposes of comparison.[40] Needless to say, the 25·6% investment in petroleum for Europe is not in oil wells, but in refineries and distribution.

As the table shows, investment expenditure in mining and petroleum greatly exceeds that in manufacturing for both Africa and Asia. Only in Latin America is such extractive investment less than 50% of the total. Furthermore, we should note that of the $4 billion or so invested in manufacturing abroad by United States companies in 1950, $1·9 billion were invested in Canada and $0·9 billion in Europe, i.e. some 74% of the total. Brazil was then third with 7% of the total investment. However, in 1966, American manufacturing investment abroad had jumped to $22 billion, of which Canada received $8 billion, Europe $9 billion and Australia, now in third place, $1 billion.[41] Why should the United States companies *concentrate* manufacturing investment in underdeveloped countries where poverty is so widespread? The developed countries of Canada, Europe and Australia are a much better market. To be sure, this investment is not altogether welcomed by the recipients who fear American control over their economies. But this only raises another question: if the advanced capitalist nations of Western Europe fear the economic power of the United States, what must be the position of the relatively defenceless underdeveloped countries?

Their position, according to the Director of the International Trade Centre in Geneva, is that although 'it is now universally accepted that a rapid increase in the export earning of the developing countries is an essential prerequisite of their economic development', it is an unfortunate fact that 'exports from the developing countries have not been keeping pace with world exports'.[42] We are told that between 1948 and 1968 their share of world trade decreased from 30 to 18·2%; that between 1963 and 1968 their trade deficit worsened at an average annual rate of 23%, reaching $2 billion for 1968. According to the Director, protectionism in the industrialized markets is one important reason for this 'sad situation'. The principal reason, however, is that 'four-fifths of the developing countries' exports are primary commodities while the developed countries are mainly selling manufactured goods. Not only is demand for manufactured goods rising faster than for primary commodities, but their prices are also going up, while the prices of primary goods, with some exceptions, are tending to stagnate or to drop, sometimes severely.'

Let us now observe some of the effects of American investment in Latin America, a continent where American manufacturing investment is not inappreciable (24% for example, of the total American investment in that continent in 1964). In particular, we note that whereas the exports of American-owned *mining* firms in Latin America in 1965 amounted to 82% of the value of all sales, the exports of American-owned *manufacturing* firms amounted to only 8%.[43] Clearly American manufacturing investment will increase the rate of production of commodity goods in underdeveloped countries but equally clearly American owned manufacturing firms have no intention of competing with parent firms in the advanced capitalist countries. Thus private manufacturing investment in underdeveloped countries considerably limits their possibility of exporting manufactured goods in order to get the foreign currency much needed both for the purchase of foreign equipment as well as for payment of debt on past loans.

Debt repayment

In Latin America between 1962 and 1966 the average annual debt repayment amounted to $1·6 billion whereas the average annual assistance from the United States to these countries amounted to $1·2 billion.[44] Hence economic aid from the United States did not even enable the Latin American countries to service their external public debt! The President of the World Bank was himself moved to write in 1966 that

> when all [debts] are taken into account, the backflow of some $6 billion from the developing countries offsets about one half the gross capital inflow which these countries receive. These payments are continuing to rise at an accelerating rate and in a little more than 15 years, on present form, would offset the inflow completely. In short, to go on doing what the capital export countries are now doing will, in the not too long run, amount to doing nothing at all.[45]

Who is 'aiding' whom?

Finally it is necessary to point out that whereas the net flow of income during the years 1960–5 from the United States to the minor capitalist nations was positive, from the United States to the underdeveloped countries the net flow was negative. In fact, as Magdoff has pointed out, nearly three times as much money was transferred from the underdeveloped countries to the United States as in the reverse direction (see Table 8.3).[46] To argue, therefore, that United States investment in the underdeveloped countries assists the development of these countries is, to say the least, problematic.

Table 8.3

	Europe	Canada	Latin America	All other areas
Flow of direct investments from U S	$8·1	$6·8	$3·8	$5·2 billion
Income on this capital transferred to U S	5·5	5·9	11·3	14·3 billion
NET	+$2·6	+$0·9	−$7·5	−$9·1 billion

4. On Social Reform in the Underdeveloped Countries

We have already mentioned that social unrest is not to the advantage of the Western nations who would undoubtedly prefer to see governments in power in the underdeveloped countries that would introduce meaningful land reform (provided American-owned land is not expropriated) and eliminate certain other flagrant injustices which provoke revolt and hence threaten the stability of the Western Empire. Such was the purpose of the Alliance for Progress; United States aid to, and progressive reforms in, Latin America were to make that rich continent safe for United States mining and other interests. But the United States is asking the impossible. John Moors Cabot appreciates very sensitively the United States dilemma. In November, 1963 he declared: 'Whereas our policy seeks to promote reform and social justice in Latin America, the need to protect our large economic stake inevitably injects a conservative note into our policies.'[47] Edmundo Flores explains further: 'The position of the United States government is tragic, and perhaps absurd: it wishes to entrust what is nothing less than a revolution to the very group—the safe conservative elements—which in its own interest must block it, as it always has . . .* Subsidizing and arming the anti-revolutionary and dictatorial groups in power, so that they will be in a better position to persecute

---

* This was not President Kennedy's view. At Bogota he declared that 'the leaders of Latin America, the industrialists and the landowners are, I am sure, also ready to admit past mistakes and accept new responsibilities'. M. Harrington in *Dissent*, Sept.–Oct. 1967, comments: 'In retrospect, the President's confidence was either naïve or ceremonial. The Latin-American *status quo* is not even prepared to take a position of enlightened self-interest.'

and kill the opposition, only adds to the popularity and power of the Communists. If, following current misconceptions, the United States backs the quasi-feudal and militaristic governments in power, there will be a pretence of economic development and Allianza funds will be misallocated and wasted without changing the conditions responsible for political unrest and economical stagnation. This will eventually lead to the establishment of military dictatorships of the extreme right.'[48] One year after Flores' article was published, the constitutional president of Brazil was overthrown by a military coup and Brazil's 80 million people were placed under the right-wing dictatorship of General Castelo Branco. While President Goulart was still resisting on Brazilian soil, the military junta received the recognition of the Government of the United States.

The case of Brazil

Since Brazil is potentially one of the world's richest countries, it will be instructive to look briefly at the initial course of events since the military assumed control.

That the United States government approved and aided the *coup d'état* is not denied. Indeed, in 1965, a year after the coup, the Chairman of the House of Foreign Affairs Committee declared that

> every critic of foreign aid [sic] is confronted with the fact that the Armed Forces of Brazil threw out the Goulart government and that US military aid was a major factor in giving these forces an indoctrination in the principles of democracy and a pro-US orientation. Many of these officers were trained in the United States under the AID [Agency for International Development] programme. They knew that democracy was better than communism.[49]

(Between 1949 and 1964 more than 16 000 Latin American military officers had undergone 'internal security' training courses at the US Army School of the Americas in the Panama Canal Zone.)[50]

It has become abundantly clear what the Western powers mean by the defence of democracy against communism; the case of Brazil is but another example. Because of his advocacy of social and political reform, United States-controlled financial organizations had imposed great pressure on the constitutional President. As noted by the US Senate Committee on Foreign Relations, the World Bank had 'refused to make any loans to Brazil for several years prior to 1964, mainly because of the unsound financial policies of the government preceding the Branco administration.'[51] The figures for AID expenditures in Brazil[52] given in Table 8.4 also tell the same story. Specifically Goulart's 'unsound' financial policies included an attempt to limit the scope of private enterprise and the amount of profit that could be remitted back to the United States by American investors, a policy quickly

PROBLEMS OF UNDERDEVELOPED COUNTRIES 221

reversed by the military whose 'sound' financial and economic measures were summarized thus in a 1966 AID report:

> The Castelo Branco administration has conducted an effectively tough economic programme of stabilization, development and reform. . . . Private enterprise has been encouraged by policies halting the previous trend towards state ownership. New incentives have been created and old obstacles removed in an effort to increase the participation of private enterprise, both foreign and domestic. New foreign investment is being sought for development of minerals and petrochemicals, and an Investment Guaranty agreement has been signed with the United States.[53]

Table 8.4

Fiscal year ending June 30	Expenditure in millions of dollars
1962	$81·8
1963	38·7
1964	15·1
1965	122·1
1966	129·3

In this agreement, the military government committed itself (1) never to expropriate an American firm except with that firm's 'full and complete agreement'; (2) to pay any damages caused to American enterprises by 'war, revolution, insurrection, strikes and sabotage'; and (3) to allow American firms to invest in Brazil under the regulations of American laws.[54]

A single example of Brazilian development under military rule: in 1958 the Hanna Mining Company of Cleveland had had certain concessions on rich iron ore reserves in Brazil cancelled by the Kubitschek administration (Kubitschek was but one of three former Presidents of Brazil whose political rights were cancelled for ten years by the military dictatorship). In December, 1964, however (just eight months after the coup), Castelo Branco promulgated a Presidential decree which 'called . . . for private competitive development of Brazil's vast iron ore reserves and ordered discouragement of any monopoly by the state or other enterprises'. The following year, the Brazilian 'supreme court' re-granted Hanna's concessions, and in addition the right to build an iron-ore shipping port just south of Rio.[55]

Clearly the American corporations know the advantages to be gained in defending 'democracy' against 'communism'. The underdeveloped countries are not to be allowed to choose their own path to development. For, as a US House Committee on Foreign Affairs has euphemistically put it, 'Only by participation in that process will we have an opportunity to direct their development along lines that will best serve our interests.'[56]

Sale of arms to Latin America

The rapid rate of obsolescence of modern weapons not only helps to ensure permanent investment outlets for the capital goods industries of the advanced capitalist countries but also ensures an adequate supply of obsolete weapons that can be sold to the military of underdeveloped countries. Latin America, for example, is a continent surrounded by oceans; it is a little difficult to conceive of invasion-planning 'communist' enemies. Nevertheless, the military establishments of Latin America compete among themselves for possession of the most modern weapons available. Although the people of Latin America live in poverty, their rulers prepare to defend themselves against an external attack no one is planning. More to the point, of course, they are also arming themselves against possible uprisings within their own countries. From the point of view of the advanced capitalist countries this situation is very satisfactory, giving them the opportunity of selling both obsolete and even modern arms to the military governments and privileged élites of Latin America, and, at the same time, of improving their balance of payments accounts.* We give a few items of such commerce and 'aid' to the underdeveloped countries of Latin America.[57] In 1966 the United States sold 25 Skyhawk attack jets to the military government of Argentina as part of a commercial jet sale. Chile then purchased 21 reconditioned British Hunter jets for $20 million, after which the United States (more concerned about internal security in Latin America than about non-existent external threats) responded to European competition by offering the $1·1 million Northrop F-5 Mark 1 'Freedom Fighter'. By February, 1968, the United States and France had each sold $30 million worth of F-5's and Mirage V's to Peru.

An example from India

The judicious use of foreign aid is clearly one way in which the advanced capitalist nations can ensure the 'development' of the countries of Latin America, Africa and Asia along lines that best serve the interests of the advanced capitalist nations. The latter make no secret of their manipulation of aid to achieve this end. According, for example, to *Forbes Magazine*:

> For a long time India insisted that it handle all the distribution of fertilizer produced in that country by US companies and that it also set the price. Standard [Oil] of Indiana understandably refused to accept these conditions. AID put food shipments to India on a month-to-month basis until the Indian government let Standard of Indiana market its fertilizer at its own prices.'[58]

AID is proud of its achievements in India. Indeed, one of its administrators declared in 1968:

> Private enterprise has greater opportunities in India than it did a few years ago. . . . There are still more sectors of the economy in which

---

* A major problem, of course, confronting the advanced capitalist countries.

private enterprise has a hard time, but fertilizer is an example of a field which is now open to the private sector, and was not in the past. This is largely a result of the efforts which we have made, the persuasion that we, along with other members of the consortium, have exerted on the Indian government. We feel that conditions in India are improving steadily. They still have not gone as far as we would like to see them go.[59]

No doubt.

5. Some Concluding Comments

The plight of the underdeveloped countries is desperate. The peoples of these countries are not to be allowed by the capitalist nations to make the radical, social and political changes that would make economic and human progress possible. They are to remain economic dependencies. Whenever necessary and wherever possible the United States with the approval of the minor capitalist nations will intervene in the underdeveloped countries in order to defend 'freedom against the communist threat'. Noam Chomsky has spelt out the harsh reality as he sees it:

> If we consider governments maintained in power by force or overthrown through subversion or intrigue, or the willingness to use the most awesome killing machine in history to enforce our rule, or the means employed—saturation bombing, free-strike zones, napalm and antipersonnel weapons, chemical warfare—there seems to me no other conclusion: we are simply without a rival today as an agent of international criminal violence.[60]

It is not easy to disagree with this verdict. Although this chapter has but touched the surface of problems confronting the 'third world' countries, about the overall Western objective there can be little doubt. From 'White Man's Burden' to 'Defence of the Free World', this objective has been basically the same: to make the world safe for private enterprise and private profit. Clearly 'communism' became a threat to this overall objective. Hence Western intervention in both the Russian and Chinese civil wars; hence the terrible additional burden imposed on the successful 'communist' governments. A sketch of the development of the Cold War and of the mistakes and crimes committed by Stalin will be given in the next chapter. But this much is already clear: it is the social structure of capitalism that is the principal source of world violence.

Chapter 9

SOCIALISM IN CRISIS: THE RUSSIAN REVOLUTION AND THE ORIGINS OF THE COLD WAR

> We are involved in one of the great ideological struggles of all times. We are in it so deep that it is hard to see it in perspective. Essentially it is a contest between two quite basic concepts. One is that men are capable of faith in ideas that lift their minds and hearts, ideas that raise their sights and give them hope, energy and enthusiasm. Opposing this is the belief that the pursuit of material ends is all that life on this earth is about.
>
> F. R. KAPPEL, President of the American Telephone and Telegraph Company in *Vitality in a Business Enterprise*, 1960

IN preceding chapters it has been argued that capitalism is essentially a social system that—no matter how progressive it has been in the past—is now not merely preventing human development and perpetuating violence on a global scale but which will—if unchecked—take the world increasingly nearer to ecological or nuclear disaster. Yet scientific knowledge and available resources are now such that there appears to be no reason why, in principle, every human being should not only be able to enjoy the basic necessities of life but also the possibility of developing to the full his most valued potentialities. Clearly, however, the construction of such a humane society is not possible within a framework of competing corporations where production is *necessarily* undertaken not for the purpose of easing pain and hardship but for the purpose of maintaining and, if possible, increasing the profit and growth rates of corporations. Indeed, it is worthwhile to note explicitly that the continued existence of poverty, illiteracy, disease and violence does not result from the actions of men who *deliberately* desire this condition of man but rather from the actions of men who are trapped within the socio-economic capitalist order, who can see only this order, who can attempt to resolve problems only within this order thereby making the existing problems even more grave or causing new problems of even greater gravity than the 'solved' ones, who in the last analysis have benefited from the existing social order and fear the loss of their privileges should the social order be radically changed. But transcended the capitalist system must be if there is to be a chance of solving the present world crisis in a way that will allow people to live in dignity and comradeship together.

However, although some of the general features of a social system preferable to capitalism would appear to be obvious enough—namely public ownership of all means of production and cooperative planning for the attainment of social goals—nevertheless it is often argued that such a system has been tested and has failed to prove itself to be either more just or more humane than its rival capitalist system. For (it is asked) has not the Union of Socialist Soviet Republics witnessed the extermination of several million 'rich' peasants, the setting up of concentration camps, prolonged purges characterized by forced confessions and mass executions, and more recently the invasion of fellow socialist countries, Hungary and Czechoslovakia? And so, the argument usually concludes, since socialism has given birth to atrocities comparable with those of capitalism, the fault must lie in man himself and not in the social system. Hence if man is to be saved from his own innate violence and greed, it is research into the biological foundations of the human species that must be given first priority. And until the desired genetic breakthrough is achieved, the only hope is to stall for time by attempting to make whatever slight social improvements man's perverted nature permits.

The argument is exceedingly dangerous. For commitment to such a research programme on the part of leading scientists combined with a general belief that man's intrinsic nature does not permit the construction of a fundamentally better world society would almost certainly guarantee its non-construction. (We shall return in the last chapter to touch on the subject of man's 'innate' aggressive nature.) However, the argument is not only dangerous but all its premises are, I believe, false. In particular, socialism has not been tried out, and could not have been tried out, in the Soviet Union. Indeed socialism (as I understand the meaning of the concept) is not possible in any country while the Western world remains capitalist. To understand the basis underlying this assertion it is necessary only to analyse the development of the Soviet Union since the October revolution (or that of China and Cuba since their revolutions), an analysis that we therefore undertake in this chapter.

Yet it is not only the failures of the Soviet Union with which we must be concerned. For the Soviet Union has made very great material progress in its fifty or so years of existence—it remains the only peasant country in this century to have industrialized and to have raised the standard of living of all citizens to well above the pre-revolution subsistence level. And it is significant that while the health of the Chinese people has dramatically improved* (as has that of the Cuban people), the same cannot be said with respect to the peoples of such potentially wealthy countries as India and Brazil, countries

---

* It is worth noting the observations of two American biologists who visited China in the Spring of 1971: 'Our immediate and lasting impression was that the Chinese people today are well fed, healthy, adequately clothed and housed, extremely hardworking, and loyal to the present government. The streets are clean and orderly, and the cities, despite their crowded and busy condition, appear peaceful and free of tension. . . . It is worth recalling for comparison that starvation, disease, alternating flood and drought, crime, drug addiction, prostitution, and sale of children were frequently described as standard features of life in China before 1949 when the present government took power.'[1n]

which have not achieved 'socialist' revolutions. The assertion, therefore, that socialism is not possible while the United States and Western Europe remain capitalist is not meant to imply that the peoples of the underdeveloped countries achieve nothing by fighting for liberation and 'socialism'. On the contrary, their struggle against Western domination is a necessary part of the struggle to achieve the only socialism possible, *world* socialism. But this goal will be realizable only after capitalism has been transcended in the West in favour of a social system valuing beauty and harmony rather than competition and commercial success. We shall return to this theme in the last chapter. For the remainder of this chapter we sketch the development of, and difficulties that faced, the new social order of the Soviet Union. Clearly only the roughest of outlines will be possible. Nevertheless, such an outline should serve to provoke at least doubt in the mind of any reader who believes that the Soviet Union has been a test case of socialist practice.

1. The Russian Revolution

Marxism and the Russian Revolution

Marx and Engels believed passionately that capitalist societies would inevitably give birth to socialist societies. However, although force would invariably act as midwife at these births, nevertheless the successful completion of the bourgeois stage in the accumulation of capital would obviate the necessity of prolonged violence during the early years of the new society. For bourgeois society would necessarily give rise to that level of scientific, technical and administrative achievement on which the proletariat could securely build first socialist, and then communist, society.

To the communist revolutionaries in late 19th-century England and Germany, Marx's theory meant the immediate practice of such revolutionary tactics as would ensure the 'negation' of their developed bourgeois societies. But to the communist revolutionaries of underdeveloped Russia, Marx's theory gave rise to much debate as to whether they ought to assist the emerging bourgeoisie to establish the social order which would *later* give birth to a communist society or whether, given the special conditions in Russia, they could hope to by-pass through an immediate revolution 'the same process of dissolution as constitutes the historical evolution of the West'. In the preface to the Russian edition of the Communist Manifesto, Marx and Engels replied as follows: 'If the Russian revolution becomes the signal for a proletarian revolution in the West, so that both complement each other, the present Russian common ownership of land may serve as the starting point for a communist development'. The Russians could therefore avoid the horrors of the bourgeois stage of development only if the proletariats of the advanced industrialized societies had already achieved power (or very quickly achieved power after a successful Russian revolution), and actively assisted their Russian comrades in overcoming the backwardness of their country.

This was the general opinion of Russian communists. If the Russian proletariat achieves power, Trotsky wrote in 1905, 'only as the result of a temporary conjuncture of circumstances in our bourgeois revolution', and if no aid arrives as a result of socialist revolutions in the advanced countries, 'the working class of Russia will inevitably be crushed by the counter-revolution the moment the peasantry turns its back on it.'[2] There were to be no supporting revolutions and the peasantry did turn its back on the proletariat.

The Russian Revolution and Western action

In 1917 Lenin and the Bolshevik Party seized power in Russia, negotiated a humiliating peace treaty with Germany, and looked anxiously westwards. However, the hoped-for revolutions in England and Germany did not occur. In Germany the social democrats used the Freikorps, financed by German capital, to put down uprisings by workers and in one of these Rosa Luxemburg and Karl Liebknecht, the two leading communist revolutionaries, were both murdered. It was not Western aid which followed the Russian revolution but Western intervention! British, French, American and also Japanese troops soon found themselves fighting in the Russian civil war in support of the anti-Bolshevik forces. As Lloyd George explained: 'Bolshevik imperialism does not merely menace the states on Russia's borders, it threatens the whole of Asia and is as near to America as it is to France.'[3] (The Russians might be forgiven for viewing the matter somewhat differently: before the Revolution Western shareholders owned some 90% of Russia's mines, 80% of her chemical industry and more than 40% of both engineering plant and banking stock.[4]) For Churchill, the architect behind the British offensive in Russia, the intervention was a great success; it had gained for the West, he declared, 'a breathing space of inestimable importance'.[5] Undoubtedly so. The hard-pressed Bolsheviks, encircled by Western-armed forces, offered in February, 1919, to pay the Tsarist debts which they had earlier repudiated and also to give up propaganda if the Allies would end intervention (although Lloyd George declared two weeks later in the House of Commons that 'we have had no approaches of any kind'). The Bolsheviks, however, under Trotsky's war leadership, eventually succeeded in defeating the White forces; Allied troops were withdrawn and in January, 1920, the Allied Supreme War Council's blockade of the Soviet Union was lifted.

But the cost had been tremendous: Russia was devastated throughout vast expanses. The national income was only one-third of its level in 1913. Russian industries had been either destroyed or run down; the coal mines yielded only one-tenth of their normal output and the iron foundries only one-fortieth. The railways were also destroyed and all stocks and reserves exhausted. Furthermore, Russia's cities and towns, originally the centres of Bolshevik strength, had become seriously depopulated; indeed, in 1921 the populations of Moscow and Petrograd were, respectively, only one-half and one-third their pre-war levels.[6] The all-important proletariat, the backbone of the revolutionary movement, had been decimated by the war and the survivors dispersed over the country. In the fourth year of the revolution the Russian people faced famine on a terrible scale. Such were the initial conditions which faced Lenin and the Bolshevik party in their gigantic task to ensure the survival of the Soviet Union—a far, far cry from the expropriation of the bourgeoisie's accumulated capital which, according to Marx, would herald the advent of communism in the industrially advanced countries of Western Europe.

Socialism in one country?

Lenin faced the unpleasant reality. There could be no question of the Soviet Union's achieving socialism alone: 'We have always proclaimed and repeated this elementary truth of Marxism,' he wrote in 1922, 'that the victory of socialism requires the joint efforts of workers in a number of advanced countries.' It was in many ways a terrible misfortune that the privilege of being the first country to overthrow bourgeois power should have fallen to Russia. Not only was there no material base for the establishment of socialism but the requisite culture was also missing. 'Is it true,' asked Lenin in 1922, 'that one might have the impression that the culture of the vanquished is of a high level? Not so: it is wretched and insignificant. But it is still superior to ours.'[7] 'We took over the old machinery of state, and *that was our misfortune*', he wrote. 'We now have a vast army of government employees, but lack sufficiently educated forces *to exercise real control over them*.' In short, Lenin declared in 1923, we 'lack enough civilization to enable us to pass straight on to socialism.' And thus he emphasized that it would be a mistake 'to lose sight of the fact that after the victory of the proletarian revolution in at least one of the advanced countries. . . . Russia will soon cease to be the model country and once again become a backward country (in the "Soviet" and the socialist sense).'[8]

Not least among Lenin's worries was the threat of renewed capitalist aggression. 'Can we save ourselves from the impending conflict with these imperialist countries?' he asked in a *Pravda* article in 1923. 'In the last analysis,' he wrote, 'the outcome of the struggle will be determined by the fact that Russia, India, China, etc., account for the overwhelming majority of the population of the globe. . . . In this sense, the complete victory of socialism is fully and absolutely assured. But what interests us', he went on, 'is not the inevitability of this complete victory of socialism, but the tactics which we . . . should pursue to prevent the West European counter-revolutionary states from crushing us', to ensure the survival of the Soviet Union 'until the next military conflict between the counter-revolutionary imperialist West and the revolutionary and nationalist East . . .'[9]

The immediate concern, however, was sheer physical survival against the threat of mass starvation. Lenin recognized that the Party had been too hasty in decreeing the nationalization of all small-scale industry and in 1921 the decree was revoked. While the state thereafter retained control of banking, foreign trade and large-scale industry, Lenin's 'New Economic Policy' encouraged private trading and likewise the private development of small-scale industrial enterprises. The role of the market grew in importance. Private traders, the so-called Nepmen, bought from and sold freely to the peasants and also from and to the various state enterprises, so much so that by 1922–3 some 75% of all retail trade was in private hands. Lenin could not but be concerned over the new direction taken by the economy. While he believed that no alternative policy was possible, nevertheless the rise of a new bourgeoisie could only exacerbate relations between the state and the peasantry. 'In the final analysis', Lenin wrote, 'the fate of our Republic will depend on whether the peasant masses will stand by the working class, loyal to their alliance, or whether they will permit the "Nepmen", i.e. the new bourgeoisie, to drive a wedge between them and the working class, to split them off from the working class.'[10] Although Lenin believed that the peasantry had to be persuaded to cooperate with the state in laying the material basis for a communist society he strongly advised against the use of communist propaganda: 'Under no circumstances must this be understood to mean that we should immediately propagate purely and strictly communist ideas in the countryside. As long as our countryside lacks the material basis for communism, it will be, I should say, harmful, in fact, I should say, fatal for communism to do so.' *Lenin at least was well aware of the danger of allowing the idea of socialism and communism to be equated with the desperate situation in which the Soviet Union found itself.*

There were also other dangers. As the revolutionary wave in Western Europe subsided and Soviet policy began to adapt itself to the prospect of securing the Russian Revolution against its capitalist enemies, so Lenin became increasingly alarmed over the many dangers such a psychological readjustment would entail. There would be the temptation to put the interests of the Soviet Union above those of the communist parties of other countries, indeed to regard them solely as the first line of defence of the Soviet Union. Furthermore—all in the interests of Soviet security—there would be a strong temptation to take an uncompromising position with respect to minority groups and nationalities within the Soviet Union. 'Scratch a Bolshevik,' Lenin warned, 'and you will find a Great Russian chauvinist.'[11] In his last message to Trotsky, Lenin saw only one course of action, 'I am declaring war on Great Russian chauvinism.' The arrogant behaviour of highly placed Party leaders towards minority nationalities greatly alarmed Lenin. 'The necessity of solidarity of forces against the international West which defends the capitalist world is one thing. . . . It is another thing when we ourselves fall into something like imperialistic relations towards the oppressed nationalities.' Such actions, Lenin believed, made nonsense of 'all the sincerity of principle, all the defence of the principle of the struggle against imperialism', and were especially dangerous since 'the morrow of world history will be a day when the awakening peoples oppressed by imperialism are fully aroused and the decisive long and hard struggle for their liberation begins'.[12]

It was against Stalin that Lenin's criticism was chiefly directed. Although a Georgian himself, Stalin had first risen to prominence through his overthrow in 1921 of the Menshevik government in his native state of Georgia and by his purge of those members of the ensuing Bolshevik government who resisted the proposal for a Transcaucasian regional republic in which Georgia would be submerged. Although Lenin at first rejected Trotsky's protest at Stalin's high-handedness, accepting Stalin's explanation, later events made him change his mind and send the Georgian Bolsheviks a pledge of support. Stalin, he decided, should be deprived of his post; the time had come to put an end to the growing bureaucratization of the Party and to 'Great Russian chauvinism'. But shortly afterwards, in the spring of 1923, the ailing Lenin suffered an attack which deprived him of the power of speech and left half his body paralysed. His inactivity and death eleven months later left Stalin firmly seated in the commanding position of General Secretary of the Central Committee of the Party, a post he had occupied since April 1922. In the postscript to his 'testament', however, written shortly before his final stroke, Lenin had attempted to warn the Party against Stalin's ruthlessness: 'Stalin is too rude and such a shortcoming is unsupportable in the office of General Secretary. Hence I propose to our comrades that a way be found to remove Stalin from that position and appoint another man . . . more patient, more loyal, politer and more attentive to other comrades, less capricious and so on. This is not a trifle—or at least it is one that may acquire decisive significance.'[13] Those of Stalin's colleagues who agreed after Lenin's death that in the public interest the testament should not be published were to pay for their mistake with their lives.

All this is not to suggest that the violence inflicted on the Soviet people in the 1930s could have been avoided if Lenin had lived. For Lenin, too, would have had to face the problem of how to create the conditions for the successful industrialization of the Soviet Union. Should he (or his successor) not have laid the foundations for heavy industry it is doubtful whether the Soviet Union could have resisted the Nazi invasion in 1941 (although almost certainly Lenin would have adopted a more realistic policy towards the fascist threat in Germany than had Stalin in the 1930s). And he would have had to take action against the peasants when they in effect threatened to starve the country by refusing to sell their grain because of the scarcity of consumer goods. He too would have had to break out of the vicious circle of no grain surplus therefore no industry, no industry therefore no grain surplus. But no matter how Lenin would have acted, one thing is certain: he would have strongly resisted any tendency on the part of Party members to identify the birth pangs of an underdeveloped society in the throes of rapid industrialization with the successful attainment of a socialist society. And this would have meant a great deal.

The industrialization of the Soviet Union

The famine and scarcity caused by the Civil War and Western intervention were slowly overcome in the 1920s. In particular American famine relief during the terrible famine of 1921–3 mitigated its impact to some extent. But American political recognition of the Soviet Union was not forthcoming. On the other hand, the British Labour government recognized the Soviet Union in 1924 and offered to pay compensation for British intervention in the Civil War, an offer immediately repudiated by the next Conservative government. The truce between the capitalist powers and the Soviet Union seemed temporary and in 1927, the year in which the British government broke off diplomatic relations with the USSR, Stalin characterized the international situation as one in which a new wave of Western aggression against the Soviet Union was inevitable. From then on, making a virtue out of necessity, Stalin pursued his campaign of 'socialism in one country' and the world communist movement effectively ceased to exist. The European communist parties were to serve as Stalin's first line of defence against the capitalist powers. The Chinese communist party, acting on Stalin's advice, had been all but destroyed and the proletariat of Shanghai wiped out. In 1928 Stalin launched his first 5-year plan to accomplish the Soviet Union's rapid industrialization. Exactly ten years before Hitler was to launch his massive attack on the Soviet Union Stalin made his justly famous prophecy, 'We are fifty or a hundred years behind the advanced countries. We must make good this lag in ten years. Either we do it, or they crush us.'[14] Rapid industrialization was a military necessity. For example, the project undertaken in 1930 to link the iron ore of the Urals with the coking coal of the Kuzbas, 1 000 miles away in Central Siberia, required a great deal of capital and yet, as Nove has pointedly asked, 'Where would the Russian army have been in 1942 without a Urals-Siberian metallurgical base?'[15] But by 1928 the national product was only just at the level of that of 1913 and neither Stalin nor any of the supporters of rapid industrialization could have had any illusions over the initial suffering that implementation of their policy would inflict on the Soviet people.

The big problem was the peasantry. The Soviet government needed a large surplus of grain and other foodstuffs from the peasants without being able to offer industrial goods in return. Without a large food surplus a growing urban population cannot be fed, nor machinery imported for the industrialization programme. The peasants, however, refused to sell food at the prices offered by the Soviet government and instead waited for the prices to go up. The government refused to increase prices and instead sent troops to confiscate grain. Clearly it was no longer in the interests of the peasant to plant for surplus when the state would simply seize the produce at a price of its own choosing and imprison those peasants who attempted to conceal grain. The five-year plan faced disaster. If the peasants would not willingly produce a surplus then, decided Stalin, they would be compelled to do so. The rich peasants, the kulaks, were to be deprived of their property and the remaining peasants forced into collective farms. The collectivization programme, launched by Stalin in 1929, has appropriately been called by Nove 'one of the great dramas of history'. The peasants resisted collectivization so bitterly as to slaughter their livestock rather than allow ownership to pass to the Soviet government. Indeed, by 1934 the livestock population was reduced to a half of its 1928 level.* Coercion and censorship increased, free discussion was suppressed and 'toughness in executing unpopular orders became the highest qualification for Party office'. Although the urban population did not starve, the priority given to heavy industry meant a fall in living standards. Stalin, however, was of a different opinion, announcing in 1933 that 'we have unquestionably attained a position where the material conditions of the workers and peasants are improving from year to year', while adding ominously that 'the only ones who can have doubts on this score are the sworn enemies of the Soviet régime.'[16] But, according to Nove, the year 1933 'was the culmination of the most precipitous peacetime decline in living standards known in recorded history.' Not surprisingly the powers of the secret police increased for 'sworn enemies of the Soviet régime' were everywhere. In 1934 Stalin massacred the majority of the central committee who thought relaxation desirable. More purges followed. But, at the cost of millions dead or in concentration camps, at the cost of the most terrible initial suffering, the Soviet Union achieved the industrial base that was to serve the Soviet people so well in the equally bitter decade to follow. Neither, of course, should it be overlooked that the medical and health services had enormously expanded; by 1940 there were more doctors per thousand of population than in the United States, Britain, Germany or France. Furthermore, it should be remembered that Western capitalist countries were in the throes of depression during the 1930s. While suffering in the Soviet Union was caused by Stalin's determination to develop industry as quickly as possible, in the West the suffering resulted from the inability of the capitalist social order to use the industry already available. Thus, to a very considerable extent, much of the suffering of the Soviet Union in the 1930s was unavoidable, unlike that caused by unemployment in the Western countries. The *rapid* transition from a peasant economy to an industrial society cannot but be painful, unless the transition is accompanied by much aid from previously industrialized societies.

For Marx, let us repeat, the future socialist society would emerge out of the womb of an already developed industrial society. But the 'socialist' child had been born elsewhere. It was therefore in a hostile environment that the

* It is not surprising, therefore, that one economist has declared that the decision to collectivize livestock was probably the single most important mistake made in the Soviet Union during the 1930s. No 'communist' country has since repeated this mistake. In China during the collectivization process the peasants were allowed to keep their pigs, the principal domestic animal.[15n]

Soviet Union embarked on its industrialization programme. However, the first military attempt to kill off the emerging society had failed. We now turn our attention to the second attempt.

The Soviet Union and the West: the years of 'appeasement'

In 1928 the German Nazi party received only 800 000 votes in the Reichstag election; in 1930 it polled 6 400 000 votes and in 1932 obtained nearly 32% of the votes, 13 700 000 in all. On January 30, 1933, Hitler was constitutionally named Chancellor of the Reich. The very considerable left-wing opposition was, however, bitterly divided: the Social Democrats saw the Communists as the major threat to democracy and the Communists the Social Democrats as the major threat to their own success. The failure of these two parties to form a united anti-fascist front left the way clear for the success of the Nazi party. In *Mein Kampf* Hitler had written: 'In Europe there will within the foreseeable future only be two allies for Germany: England and Italy'. A few pages further on he wrote: 'When we talk today about new land and soil we can think in the first place only of Russia and the states bordering upon it and subject to it.'[17] According to Lord Rothermere in the *Daily Mail* of November 18, 1933: 'The sturdy young Nazis of Germany are Europe's guardians against the Communist danger. Once Germany has acquired the additional territory she needs in Western Russia, her need for expansion would be satisfied.'[18] In 1933 Germany, as did Japan, resigned from the League of Nations. In the following year, when the Soviet Union joined the League, Hitler concluded a non-aggression pact with Poland and a year later concluded a Naval Treaty with Britain which gave the German fleet the right to dominate the Baltic (of strategic importance to the Soviet Union) and to build up the strength of the German submarine fleet to that of Britain's. In 1936 Hitler occupied the Rhineland unopposed. In the same year Franco's fascists, aided by troops from Germany and Italy, overcame Republican resistance. Guernica was the foretaste of things to come. It took the United States government only four days after the fall of Madrid to recognize the Franco régime; it had taken the American government sixteen years to recognize the new government of the Soviet Union.

Not all of Hitler's ambitions were entirely unfathomable. In September, 1936, Hitler publicly declared at Nuremberg that 'if I had the Ural Mountains with their incalculable store of treasures in raw materials, Siberia, with its vast forests, and the Ukraine with its tremendous wheat fields, Germany and the National Socialist leadership would swim in plenty.'[19] In November, 1937, Lord Halifax, one of the principal Munich appeasers, told Hitler that 'he and other members of the British government were fully aware that the Führer had not only achieved a great deal inside Germany herself, but that, by destroying Communism in his country, he had barred its road to Western Europe, and that Germany could rightly be regarded as a bulwark of the West against Bolshevism.'[20] English, French, American, Czech and Dutch armament firms all sought to assist the Nazi rearmament programme. According to Brockway and Mullally, 'In an exhaustive, painstaking study of all the relevant evidence over the years 1933-8, [we] have been unable to find a single instance of an Allied armament manufacturer refusing to supply Germany on political grounds.'[21]

Stalin's objective at this time must have been to try to prevent an alliance between England, France, Germany and Mussolini's Italy, which he saw would have only one end—the destruction of the Soviet Union. However, the Soviet Foreign Minister's plea to save Austria went unheeded, as did Litvinov's proposal for a conference between Russia, the UK, France and the United States to guarantee peace in Europe. The Soviet Union did have, however, a treaty with France to guarantee the frontiers of Czechoslovakia against a fascist invasion. If France declared war on an aggressor the Soviet Union would then support France and come to the aid of Czechoslovakia. Chamberlain's response to this was to instruct the Czech government to cede the German-speaking part of Czechoslovakia to Hitler, after which Britain and France together would guarantee the new—and indefensible—frontiers on condition that Czechoslovakia broke her alliance with the Soviet Union. Upon Czechoslovakia's reluctance, Britain and France issued an ultimatum to Czechoslovakia to accept their plan immediately and unconditionally, and through their ambassadors threatened the well-armed and well-defended Czechoslovakia with an attack from all sides, an attack in which Poland and Hungary would be invited to participate. The Czech government then yielded but a massive demonstration in Prague forced the resignation of the government and the Czech nation prepared to resist. After some hesitation Chamberlain appealed to Hitler for an international conference to give him what he wanted without war and replied to one of Hitler's messages with the reassurance, 'However much you distrust the Prague government's intentions, you cannot doubt the powers of the British and French governments to see that the promises are carried out fairly and fully and forthwith.'[22] The 'international' conference at Munich, to which Russia was not invited, ceded all to Hitler. Poland and Hungary joined Germany in the seizure of Czechoslovakia. 'The British government', wrote Fleming, 'had at last achieved its long pursued dream of achieving a 4-power union with the fascist dictatorship.'[23] Surely, however, this is too extreme a statement. The British government undoubtedly wished to see neither a triumphant Germany nor a triumphant Russia. The hope must rather have been that the two countries would eventually destroy each other (a hope later expressed quite explicitly by Senator Truman, President-to-be). In the League of Nations, however, Litvinov, excluded from the Munich talks, made an impassioned defence of the Soviet Union's attempted defiance of fascism:

> ... at a moment when there is being drawn up a further list of sacrifices to the god of aggression, and a line is being drawn under the annals of

all post-war international history, with the sole conclusion that nothing succeeds like aggression—at such a moment, every State must define its role and its responsibility before its contemporaries and before history. That is why I must plainly declare here that the Soviet government bears no responsibility whatsoever for the events now taking place, and for the fatal consequences which may inexorably ensue.'[24]

From the West's point of view everything had gone as well as could be expected in a difficult situation. Britain and France had given proof of their 'friendship' for Hitler; in return Hitler was to march east. However, on March 16, 1939, Hitler gave a possible indication of a somewhat different future policy by consenting to the Hungarian annexation of the extreme tip of Czechoslovakia and the ideal jumping-off point for an invasion of the Ukraine. It appeared that Hitler, after absorbing the resources of Poland and Rumania, might well decide to deal with the West before invading the Soviet Union. Even so, in March, 1939, the British Federation of Industries concluded a series of cartel agreements with its Nazi counterpart. Two months later (Fleming notes), while economic advisers to Chamberlain and Hitler were negotiating in London over a possible British loan of $5 billion to the Reich, the British government permitted $25 million of Czech gold to be transferred from London to Berlin. And in Berlin, Fleming further notes, the British ambassador to Germany 'was still offering British friendship'.[25]

Yet it seemed that Hitler could not be persuaded to attack the Soviet Union. On March 31, Chamberlain guaranteed Polish, Greek and Rumanian independence and called on Russia to give a unilateral guarantee to Poland and Rumania. Russia promptly proposed a mutual guarantee by Britain, France and Russia of all the border states from the Baltic Sea to the Black Sea. Chamberlain resisted the idea of an allegiance with the Soviet Union. On April 28, Hitler made a speech in which he revoked his Non-Aggression Pact with Poland but at no point made an attack on Russia or communism. Stalin took the hint. The tables were to be turned on Chamberlain. On May 7, the French Ambassador in Berlin reported that Hitler would come to an understanding with Russia. At the same time a Gallup poll showed that 92% of the British people favoured an alliance with Russia while even Churchill declared that the 'brutal truths' of the matter were that 'without an Eastern Front, there would be no satisfactory defence of our interests in the West, and without Russia there can be no effective Eastern Front!' Yet only a minor Foreign Office official, who had been a member of the British Munich team, was sent to Moscow. On June 29, a *Pravda* editorial complained that 'the British and French governments are not out for a real agreement acceptable to the USSR but only for talks about an agreement'.

Sooner or later, it must have been clear, Stalin would be forced to play Chamberlain's own game. His preferred 'allies' had rejected him. The only hope remaining for Stalin was to conclude a pact with Nazi Germany by which Hitler would be able to invade westwards without fearing a Soviet attack on his eastern flanks. Nevertheless, Chamberlain continued to ignore warnings from Moscow. Fleming writes that not until July 31, did he agree to naming a military mission to be sent to Moscow (by boat through the Baltic and not by air). Furthermore, although the Russian delegation was headed by the Soviet Chief-of-Staff, 'the British-French delegation was headed by an obscure British admiral, Sir Reginald Plunkett-Ernle-Erle-Drax, and by a French general of comparable obscurity.' The talks stalled: the colonels who ruled Poland did not favour an alliance with the Soviet Union.

The Soviet-German pact

On August 23, Stalin accepted the pact being offered him by Hitler and a non-aggression treaty between Germany and the Soviet Union was signed in Moscow. While Hitler's way was apparently clear for an invasion of Western Europe, the Soviet Union had achieved frontiers many miles further westwards and, as it turned out, a two-year grace to prepare for the inevitable invasion. The Western leaders feigned surprise at Stalin's act and denounced his 'treachery'. Although they could not prevent the partition of Poland, Britain and France declared war on Germany as Hitler and Stalin divided Poland between them. When the Soviet Union invaded Finland as a result of the Finnish refusal to cede territory strategically important for Leningrad's defence (a refusal reasonable enough from the Finnish point of view), the League of Nations acted for the first time and expelled the Soviet Union for aggression.

Despite the fact that Britain and France were at war with Germany, the Allied Supreme War Council was in favour of sending troops to Finland. The two governments were saved from fighting both Germany and the Soviet Union simultaneously, writes Fleming, only by the refusal of Norway, Sweden and Turkey to allow foreign troops to cross their territories.[26]

2. The Soviet Union and the West 1941–5

The Soviet-Western partnership

Continental Europe fell quickly to Hitler's armies, only the Channel saving England from immediate invasion. In such desperate circumstances a very different English policy towards the Soviet Union became obviously necessary. In June, 1941, when Hitler launched his invasion of the Soviet Union, Churchill declared that if Hitler invaded Hell he would assist the Devil and offered help to the Soviet Union. Senator Truman, to become President of the United States in 1945, proclaimed: 'If we see that Germany is winning the war we ought to help Russia, and if Russia is winning we ought to help Germany, and in that way let them kill as many as possible. . . .'[27] The US War Department intelligence officers estimated that Russia could hold out only between one to three months. But at the end of 1941 the Soviet Union still existed, although the initial losses had been tremendous. By December 1941, 'the Soviets in their retreat had lost vast territories which contained 63% of all coal production, 68% of pig iron, 58% of steel, 60% of aluminium, 41% of railway lines, 84% of sugar, 38% of grain, 60% of pigs.'[28] Yet the Soviet Union not only survived but began to fight back. In 1940 some 15% of the national income had been devoted to military purposes; in 1942 the figure had reached 55%—according to Nove, perhaps the highest ever reached anywhere—but it was this priority given to the surviving arms industry that allowed the Soviet Union to produce the bulk of the guns, planes and tanks used against the Nazi invader. The Allied help the Russians most wanted in 1942 was an invasion of France in order to divert at least some of the German divisions from the Eastern front. Although Stimson, the American Secretary of War, and the American Chiefs of Staff strongly argued for a European second front, Churchill categorically refused. His alternative plan was for an invasion of North Africa, far from the heart of Germany. In the face of Churchill's refusal to agree to an invasion of France, the Americans could do nothing but accept the alternative plan. Once again the Russians were left alone on the European Continent to bear the full brunt of the German armies. Stalin could not but be suspicious of American and British intentions. To be sure, the Allies wished to see Nazi Germany defeated but at the same time, Stalin must have thought, to see the Soviet Union weakened beyond repair. As it turned out, of course, it was the massive Soviet victory at Stalingrad, perhaps the turning point of the war, that facilitated the Allied campaign in North Africa, not the other way round. There was also to be no second front in 1943. Although the tide of the war turned on the eastern front, nevertheless every day the Russians were suffering great losses. By the time the second front opened in 1944 the Russians had all but won the war. But never had victory been so dearly won. As President Kennedy himself explained to the American people in 1963, 'No nation in the history of battle ever suffered more than the Soviet Union in the Second World War. At least 20 million lost their lives. Countless millions of homes and families were burned or sacked. A third of the nation's territory, including two-thirds of its industrial base, were turned into a wasteland—a loss equivalent to the destruction of this country east of Chicago.'[29] And it was this crippled giant that was supposed to be a mortal threat to Western Europe—despite the fact that the United States had emerged from the War stronger than ever before, and armed with the most powerful weapon ever known to man.

Socialism in one country?

We must, however, not lose sight of the central theme of this chapter—the examination of the view that socialism has not only been tried out in the Soviet Union but has there proved itself an inefficient and unworthy alternative to its rival capitalist social order. We have, of course, already noted that the first 'socialist' revolution occurred in an underdeveloped country, not an highly industrialized one; that Western intervention and civil war destroyed the incipient industrial base and decimated the proletariat; that rapid industrialization led to terror and the emergence of a brutalized bureaucracy; and now we note the massive destruction inflicted on the Soviet Union and its people during the Second World War—so much so that it was not until 1953 that the level of pre-war industrial production was passed.[30] But above all we must now describe the next determined effort of the Western powers to destroy that social order which, despite its many shortcomings, nevertheless, presented the capitalist world with a challenge which it was not prepared to face: if the Soviet Union could achieve so much under the most adverse conditions, what could the industrialized West achieve were it to become socialist? It was a question the capitalist élites of the Western world did not want asked.

The development of the Cold War

We have seen that hostility between the Soviet Union and the capitalist world did not begin in 1944 or 1945 or two years later but from that year in which the Bolsheviks seized power in Russia. What we are concerned with now, however, is the resumption of open hostility between the Soviet Union and the West after the enforced 'truce' of 1941-5.*

Stalin, it appeared, perhaps out of economic necessity, initially wanted to continue and even improve on the war-time alliance. In 1943, to convince the Allies of his good faith, he dissolved the Comintern and even appeared willing to have British and American troops fighting on Russian soil. But the main objective of Stalin's foreign policy was not only to maintain friendly relations with the Western powers, but also to ensure that those countries sharing a border with the Soviet Union should have governments favourably disposed to the Soviet Union (Hungary, Bulgaria, Rumania and the Baltic States had all participated in the Nazi invasion of the Soviet Union). During October, 1944, Churchill and Stalin, in dividing the Balkans between them, agreed that Britain and Russia should each have 50% influence in Hungary and Yugoslavia, that Russia should have 90% influence in Rumania and 75% in Bulgaria, while Britain was to have 90% influence in Greece.[31] Subsequently, in November, and December, 1944, British troops under Churchill's orders attacked those Greek guerrilla organizations which were not only leftist inspired but which, as Anthony Eden admitted in the House of Commons in April, 1944, controlled 75% of the resistance forces against the Nazis. The upper classes had collaborated with the Nazi invader, but it was Churchill's aim to restore the monarchy in Greece. The British suppression

---

* It is of interest to note that on October 11, 1939, the Assistant Secretary of State, Breckinridge Long, noted in his diary that the 'eventual enemy' of the United States is the Soviet Union and that if the 'responsible' Nazis would replace Hitler with Goering then the Western powers might be able to reach rapid agreement with Germany.

of Greek resistance was bloody. But, as Churchill recorded in volume 6 of his War Memoirs, Stalin 'adhered strictly and faithfully to our agreement of October, and during all the long weeks of fighting the Communists in the streets of Athens, not one word of reproach came from *Pravda* or *Isvestia*.' Thus, as Fleming quite rightly remarks, 'It was Churchill who acted first and Stalin who followed his example, in Bulgaria and then in Rumania, though with less bloodshed.'[32] (The British pursued a similar policy in Vietnam. Although welcomed by the Vietnamese and the Viet-Minh guerrillas upon Japan's surrender, the British not only did not disarm the Japanese, but used them instead for 'security duty'. At the same time they freed French and Foreign Legion soldiers who in September 1945 attacked the Vietnamese in Saigon with the intention of regaining French control of the city. The colonial war began in all its savagery when the French bombarded the Vietnamese quarter of Haiphong, killing thousands of civilians.)

The question of Eastern Europe

Despite the fact that capitalist élites in both Britain and the United States loathed the idea of Russian domination of Eastern Europe, there seemed nothing they could do to force *immediate* Soviet withdrawal from the territory Stalin considered to be of strategic importance to the Soviet Union. However, after Roosevelt's death in April, 1945, talk of an inevitable war with the Soviet Union became commonplace in the United States, so much so that in June of that year eleven faculty members of Yale University saw fit to state their concern about 'the deplorable and dangerous state' of American relations with Russia. In their letter to the *New York Times* they warned against the irresponsible 'talk in some quarters of the "inevitable" war with Russia—the fulfilment of Hitler's dream', pointing out that the fear of such a conflict could become 'the premise of future policy on both sides'.[33] On July 16, the atomic bomb was successfully tested in New Mexico. Allied policy towards the Soviet Union could now change. As Churchill put it: 'We now had something in our hands which would redress the balance with the Russians', declaring in the House of Commons that 'We were ... possessed of powers which were irresistible.'[34] And as President Truman was advised by his Secretary of State, Byrnes: 'The bomb might well put us in a position to dictate our own terms.'[35] Truman's policy was clear. On the same day in which a second atomic weapon was dropped on a Japanese city, President Truman declared that the East European countries were 'not to be spheres of influence of any one power'.[36]

Atomic Diplomacy

On August 6, 1945, Hiroshima was destroyed by atomic attack; 80 000 people were killed and as many injured by the explosion of just one bomb. Three days later Nagasaki was similarly destroyed. On August 14, Japan accepted United States terms for surrender on the assurance that the institution of the

Emperor would remain (the Western powers had originally insisted on unconditional surrender). The invasion of Kyushu planned for November 1, 1945, would now no longer take place. On August 9, President Truman reported that he had ordered the use of atomic weapons 'to shorten the agony of war in order to save the lives of thousands and thousands of young Americans'.[37] Yet this opinion was not shared by all American Chiefs of Staff. When the Secretary of War, Stimson, told General Eisenhower that the atomic bomb would be used against Japan, Eisenhower

> voiced to him my grave misgivings, first on the basis of my belief that Japan was already defeated and that dropping the bomb was completely unnecessary, and secondly because I thought that our country should avoid shocking world opinion by the use of a weapon whose employment was, I thought, no longer mandatory as a measure to save American lives. It was my belief that Japan was, at that very moment, seeking some way to surrender with a minimum loss of 'face'.

Eisenhower concluded that 'it wasn't necessary to hit them with that awful thing'. Indeed as Alperovitz has pointed out: 'Before the atomic bomb was dropped each of the Joint Chiefs of Staff advised that it was highly likely that Japan could be forced to surrender "unconditionally" without use of the bomb and without an invasion.'[38] It was Admiral Leahy's opinion, for example, that 'the use of this barbarous weapon at Hiroshima and Nagasaki was of no material assistance in our war against Japan. The Japanese were already defeated and ready to surrender...'.[38] According to the post-surrender study of the United States Strategic Bombing Survey: '... certainly prior to December 31, 1945, Japan would have surrendered, even if the atomic bombs had not been dropped, even if Russia had not entered the war, and even if no invasion had been planned or contemplated'.[39]

The evidence is therefore considerable (and has been documented by Alperovitz) that the atomic bomb was not primarily used against Japan in order 'to save American lives' but in order to intimidate the Russians into accepting American objectives in Europe. According to Szilard, one of the scientists responsible for Einstein's 1939 letter informing President Roosevelt about the military consequences of nuclear fission,* the American Secretary of State 'did not argue that it was necessary to use the bomb against the cities of Japan in order to win the war. ... Mr. Byrnes' ... view [was] that our possessing and demonstrating the bomb would make Russia more manageable in Europe....'[40] To his credit, however, the American Secretary of War, Stimson, had concluded by early September that he had been 'wrong' in attempting to gain political objectives by means of atomic diplomacy; it was, he wrote, 'by far the more dangerous course'. He attempted to convince President Truman of his error: American relations with the Soviet Union

---

* The role of the scientists in this 'atomic diplomacy' will be discussed in chapter 11. It should be noted that Byrnes interpreted the gist of the conversation rather differently.

might be 'perhaps irretrievably embittered by the way in which we approach the solution of the bomb with Russia. For if we fail to approach them now and merely continue to negotiate with them, having this weapon rather ostentatiously on our hip, their suspicions and their distrust of our purposes and motives will increase'. Stimson warned the President that 'unless the Soviets are voluntarily invited into the partnership [between Britain and the United States] upon a basis of co-operation and trust', there could be no avoiding 'a secret armament race of a rather desperate character'.[41] Stimson's advice was not accepted; the Joint Chiefs of Staff were all opposed to the idea of sharing atomic secrets with the Soviet Union.

3. The Cold War Years

Formalization of the Cold War

The Cold War was now on in earnest. As the arms race developed, so the Soviet grip on Eastern Europe tightened. And as Schlesinger has explained, 'American liberals' could not 'watch with equanimity while the police state spread into countries which, if they had not been real democracies, had mostly not been tyrannies either.'[42] He did not, however, see fit to explain why American liberals had tolerated, and continued to tolerate, dictatorships of the most brutal kind in Latin America, a continent of strategic importance to the United States. America and Britain could do as they wished in their 'spheres of influence'. Russia, however, was not to be allowed to have a sphere of influence and thereby to achieve effective control of those countries through which Western invaders had so often passed. On March 5, 1946, Churchill, with President Truman at his side, declared in his famous Fulton speech that an 'iron curtain' from Stettin to Trieste now separated the free countries of Western Europe from the 'police governments' in the East.[43] Churchill demanded the 'establishment of conditions of freedom and democracy as rapidly as possible in all countries'; the Anglo-Saxon countries must negotiate with Russia from a position of overwhelming strength, not of weakness.

From then on events assumed the course that Stimson had feared. The Baruch plan for the international control of atomic energy of June, 1946, was rejected by the Russians since the first stage of the proposed plan would give the West detailed information about the military preparedness of the Soviet Union while only in the last stage—should it ever be reached—would American manufacture of atomic weapons cease and destruction of stock piles begin. The Soviet Union was suspicious of the West's intentions. Speeches expressing hatred of the Soviet Union and readiness to subject Russia to atomic blitz were common. Fleming writes that in February, 1946, the President of the National Industrial Conference Board, an organization supported by large corporations, made a speech to a 'distinguished audience', including many prominent American industrialists.[44] In it he declared that Russia is a 'primitive, impoverished, predatory Asiatic despotism', resting on 'a vast pyramid of human skulls', resourceful only in 'insolence, intrigue, treachery and terrorism' which she had used to the utmost 'in the forced march of communist imperialism'. To counter the communist menace, as he saw it, Jordan proposed that America should offer economic aid to all nations who would unilaterally disarm, demand the unlimited right of continuous inspection and control of all industrial processes which could, however remotely, be used for armament manufacture, and finally 'suspend in principle' improved atomic weapons 'over every place in the world where we have any reason to suspect evasion or conspiracy against this purpose; and let us drop them, in fact, promptly and without compunction wherever it is defied.'

The war hysteria building up or, rather, being built up in the United States was, however, repudiated by a member of President Truman's Cabinet, Secretary of Commerce Wallace. In September, 1946, he declared that 'the real peace treaty we now need is between the United States and Russia. On our part, we should recognize that we have no more business in the political affairs of Eastern Europe than Russia has in the political affairs of Latin America, Western Europe and the United States.'[45] At the same time, Wallace made public a letter he had sent to the President. In it he asked how recent American actions must appear to other nations:

> I mean by actions the concrete things like $13 billion for the War and Navy Departments, the Bikini tests of the atomic bomb and continued production of bombs, the plan to arm Latin America with our weapons, production of B-29's and planned production of B-36's and the effort to secure air bases spread over half the globe from which the other half of the globe can be bombed. I cannot but feel [he continued] that these actions must make it look to the rest of the world as if we were only paying lip service to peace at the conference table. These facts rather make it appear either (1) that we are preparing ourselves to win the war which we regard as inevitable; or (2) that we are trying to build up a predominance of force to intimidate the rest of mankind. How would it look to us if Russia had the atomic bomb and we did not, if Russia had 10 000 mile bombers and air bases within 1 000 miles of our coastlines, and we did not?

His conclusions were that:

> We should make an effort to counteract the irrational fear of Russia which is being systematically built up in the American people by certain individuals and publications. . . . We should not act as if we, too, felt that we were threatened in today's world. We are by far the most powerful nation in the world, the only Allied nation which came out of

the war without devastation and much stronger than before the war. Any talk on our part about the need for strengthening our defences further is bound to appear hypocritical. *

Three days later Wallace was dismissed. The President declared complete support for his Secretary of State, Byrnes. That evening, Wallace said in a radio talk that he was against all types of imperialism and aggression, whether they were of Russian, British or American origin. Such a man was unwelcome in President Truman's cabinet.

The Truman Doctrine

On March 17, 1947, President Truman launched the military campaign to 'quarantine' the Soviet Union. In a speech to the two Houses of Congress, the President declared the time had come when 'nearly every nation must choose between alternative ways of life', between freedom and tyranny. The President was convinced 'that it must be the policy of the United States to support free peoples who are resisting attempted subjugation by armed minorities or by outside pressure.'[46] From now on the world would be divided: those countries in favour of the United States and those against. From now on, under the banner of 'defence of the free world', the United States would openly oppose all revolutionary change.

Containment and NATO

There could, it seemed, be no turning back from the perilous path recommended by Churchill and Truman. From the famous July, 1947, article by George Kennan in *Foreign Affairs*, in which the Director of the Policy Planning Staff of the State Department recommended 'containment' of the Soviet Union, through the Soviet-inspired coup in Czechoslovakia (the gateway to the Soviet Union), to the creation of NATO in 1949, the international situation inexorably deteriorated. It was very late in the day when Kennan admitted in a 1965 lecture that 'it was perfectly clear to anyone with even a rudimentary knowledge of the Russia of that day, that the Soviet leaders had no intention of attempting to advance their cause by launching military attacks with their own armed forces across frontiers'.[47] They would not do this, he explained, because it 'fitted neither with the requirements of the Marxist doctrine, nor with Russia's own urgent need for recovery from the devastations of a long and exhausting war, nor with what was known about the temperament of the Soviet dictator himself'. He further explained that he had privately criticized the formation of NATO as a 'military

* It is of interest that on September 18, Einstein wrote to Wallace: 'I cannot refrain from expressing to you my high and unqualified admiration for your letter to the President. You have shown deep appreciation of the factual and psychological situation and real understanding of the fateful consequences of present American foreign policy. Your courageous intervention deserves the gratitude of all of us who regard the present position of our government with grave concern.'[45n] (An account of Einstein's analysis of the Cold War is given on pp. 342–8.)

defence against an attack no one was planning'. In fact, he thought NATO could have been planned only by 'people capable of envisaging a favourable future for Europe only along the lines of a total military defeat of the Soviet Union or of some spectacular, inexplicable and wholly improbable collapse of the political will of its leaders'. From the moment that NATO was formed, 'the peaceful solution of Europe's greatest problems on any basis other than that of the permanent division of Germany and the continent, with the implied consignment of the Eastern European peoples to inclusion for an indefinite period in the Soviet sphere of power, became theoretically almost inconceivable . . .'.

Thus Kennan publicly spoke in 1965; he did not so speak in 1948–9 as American military officers advocated an atomic blitz on the Soviet Union. In a public speech in May, 1949, Lt-General Doolittle warned that Americans had to 'be prepared, physically, mentally and morally, to drop atom bombs on Russian centres of industry at the first sign of aggression. She must be made to realize that we will do so, and our own people must be conditioned to the necessity for this type of retaliation.'[48] In the previous year, however, the atomic blitz theory had been strongly attacked by the English Nobel prize-winning physicist, P. M. S. Blackett. Deploring the fact 'that American opinion, both military and civilian, in marked contrast with Russian, seems to have accepted the use of tactics of mass destruction as a normal operation of war', Blackett warned the American military planners that even if they could achieve the mass destruction of Russian cities, this would certainly not prevent Soviet armies in retaliation from 'over-running Europe, the Middle East and much of the Far East'. In any case, his opinion was that an atomic attack on the Soviet Union using 'existing heavy bombers against the fighter defence that the Soviet Air Force may reasonably be expected to possess', was by no means certain of achieving its insane objective.[49]

On September 23, 1949, President Truman announced that the Russians had exploded their first atomic weapon.

Socialism in one country?

Through the 1930s the Soviet people had laboured under the direst conditions to lay an industrial base that would not only ensure the defence of the Soviet Union but would also guarantee in the future a material standard of living on which they believed socialist society could be constructed. At the end of the war, however, although the Soviet people had defeated the Nazi armies, the Soviet Union once again lay in ruins. The Nazis had occupied Soviet territory in which nearly 90 million people lived; over 15 million of them were killed.[50] The Nazis destroyed, completely or partially, 15 large cities, 1 710 towns and 70 000 villages; 60 million buildings were destroyed and 25 million people made homeless. Nearly 32 000 industrial enterprises were demolished, 10 000 power stations, 3 000 oil wells, over 1 000 coal mines, 65 000 kilometres of railway track and 86 000 miles of main highway. The Nazis transported to

Germany 14 000 steam boilers, 1 400 turbines and 11 300 electric generators. Nearly 100 000 collective farms were sacked. The Germans either slaughtered or carried with them 7 million horses, 17 million cattle, 20 million hogs, 27 million sheep and goats, 110 million poultry. Some 40 000 hospitals and medical centres were destroyed, 64 000 schools and colleges, 43 000 public libraries with over 100 million books, 44 000 theatres, nearly 3 000 churches and over 400 museums. In the first population census after the war there were, over the age of 18, only 31 million men compared with 53 million women. As Isaac Deutscher has remarked, 'For many, many years only old men, cripples, children and women tilled the fields in the Russian countryside.'[51] It did not need Kennan's expert insight to realize that this crippled nation did not seek war with that nation which emerged from the conflict stronger than ever before, armed with atomic weapons at that. The Russians first and foremost sought security. The most important need was to rebuild the Soviet Union. From the end of the war until the proclamation of the Truman Doctrine in 1947, the Russians had demobilized their armies to such an extent that by 1947 less than 3 million men remained in uniform out of an army of 11·5 million. But once again there could be no thought of building socialism; at most of rebuilding that minimum industrial base which would make socialism *possible*—provided, of course, that the Soviet Union would be left in peace by its wartime capitalist allies.

It was, of course, not to be. Once again a crippled, devastated Soviet Union had to try to match the world's most advanced capitalist nations in the production of scientifically sophisticated armaments. Until the Soviet Union possessed a stockpile of atomic weapons, the main defence against a Western atomic blitz was to remobilize armies and mass them on the frontiers of the West. In a country deprived of the basic necessities of life, once again priority had to be given to military defence. Once again terror was inflicted on millions of Soviet people; the bureaucrats had ever more reason to try to ensure the preservation of their very substantial privileges. Yet, on the other hand, in the United States the rearmament programme dispelled all fears of a major depression. The Cold War would sustain prosperity. Indeed in 1949, for example, the Harvard economist, Professor Slichter, was 'warmly received' when he told a bankers' convention that provided the Cold War persisted, a severe depression was 'difficult to conceive'.[52] For the Cold War, he explained, 'increases the demand for goods, helps sustain a high level of employment, accelerates technical progress and thus helps the country to raise its standard of living. . . . So we may thank the Russians for helping capitalism in the United States work better than ever.' Nearly ten years later (*The Nation* relates) the former Deputy Secretary of Defense under Truman, W. C. Foster, would advocate an increase in America's military budget from 10 to 20% of GNP, arguing (in the *General Electric Defense Quarterly*) that such an increase would compel the Russian leaders to do the same and thus would force them to 'take away from their people one-third of the already sparse good things of life they have.'[53] In 1961 President Kennedy appointed W. C. Foster to be the Director of the new United States Arms Control and Disarmament Agency.

4. Some Concluding Comments

Our historical sketch of the development of the Soviet Union does not suggest that socialism has been tried out and has failed. On the contrary, it suggests that the Soviet Union has never been in a position to develop socialism. For socialism (as we earlier emphasized) does not merely mean the successful laying of an industrial base—although to be sure science and technology must be developed to the point where all citizens can enjoy good health and the basic necessities of life. It certainly does not mean that all aspects of life and industrial production must be planned by a bureaucratic élite. Socialism (as I am using the term) means above all, not merely public ownership of the means of production, essential though this is, but a way of life, an ethos, in which cooperation for the common good allows a *diversity* of life-styles to develop and flourish, an ethos that encourages a richness of emotional, intellectual and spiritual life rather than an emphasis on ever-increasing material welfare and achievement.

Starting, therefore, from an industrially underdeveloped position and forced into economic and military competition with the West, the Soviet Union is far from the achievement of such a humanistic socialist society. Among other shortcomings, the society is far from non-exploitative, dissenters to Soviet orthodoxy are cruelly repressed, and science and technology appear to be worshipped for their own sake rather than for the benefits they can bring to a suffering humanity. (Despite the high prestige of science in the Soviet Union scientific dissenters are not exempt from repressive measures; Medvedev, for example, was temporarily committed to a psychiatric hospital because of his criticism of Soviet bureaucracy in general and Soviet administration of science in particular.)

What our historical sketch does suggest with respect to the *relative* failure of the Soviet Union is that the Western nations have been primarily responsible for the very unfavourable environment in which Soviet development has occurred. Thus combined with the sketches of previous chapters, our analysis suggests not only that the capitalist nations have been and still are *primarily* responsible for an increasingly troubled world situation but that a more just and humane world cannot be built until the social structure of capitalism and its intrinsically exploitative ethos have been transcended. While Western capitalism still exists countries aspiring to socialism are forced to compete in an arms race with the West either with the aim of achieving a 'detente' with the Western powers or with the aim of more effectively opposing Western imperialism. In neither case is the realization of humanistic socialism possible. Apart from the corruption of socialist ideals, 'socialist' countries with different policies towards Western imperialism inevitably become suspicious of each other's intentions. (The Russian refusal to supply China with atomic weapons is certainly one of the reasons for the present enmity between the two 'socialist' countries.) In any case an environment of atomic, bacteriological and chemical weapons is not recommended by most 'utopian' thinkers as particularly conducive to development of the 'good life'.

It is therefore our turn to act; instead of producing waste at home, and perpetuating oppression and hunger in the underdeveloped countries, we could attempt to transcend our own capitalist societies and develop a social structure based on a cooperative ethos and non-commercial values. In doing so, not only would we be able to vastly improve the quality of our own lives but we would be able to help—not hinder—the attempts of the peoples of the underdeveloped countries to eradicate illiteracy, poverty and disease, and so bring the world nearer to the socialist unity it must eventually acquire if man is to be able to continue the only worthwhile quest he has set himself— the building of a beautiful world.

Of course, the problems of transcending the capitalist social order are formidable enough. And it is far from certain that the death throes of capitalism will not result in the death throes of the whole world. Nevertheless, if it is possible to be fairly certain that capitalism is by far the *major* cause of the problems facing mankind then a research-action programme could be articulated based on such a conviction. Central to such a programme would be the thesis that if science is to serve mankind then the social structure of capitalism must be transcended and a socialist society constructed in which, no matter what the difficulties of planning (and they would be considerable), science would be deliberately put at the service of people rather than (as now) at the service of autonomous large-scale corporations whose sole concern is to achieve the maximization of their own growth and profit rates.

However, such a programme, complicated and uncertain of success though it is, has recently come under attack by radical critics of advanced industrial society. These critics argue that science by its very nature cannot be put at the service of mankind and that any attempt to build a society that has the scientific enterprise as a central institution will inevitably become a dehumanized and depersonalized society of *thing*-beings rather than *human* beings. These critics argue that not only must capitalist society be rejected (together with bureaucratic 'socialism') but, and even more important, the scientific enterprise itself must be brought to a speedy end. Such an attack thus comes as almost a mortal blow to humanistic socialist thinkers who for long have put their faith in the possibility of building a society in which 'science will serve the people'. In the next chapter we therefore look in considerable detail at the ethos of science and at the arguments of some of its principal critics.

Chapter 10

ON THE ETHICAL NEUTRALITY OF SCIENCE — THE INDICTMENT OF THE SCIENTIFIC MENTALITY

> There is but one interest common to the whole of humanity, the progress of the sciences.
> SAINT-SIMON *Letters from an Inhabitant of Geneva to his Contemporaries*, 1803

> I hate and fear 'science' because of my conviction that, for long to come if not for ever, it will be the remorseless enemy of mankind. I see it destroying all simplicity and gentleness of life, all the beauty of the world; I see it restoring barbarism under a mask of civilization; I see it darkening men's minds and hardening their hearts; I see it bringing a time of vast conflicts, which will pale into insignificance 'the thousand wars of old', and, as likely as not, will whelm all the laborious advances of mankind in blood-drenched chaos.
> GEORGE GISSING *The Private Papers of Henry Ryecroft*, 1903

1. The Rejection of Science

AT the end of chapter 3 we suggested that in a civilized world physics could well become a kind of delightful play, providing men with nourishment for their intellectual and aesthetic faculties. But in this far from civilized world, such a statement appears almost grotesque in the apparent lack of comprehension it betrays concerning the nature of the scientific quest. For scientists, and physicists in particular, have come to find themselves regarded as indulging not in a delightful play but in a degrading and life-reducing activity whose consequences are the appalling dangers to which mankind is now exposed. The nuclear arms race, it is unkindly pointed out, does not result from the work of village craftsmen and farmers. On the other hand, displeased by, as they see it, such irrelevant and politically naïve criticism, leading spokesmen for the scientific community reply that in a world of nation states in conflict, it is inevitable—and proper—that the results of scientific research should be applied by scientists in the defence of their own countries. It would, they claim, be grossly irresponsible to do otherwise. It is the responsibility of politicians—of governments—to end the present state of conflict and bring about those social conditions in which military applications of science would be no longer necessary.

At a more mundane level it is also pointed out that, like everyone else, scientists have to eat and must therefore earn their daily bread by placing their science at the service of the appropriate ruling élite. If the resource-allocating élite has military needs, so be it—the would-be scientist will be able to earn his bread and butter, together with his equipment, only if he is prepared to satisfy those needs. A cynic could well attempt to argue that Galileo's claim to the title of the Father of Modern Science could be defended more firmly and less controversially on the grounds of his prompt and lucrative sale of the telescope as a war instrument to the Venetian Senate rather than by pointing to the scientific discoveries he made with it. Galileo certainly makes the dependent status of scientists very explicit in his employment-seeking letter to the Grand Duke of Tuscany's Secretary of State: 'Great and remarkable things are mine but I can only serve (or rather, be put to work by) princes, for it is they who carry on wars, build and defend fortresses, and in their royal diversions make those great expenditures which neither I nor other private persons may.'[1] To back up his case, Galileo tells the Secretary of State that he has

> in mind the writing of some books about military matters, setting these forth not merely theoretically but showing by very elegant rules everything in that science which depends upon mathematics, such as the practice of fortification, ordnance, assaults, sieges, estimation of distances, artillery matters, and the uses of various instruments and so on. Of course I must also reprint my instructions for the use of the military compass . . .

We see here, of course, emphasis on science to satisfy military needs rather than manufacturers'. Italy however was not the first country to become capitalist; that honour fell to England and it was in England that the new science was to flourish most. Significantly enough, Francis Bacon, for whom science was to achieve 'the effecting all things possible' and whose goal was 'the relief of man's estate', did not argue in his scientific utopia, as More had done a century before him and as had his contemporaries, Andreae and Campanella, that private property should be abolished. For science was to be done either by or under the patronage of gentlemen of independent means or in the service of commercial interests, a feature of the new science we noted in chapter 4. It surely cannot come as a surprise that in an emerging capitalist society scientists recognized that their own interests lay in serving the interests of the emerging bourgeoisie and thus in allowing their exploitable new science to be exploited in the interests of private profit.

If this were substantially the whole story then there could be reasonable hope that as the awareness increases that capitalist society is inherently an exploitative system, manipulative of people in particular and nature in general, but that as the means develop (thanks to both capitalism and science) whereby material welfare for all becomes in principle possible, so making the continued existence of scarcity—the very foundation stone of a repressive, exploitative society—a scientific absurdity, so scientists would join forces with other groups in the attempt to transcend capitalist society and construct a non-exploitative, non-manipulative human society. Scientific knowledge is ethically and politically neutral, so the conventional wisdom runs. While in a capitalist society science must inevitably be used for exploitative ends, in a society that has transcended the capitalist stage in the direction of greater human freedom the very same science could be put at the service of human love and freedom just as readily as it had been previously put at the service of private profit.

So it is comforting to believe. But in recent years very different theses have been increasingly argued, culminating in the most radical of theses that in order to build a truly human world, a world responsive and receptive to love, beauty and freedom, the practice of science would have to be discontinued. For, it is argued, the image of nature that necessarily underlies the practice of any science reinforces a pattern of mind and behaviour which automatically justifies and absolves 'even the most destructive and oppressive features' of a manipulative society. Even to expect scientists to agree to help build a society of freely creative and cooperative human beings is as unrealistic as to expect capitalists to agree to help build a society in which there would be no right to private ownership of the means of production. Such action by either group would amount to the signing of its own death warrant. Scientists, it is argued, necessarily view nature—and other people—as manipulable, controllable. It is therefore not in the interests of scientists to help create that kind of society whose immediate goal is the minimization of all forms of manipulation and control and whose ultimate goal is their total elimination.

This indictment is radical indeed. If it is believed the implications will be far reaching for the scientific community to say the least. For presumably it is not by chance that around the necks of dissenting young in advanced capitalist countries can be found beads and talismans, symbols of an anti scientific mentality. Indeed Hilary and Steven Rose claim that 'the whole hippy flower-power sub-culture represents a total rejection of science and rationality', that 'comprehension is sought instead through mystical experience supported by hallucinogenic drugs', and that 'a meta-language system is sought to describe the one-ness and unity with nature, which by-passes the scientific job of analysing it rationally.'[2] It should be noted that when America's dissenting young marched on the Pentagon some of them chose to mock this hated symbol of a scientific civilization and its 'culture of death' with a symbolic levitation attempt accompanied by magical incantations. It is obviously not by chance that Theodore Roszak, a leading spokesman for the 'counter culture'—the culture of life, openly espouses a magical vision of nature, and welcomes the much discussed increasing hostility of the young to rational and scientific thought.[3] Speaking of the student movement at Berkeley a leading molecular biologist has emphasized (as he sees it) that 'the really radical aspect of the new student mentality was not the superficially obvious and by no means novel attitude of social protest but its underlying *anti-rational* basis'.[4] Lamented a University Senate Report: 'Students who hold the belief

ON THE ETHICAL NEUTRALITY OF SCIENCE

that feeling is a surer guide to truth than is reason cannot readily appreciate the University's commitment to rational investigation.'[5] Such comments appear to be almost universal. In an address, for example, to the Society for the Study of Social Problems the 1967 President saw fit to condemn the young for, in his words, 'their deliberate and suicidal rejection of intellect and reason'.[6]

This book opened by citing some of Popper's statements that reason must rule the world, not emotion, and that since scientific thinking *is* rational thinking *par excellence*, it is science that must be the foundation stone of civilized society. Many of the dissenting young appear to think otherwise. If life is to be preserved on this planet, if a world society based on freedom, love and tenderness is to be achievable, then an essential prerequisite must be the rejection of the scientific image of nature and of any practice based on this image. For in some very deep sense, it is implied, a society that practises science in any form will eventually become—if that science is not rejected—a society hostile to life itself.

Such a thesis is an astonishing one. How could it possibly stand up to elaboration and criticism? Surely it amounts to nothing more than gross exaggeration of a few well-founded complaints about unethical scientific practice? The thesis is all the more astonishing when one recalls Francis Bacon's declaration that the purpose of science is 'the relief of man's estate'. Indeed did not all the founders of the 'new science' similarly proclaim their belief that science would at long last enable man to free himself from a life of poverty, disease and toil. They at least did not suppose that science is in its very nature hostile to the life of man. On the contrary; but was this not their very great error?

The questions raised by such attitudes towards science, even in their weakest form—that science is not ethically neutral—are complicated indeed. Nevertheless, they are ones with which we must attempt to come to terms if only tentatively. Several methods of approach are possible. The one adopted here is to look at those historical roots of the new science that not only focus attention on essential features of the new image of nature, together with the motivations of its protagonists, but which directly lead to a consideration of central issues in the 20th-century indictment of science.

Let me first, however, state some of my own beliefs more clearly.

If men are to succeed in building a world that is more just and more humane than this one, free as much as possible from pain, disease and toil, not a world characterized by exploitation and the treatment of people as things, but rather by comradeship and cooperation between people, by joy and delight in the offerings of nature and of life, then at no time in the struggle to build such a world must sight be lost of all those human values that alone make the struggle worthwhile and the present suffering bearable. Capitalist society must be transcended not in order to take revenge on those men who most manipulate and exploit, but because too many people are living in pain and deprivation, victims of capitalism and its consequences, denied the right to fulfil themselves as complete human beings, above all because no one whether exploiter or exploited is able to be truly human in a capitalist society. Nature, which not only inspires but also oppresses men, is not so much to be dominated and exploited by scientific man as understood and interacted with so that her oppressive ways can be increasingly made to yield to those of her ways that nourish life. Above all, knowledge is to be sought not in the interests of power for its own sake, but so that men in liberating themselves, will be able to live in harmony with each other—and, as much as possible, with nature as well.

Why is it that scientists seek knowledge? *Any* knowledge? Clearly not. Only *interesting* knowledge, *relevant* knowledge. But relevant to whom, for what ends? Is the scientists' goal liberation for themselves and all men? If not, what is their goal? Let us reflect on the words of Marlowe's *Tamburlaine* (bearing in mind the ambitions of the same author's *Doctor Faustus*):

> Nature, that framed us of four elements
> Warring within our breasts for regiment,
> Doth teach us all to have aspiring minds.
> Our souls, whose faculties can comprehend
> The wondrous architecture of the world,
> And measure every wandering planet's course,
> Still climbing after knowledge infinite,
> And always moving as the restless spheres,
> Will us to wear ourselves and never rest,
> Until we reach the ripest fruit of all,
> That perfect bliss and sole felicity,
> The sweet fruition of an earthly crown

If scientists are not seeking an 'earthly crown', do not see nature and other men as so much subject matter for experiments, then it is possible that their image of nature and man, and the implied action programmes, will not only reflect their desire for liberation but also their belief in its possibility. And conversely.

2. The 'New Science' and its Image of Nature

By the end of the 15th century man no longer viewed himself as merely a pious spectator of God's works but as an active participant in nature's processes—a creature who could, by gaining knowledge that would give him power over nature, make of himself what he willed. It was this transformation in man's view of himself that had helped engender the scientific revolution and that had itself been strengthened by the astonishing scientific successes of the 16th and 17th centuries. At long last it had become possible to think in terms of making brute labour unnecessary and of eradicating disease. Nature, for so long the ruthless oppressor of men, was to be mastered by means of the new science. Whereas the Aristotelian image of nature and its accompanying teleological explanations had not resulted in the gaining of power over nature, it was to be one of the central aims of the new science, in Bacon's words, to ensure 'the enlarging of the bounds of Human Empire', indeed to achieve 'the effecting of all things possible'. In Bacon's eyes, science was to regain for man that power over nature granted him 'by divine bequest' but which had been lost at the Fall. To this end, scientific knowledge became *redefined* by Bacon as knowledge that would lead to power over nature—knowledge of causes of phenomena was what was required, not knowledge of purposes and goals in nature. Aristotle's belief in purposive behaviour in nature was therefore mocked by Bacon: teleology, he wrote, 'like a virgin consecrated to God, produces no offspring'.[7] Acceptable scientific explanation was now causal, not teleological. Truly could Bacon now write: 'Human knowledge and human power meet in one; for where the cause is not known the effect cannot be produced'.[8]

The men of the new science, explicitly aware of its practical implications, greatly welcomed them. Descartes in particular was typical of this new spirit of optimism and confidence in the future. He could not, he declared, keep his discoveries secret 'without sinning against the law which lays us under obligation to promote, as far as it lies within us, the general good of mankind'. And how could this general good of mankind be brought about?

Descartes answers:
> It is possible to obtain knowledge highly useful in life, and that in place of the speculative philosophy taught in the Schools, we can have a practical philosophy by means of which, knowing the force and actions of fire, water, air, the stars, the heavens and all the other bodies that surround us as distantly as we know the various crafts of the artisans—we may in the same fashion employ them in all the uses for which they are suited, thus rendering ourselves masters and possessors of nature.[9]

Masters and possessors of nature! With the hindsight of three centuries and with the contemporary indictment of science in mind, Descartes' five dramatic words compel us to ask the following question: What moral right had the new scientists to assume such an arrogant objective? Their answer— 'every right'. Had not nature always been a ruthless oppressor of man? Had not one-third of Europe's population been wiped out by the 'black death' alone? Were not men still living in the most terrible poverty?

But above all—a feature of modern science that is of far-reaching importance—the scientists had given themselves a new image of nature that appeared to justify not only any kind of intervention in nature's affairs but also the rejection of teleological explanation as unscientific. For no longer was the world of nature conceived to be the living, feeling, thinking organism that Rennaissance alchemists and Hermeticists had believed themselves living in. For the new scientists it became the very opposite of such a world—a world of mere matter in motion, nothing more than this. The purposeful strivings of matter to achieve perfect forms (as envisaged by Aristotle) no longer took place; the Aristotelian hierarchical order was declared to have been but an illusion: matter 'in reality' was the same everywhere in the infinite universe, unfeeling matter, moving without intelligence, without purpose. Obviously a nature such as this demands no moral self-examination on the part of its users.

For the magicians and Hermeticists, on the other hand, nature was alive, permeated throughout by a world-soul which made everything, including matter, alive and sentient to some degree. The whole of nature was linked together by ties of sympathies and antipathies. And by gaining an insight into these secret properties of nature, the Hermetic magician believed he could 'marry' them together to produce works of great benefit for man.* Wrote Pico: 'As the farmer weds his elms to the vines, so the "magus" unites the earth to heaven, that is, the lower orders to the endowments and powers of the higher.'[10] By practising such magic, argued Pico, 'we shall more ardently be moved to love and worship God in his works, until finally we shall be compelled to burst into the song "The heavens, all of the earth, is filled with the majesty of your glory".' A sense of the sacred appeared to be built into the very heart of the Hermetic view of nature.

But for the men of the new science nature lay exposed—a fraud. The Hermeticists had been deceived. Campanella, in declaring all of nature to be alive, had ridiculed Greek and Roman atomists: 'Behold Lucretius the Epicurean attempting to show, with Democritus, that insentient and inert things can give rise to things with sense and feeling! Maintaining that from non-weeping, non-laughing elements, men, who weep and laugh, can come into being.'[11] But for the new scientists not only was nature not alive but not even tastes, colours and odours belonged to it any longer. As Democritus had done two thousand years before (only to be severely criticized by Aristotle), Galileo, the father of the new science, insisted that 'tastes, odours and

---

* To be sure, it was thought that knowledge of the secret properties of nature could also be used to attain great personal power over other men and nature and that to achieve this end men—and women—would be tempted to join forces with the Prince of Darkness. From Ficino to della Porta the Hermetic philosophers and natural magicians inveighed against the practice of such demonic magic. When John Webster in the middle of the 17th century advocated the teaching in English universities of Hermeticism and the works of the Paracelsian Hermeticist Robert Fludd he took care to condemn 'that impious and execrable *Magick*, that either is used for the hurt and destruction of mankind, or pretends to gain knowledge from him who is the grand enemy of all the sons of Adam.'[9n] Yet, so it was believed, men and women throughout the 16th century had turned in increasing numbers to the worship of the Devil, to the practice of demonic magic, of necromancy and sorcery; and throughout the 16th century and beyond these men and women—and even children—were hunted down, tortured until they confessed their evil practices and then burnt. By 1630 when the slaughter was at its height lawyers, judges and even clergy joined witches first in the torture chambers and then at the stake. Some forty years earlier, in 1587, the Faust book had been published informing the unwary of the terrible end of the necromancer Johanne Faustus who had made a pact with the Devil. Within a few years the book had been translated into several European languages and the Faust legend had been born.

ON THE ETHICAL NEUTRALITY OF SCIENCE

colours and so on are no more than mere names so far as the object in which we place them is concerned, and that they reside only in the consciousness'.[12] The conclusion automatically follows and Galileo made it: 'If the living creature were removed, all these qualities would be wiped away and annihilated.' What would be left would be the real world of matter in motion. Wrote Galileo: 'To excite in us tastes, odours and sounds I believe that nothing is required in external bodies except shapes, numbers, and slow or rapid movements.' It was therefore necessary to know only these 'primary qualities'. Thus wrote Descartes to his friend Mersenne, expressing in a nutshell the 'reductionist' faith: 'If anyone could know perfectly what are the small parts composing all bodies, he would know perfectly the whole of nature.'[13]

Side by side with the belief that the real world was one of matter in motion went the belief that in some sense the external world could be likened to a giant machine, a clockwork, designed by God, 'the Divine Engineer', for the convenience of human beings.[14] The later Middle Ages' 'passion for the mechanization of industry'* had produced a notable impact on the minds of the new scientists. 'At one time,' wrote Kepler, '. . . I believed that the motor causes of the planets was a soul. . . . The aim that I have set myself here is to affirm that the machine of the universe is *not* similar to a kind of divine, animated being but similar to a clock.'[15] For Robert Boyle the universe became 'a great piece of clockwork', with all phenomena explicable in terms of 'the two great and universal principles of bodies: matter and motion'.[16] Wrote Henry Power, 'the Management of this great Machine of the World must needs be the proper Office of only the Experimental and Mechanical Philosophers'.[17] The machine had so successfully replaced the organism as a model for the conception and understanding of nature that Descartes, who had rejoiced in the hope that men could become 'masters and possessors of nature', argued not only that animals possessed no souls but that they were in principle no different from extremely complicated machines. Although it is not clear that Descartes truly believed that these complicated machines were totally devoid of sentience, some of his followers appeared to experience no doubts: live animals were nailed to boards and opened up to exhibit their blood circulations. A nature of mere matter in motion has no rights. It is not a nature that rational men, after liberating themselves from its domination, will seek to commune with (as had the Hermetic philosophers and natural magicians)—will seek to love through ever-greater understanding of its ways. It is a nature that is there for men to use as they will.

There was certainly resistance to this mechanization of the world view. While the new scientists sought to liberate themselves from remnants of the

---

* 'From the 11th century to the end of the 15th', writes Cipolla, 'European technology moved ahead in almost every field—in agriculture as well as in the building industry, in navigation and in ship building, in the textile industry, in metallurgy, in carpentry, in accounting, in finance, in transportation, in the production of energy and warfare.'[14n]

Aristotelian and Hermetic images of nature, they were at the same time being attacked by the Aristotelians from within the universities and by the Hermetic philosophers and magicians from without. The resistance was to no avail. The Aristotelian paradigm had been struck a blow by Copernicus from which it was never to recover. In addition, not only did the Aristotelian paradigm appear intellectually inadequate but, as we have seen, it also appeared socially unacceptable. Bacon's verdict was harsh on Aristotle and the Schoolmen: 'Such teachings, if they be justly appraised, will be found to tend to nothing less than a wicked effort to curtail human power over nature and to produce a deliberate and artificial despair. This despair in its turn confounds the promptings of hope, cuts the springs and sinews of industry and makes men unwilling to put anything to the hazard of trial.'[18] As for the Hermetic philosophers, alchemists and magicians, Mersenne and the atomist Gassendi launched into a successful attack against their belief in an animistic universe. Even the mystical Kepler attacked the views of his even more mystical contemporary, the Hermeticist Robert Fludd.[19] Bacon wrote of the magicians that he was unsure 'whether we ought rather to laugh at them or to weep.'[20] To Webster's plea that English universities should turn to the teachings of Renaissance magicians and 'that profoundly learned man Dr. Fludd', the Oxford astronomer Seth Ward angrily replied that it would be utter foolishness to 'dwindle after the windy impostures of Magick and Astrology'.[21] The Hermetic world view crumbled, together with the Aristotelian image of nature. The mechanization of the world view had achieved its first important victory.

Thus by the end of the 17th century the belief had established itself that men were living in a world of matter in motion,* a completely purposeless world of manipulable matter, controllable and exploitable by men, and in particular, let us now remind ourselves, by the emerging bourgeoisies of Europe at whose service the men of the new science would be so willing to place their talents (a feature of the new science we noted in chapter 4). Furthermore the men of the new science had provided an image of nature totally appropriate for that emerging society in which nature — and other men — were to be exploited for the enhancement of private profit and personal power, for a society in which working-class men would become 'appendages to machines' and be referred to as 'hands'. Whether causally related or not, the change in society from feudal to capitalist was accompanied by a change

---

* To be sure, Newton's postulation of a gravitational attraction between bodies was believed by Cartesian philosophers to be a retrogressive step, the reintroduction of occult, non-mechanical properties into nature. The Newtonians replied that merely the mathematical form of the force had been accounted for, not its material nature. In addition they were able to argue that such a force clearly existed, as the successes of Newtonian theory showed. The mechanistic image of nature had therefore not been tarnished with occult forces but refined and strengthened by the inclusion of a *real* physical force whose effects could be, and were being, successfully predicted (see, however, Newton's letter to Bentley, quoted at the end of chapter 2).

ON THE ETHICAL NEUTRALITY OF SCIENCE

in the image of nature from organismic-animistic to mechanistic. As if symbolizing these two changes, we find that seven years before capital punishment was abolished in England for the crime of witchcraft (a social activity credible only within a magical world view) capital punishment had been introduced for the breaking of machines. While undoubtedly the Hermeticists and alchemists (who had usually been among the enemies of the witch-craze) would very much have approved of the former act, certainly some of them would have strongly disapproved of the type of society in which men were driven to the desperate act of destroying machines. For in the 'ideal' societies of Andreae and Campanella there was to be no private property, all men and women were to be educated, no one would have to work more than a few hours a day. In a properly organized society it would not be necessary for men to exploit each other. The capitalist society of England would have appalled the hearts and minds of Andreae and Campanella. But then they were not mechanical philosophers—they were men steeped in a totally different tradition at whose core was a very non-mechanical image of nature.

To give additional emphasis to this point, let us note how the ethical implications of adopting a mechanistic image of nature are exemplified in the words of an old woman from a culture whose image of nature is not unlike that envisaged by the Hermetic philosophers, the animistic vision to which the new scientists were so opposed. This Wintu Indian woman (from California) is not appreciative of the white man's attitude to nature:

> The white people never cared for land or deer or bear. When we Indians kill meat, we eat it all up. When we dig roots, we make little holes. . . . We shake down acorns and pine-nuts. We don't chop down the trees. We only use dead wood. But the white people plough up the ground, pull up the trees, kill everything. The tree says, 'Don't. I am sore. Don't hurt me.' But they chop it down and cut it up. The spirit of the land hates them. . . . The Indians never hurt anything but the white people destroy all. They blast rocks and scatter them on the ground. The rock says, 'Don't. You are hurting me!' But the white people pay no attention. When the Indians use rocks, they take round ones for their cooking. . . . How can the spirit of the earth like the white man? . . . Everywhere the white man has touched it is sore.[22]

In comparing the old Indian woman's attitude to Nature with Bacon's call to men to unite 'forces against the Nature of Things, to storm and occupy her castles and strongholds, and extend the bounds of human empire'[23] the point is forcefully made that a society with such a magical-animistic image of nature will have little ambition to become 'masters and possessors of nature' and that if it is to evolve into a capitalist society at least one prerequisite will be a change in its image of nature. Such a change was adopted in the 16th and 17th centuries by those European countries standing on the threshold of the capitalist mode of production.

3. The Mechanistic Imperative

Yet the triumph of the mechanistic world view had not been total. To be sure, the Devil and God had been swept out of nature's affairs and to a very considerable extent out of men's affairs as well. To be sure, nature had been left after the scientific springclean as so much mere matter in motion, so much manipulable matter. And yet when all the sweeping had been done the mind of man remained as the most obvious evidence of an 'occult quality', of an incorporeal substance that could interact with matter. But if men have minds, possess consciousness, perhaps so, to a lesser extent, have animals, perhaps so have trees, rocks, even atoms. The mechanistic world view had clearly left the door ajar to the reintroduction of occult qualities and to the resurgence of mystical visions. Within the mechanistic world view, therefore, the mind of man became a problematic phenomenon—not the gift of consciousness, not the joy of consciousness, but the *problem* of consciousness.

The question presents itself: How in practice can scientists solve the problem of consciousness, close the door to occult forces and communion with nature, eliminate the dichotomy between mind and matter, and at the same time (if they so wish) give themselves the right to manipulate matter, organisms and other men? The answer seems obvious. It is surely by declaring that consciousness does not *do* anything, that it is but a passive spectator of events,[24] that man himself is but a machine, and that therefore all his actions are determined. And if the latter is the case then the scientist-machine manipulates both nature and non-scientist-machines according to inexorable laws of nature over which he has, and can have, no control. Consequently, just as matter in motion obeying the laws of physics cannot meaningfully be held responsible for how it interacts with so much other matter, so all actions on the part of scientist-machines aimed at controlling the behaviour of non-scientist-machines similarly acquire a non-ethical nature. 'Good' and 'evil' become obsolete words of a pre-scientific vocabulary.

The mechanistic world view, embedded in and nourished by a capitalist social order, contained within its materialistic core the seed of a long term threat: that scientists would eventually be led to reject not only the profane but the sacred as well.

The man-machine

Clearly it will be impossible even to attempt to sketch the progress of, and the resistance to, the research-action programme that is to establish that man is but a machine. We therefore look only at some highlights of both campaigns, first at the claims and achievements of some of the proponents of the mechanistic programme and then at the response of some of their non-scientist detractors.

'Let us conclude boldly then that man is a machine,' wrote La Mettrie in 1747, following the example set by Hobbes, 'and that there is only one substance, differently modified, in the whole world. What', he asked, 'will all the weak reeds of divinity, metaphysic, and nonsense of the schools, avail against this firm and solid oak?'[25] Clearly a machine is not free, possesses no free-will. Twenty-three years later the conclusion was drawn explicitly by Holbach: 'Is not the least reflection enough to prove that the solids and fluids of which the body [of man] is composed, and that the hidden mechanism that he considers independent of external causes, are perpetually under the influence of these causes, and could not act without them? Does he not see that his temperament does not depend on himself, that his passions are the necessary consequences of his temperament, that his will and his actions are determined by these same passions and by ideas that he does not give to himself?' Holbach answers the question: 'In a word, everything should convince man that during every moment of his life, he is but a passive instrument in the hands of necessity.'[26]

This point of view was to receive support from French mathematical-physicists. At the beginning of the 19th century Laplace declared that could the position and velocity of every particle in the universe be known at a given instant then both the past *and* the future of the universe could in principle be known by solving the appropriate equations of motion. To Napoleon's question as to where God fitted in to such a universe, Laplace is credited with the famous reply, 'Sire, I have no need of that hypothesis.'

20th-century Physicalism

The research-action programme to prove man a machine, to interpret all phenomena in terms of the known laws of physics and chemistry, has been developed with increasing momentum in the 20th century. To get a glimpse into the ethos of this research programme—the programme that dominates(?)* 20th-century science—we turn to a popular exposition by one of the leading spokesmen for that discipline in the vanguard of biology's attempt to prove itself but a branch of the physical sciences—to Francis Crick, molecular biologist, and his essay *Of Molecules and Men*.

Nobel-prize winning Crick is, to say the least, explicit as to the nature of the research programme to which he is committed: 'The ultimate aim of the modern movement in biology is in fact to explain *all* biology in terms of

---

* Certainly according to the historian of science R. M. Young, 'The fundamental paradigm of explanation—the goal of all science—has been to reduce or explain all phenomena in physico-chemical terms. The history of science is routinely described as a progressive approximation to this goal.' He adds: 'This is the metaphysical and methodological explanation for the fact that molecular biology is the queen of the biological sciences and the basis on which other biological (including human) sciences seek, ultimately, to rest their argument.' A Nobel prize-winner and professor of biology, J. Lederberg, comments: 'A few eccentrics aside, the whole community of contemporary science shares the view that the same laws of nature apply to nonliving and living matter alike. All of us who investigate the chemistry and physics of living organisms pursue our work as if organisms were complex machines, and we find man to exhibit no tissues or functions that would except him from this way of analysing human nature.'[26n]

physics and chemistry' (emphasis is Crick's).[27] Although he admits that there is an alternative paradigm, vitalism, which 'implies that there is some special force directing the growth or the behaviour of living systems which cannot be understood by our ordinary notions of physics and chemistry',[28] and although, as he further admits, that 'much more knowledge is needed before we shall be able even to clarify the major questions we need to ask about our brains, our behaviour and our strange [sic] feeling of being conscious', nevertheless Crick promises that provided science continues on a considerable scale there will come a time 'when vitalism will not seriously be considered by educated man'. However, we are informed that just as there are still people who believe the earth to be flat, so people will continue to believe in vitalism, no matter how overwhelming the scientific evidence to the contrary. To these people Crick addresses a few kind words: 'To those of you who may be vitalists I would make this prophecy: what everyone believed yesterday, and you believe today, only cranks will believe tomorrow.'[29]

Crick is also very interested in the attainment of a future scientific society. We are told that 'once one has become adjusted to the idea that we are here because we have evolved from simple chemical compounds by a process of natural selection, it is remarkable how many of the problems of the modern world take on a completely new light. It is for this reason,' he writes, 'that it is important that science in general, and natural selection in particular, should become the basis on which we are to build the new culture.' Crick is confident of success. What we have to realize is that 'the old or literary culture, which was based originally on Christian values, is clearly dying, whereas the new culture, the scientific one, based on scientific values, is still in an early age of development, although it is growing with great rapidity.' University administrators are therefore advised 'to see that their universities become centres for the propagation of the new culture, and not merely homes for propping up an ageing and dying one'. We are reassured that although today's science has not made as yet a deep impact on people with training in the arts, 'tomorrow's science is going to knock their culture right out from under them.' Crick speculates on a fascinating experiment in which the two lobes of (someone's) brain are disconnected:

> It seems to me that (if it were ethically acceptable) one might try to train such a body to become two people. If, for long periods of time, one could prevent the two brains from communicating with one another, one could perhaps convince one brain that it was in the same body as another brain—in other words, one could make two people where there was only one before. Whether this can actually be done remains to be seen....[30]

If the ethos underlying the physicalist image of nature has not already been made sufficiently clear, perhaps the following examples will serve to illustrate it further.

ON THE ETHICAL NEUTRALITY OF SCIENCE

(1) In his book entitled *Mechanical Man*, Wooldridge's avowed intention is to convince his readers that 'biology is indeed a branch of physical science and that man is only a complex kind of machine.'[31] He tells us, however, that since there has not yet been time for all the 'critical research discoveries' to be checked and double-checked, 'the thesis has not yet been proved beyond all doubt.' This means that for the time being it is still possible to believe that the mechanistic research programme will come to grief, that some of the existing evidence in favour of mechanical man will be rejected by future investigators. 'No doubt,' writes Wooldridge, 'some—especially those who are repelled by the idea that man is a machine—will consider this the only acceptable interpretation. But others, like the author, will judge the existing evidence to be compelling.' Now why should Wooldridge commit himself to this image of man rather than to a non-reductionist rival image which he himself admits has not been disproved 'beyond all doubt'? Wooldridge is very explicit about the reasons for his paradigm commitment. For the group to which he belongs

> is strongly attracted by the idea of a lawful universe. We find great appeal in the notion that all we can observe or feel is caused by the operation of a single set of inviolable physical laws upon a single set of material particles. This seems to us to be a logical extension of the unbroken chain of brilliant successes of physical science in accounting for one aspect after another of human experience. Therefore to us, the evidence examined in this book seems right; we believe it easily, despite the fact that the resulting displacement of essentially non-physical Man by machine-like man requires us to relinquish even the small vestige of claim to human uniqueness left to us by the discoveries of Galileo and Darwin.

He reassures us that no matter how strange some of the resulting consequences appear to be, they are the acceptable price 'we must pay for a world view in which all human experience is basically lawful or orderly'. But there is no cause for alarm. On the contrary, when men finally realize that they are machines society will profit. For 'men who know they are machines should be able to bring a higher degree of objectivity to bear on their problems than machines that think they are Men'.[32]

Wooldridge describes some experiments even more fascinating than that described by Crick, experiments now currently being done, unfortunately, only on animals. In particular he contemplates experimentation on human brains that have been 'surgically removed' from their bodies and are kept alive and conscious by providing them with chemical fluids and electrical inputs. By suitable engineering, he writes, the brains will be able to see and 'give vocal expression' to their thoughts. Wooldridge does, however, feel 'a word of reassurance is in order regarding the seemingly gruesome aspects of the postulated experimental arrangement'. I let Wooldridge speak for himself:

> We do not necessarily have to feel sorry for the disembodied star of our planned production. In the first place, we will certainly control the conditions so that it [the isolated brain] feels no physical pain or discomfort. Furthermore, from the work on pleasure and punishment centres, we know that we can also control its emotional state, making it feel continually relaxed, happy, or even ecstatic simply by arranging for suitable patterns of electric current in selected regions of the brainstem. Indeed, if such experiments ever really become possible, a major problem may be the selection of lucky winners from the many who volunteer for disembodiment because of their wish to achieve a happier state of existence than that available to them by ordinary means.[33]

(2) Watson, founding father of the behavioural school of psychology, enunciated in the late 1920s the basic aim of that science: the behaviourist, he declared, 'wants to control man's reactions as physical scientists want to control and manipulate other natural phenomena'.[34] People (i.e. *other* people) were to be viewed as so much manipulable matter—an image of man explained at greater length by the behavioural psychologist Skinner:

> Man's vaunted creative powers . . . his capacity to choose and our right to hold him responsible for his choice—none of these is conspicuous in this new self-portrait (provided by science). Man, we once believed, was free to express himself in art, music, and literature, *to inquire into nature*, to seek salvation in his own way. He could initiate action and make spontaneous and capricious changes of course. . . . But science insists that action is initiated by forces impinging upon the individual and that caprice is only a name for behaviour for which we have not yet found a cause [emphasis added].[35]

Lest there be any misunderstanding, Skinner informs us that 'the hypothesis that man is not free is essential to the application of scientific method to the study of human behaviour. The free inner man who is held responsible for the behaviour of the external biological organism is only a prescientific substitute for the kinds of causes which are discovered in the course of a scientific analysis.'[36] In 1967 Skinner declared: 'the behaviouristic principle [is that] ideas, motives, and feelings have no part in determining conduct and therefore no part in explaining it. . . .'[37]

(3) In the preface to a book intriguingly entitled *Man-Machine Communication* the author C. T. Meadow states that 'the essence of the problem is to achieve effective communication between two quite different organisms or systems'. How are these 'two quite different systems' distinguished? 'One of these,' the author tells us, 'is capable of creative thinking, high-speed correlation and decision-making under uncertainty, but is slow at calculating, reading and writing and is error prone. The other has limited intellectual

powers but does high-speed calculating, reading and writing and rarely makes errors.'[38] The omissions are striking. Such qualities as love, tenderness and compassion do not apparently serve to distinguish men from machines. Indeed, in the book's closing pages the author makes the revealing observation that (in his opinion) modern man 'fears that the computer threatens not only his job but also *his very role as a thinking human being, dominating nature around him by his superior intellect*' (emphasis added).[39]

(4) As a concluding example of 20th-century science and its dominating (?) physicalist ideology let us note the extent to which modern scientific literature contains numerous descriptions of gruesome experiments performed on living animals and written up as if the experiments had been done on inert matter. Even where pain-inflicting experiments have been done on animals with the explicit aim of promoting the growth of medical knowledge, the subsequent scientific papers usually give no indication that pity was felt, or should be felt, for the animals involved. Just one example. In order to study 'pain perception in the monkey', two behavioural scientists placed fully conscious monkeys in 'specially designed restraining units' and then applied electric stimuli to electrodes previously placed in the monkeys' pain centres. The scientists reported that as they increased the electric current the monkeys showed 'facial grimacing, closure of both eyes, high-pitched vocalization, turning of the head away from the side of the stimulus, and generalized motor activity'.[40] After the experiment was over, six of the nine monkeys used were killed for autopsies to confirm that the electrodes had been correctly inserted. At no point in the paper was any pity for the monkeys expressed.

4. An Indictment of Science

I have selected the passages just described because I find them frightening. I am not, of course, claiming that such passages are necessarily typical of contemporary science writing—I have done no systematic survey. It is enough that they have been written—and some of them written by men very prominent in their respective disciplines. What I feel is characteristic of the cited passages is the lack of belief in the dignity and worth of man; the image of man underlying the passages is that of an automaton, an image of man that sanctions the violation of his basic integrity in the name of science. The question immediately presents itself: What sort of people tend to become scientists? And what are their basic values and commitments? If their goal is not to help make a more beautiful world, a world in which people are progressively able to become more self-aware, more able to determine their own lives, to become more *human*, then what is their goal? What, the non-scientist feels obliged to ask, do scientists intend to do with so much mere manipulable matter?*

* Two *social* scientists, S. Cotgrove and S. Box, recently concluded a discussion on the character of *physical* scientists with the verdict that 'scientists are people who have learned to manipulate things *rather than people* . . .' (emphasis added).[40n]

It is scarcely difficult to understand why the argument is being increasingly made that since scientists must experiment on matter and organisms in order to practise their science the image of nature adopted is one that justifies such manipulation. For very good reasons, the ideals of love, tenderness and compassion are not built into the scientific quest. Where, in the scientific quest, does one ever find respect for non-human living organisms? Indeed, how does one even explain 'a feeling of respect' in terms of the laws of physics and chemistry? Is it not better for a scientist not to suffer from such scientific embarrassments, particularly from such embarrassments that would impose ethical rules on the scientist in his 'distinterested search for truth (?)'. What role, it is asked, does tenderness play in the scientific quest? Is it not too a scientific embarrassment and handicap? Would not the acknowledgement that some things are sacred (apart from the right to do scientific research) call into question the entire scientific enterprise and its commitment to rational thinking? Is not the molecular biologist Gunther Stent almost certainly right in his statement that 'the very roots of rational thought are likely to lie in the will to have power over the events in the outer world'?[41] Was Bertrand Russell so off the mark in his assertion that 'All modern scientific thinking ... is at bottom power thinking, that is to say, the fundamental human impulse to which it appeals is the love of power'?[42] Even more unpleasant questions come to mind. Is not the dominant tendency among the new scientists to conceive of nature as lifeless and then, consciously or otherwise, attempt to make it so? And hence to 'prove' their image correct? Is not the (subconsciously) desired end product of the new science nothing less than the elimination of man himself and hence the elimination of all those embarrassing qualities that refuse to be reduced to matter in motion?

There can be little cause for surprise that the new science has witnessed periods of opposition to its progress. In what follows we consider some principal aspects of such 'anti-science' movements, both past and present.

The Romantic Reaction

As we have already noted, the rout of the magical view of nature had not been total. The mind of man remained as a troublesome reminder of the mystical view of nature that had once been one of the new science's most formidable opponents. The door had been left ajar and through the crack slipped the occasional mystical thinker to do battle with the new orthodoxy. During the horrors of the Industrial Revolution a chorus of such mystical voices cried out in anguish at what they saw as the inevitable fruit of the mechanistic seed planted in capitalist soil. The 'mechanical philosophy', Coleridge was eventually to convince himself, had substituted

> a universe of death
> For that which moves with light and life informed,
> Actual, divine and true.[43]

In the most famous of his poems, Coleridge describes how the spirits of the earth wreak terrible vengeance not only on the seaman who had so wantonly killed the friendly albatross but also on his shipmates who had by their later actions become accomplices in the crime. Nature was not for man to master and possess. It was Coleridge's opinion that the purchase of 'a few brilliant inventions at the loss of all communion with life and the spirit of nature' was a poor bargain indeed.

Blake agreed. 'Art is the Tree of Life,' he wrote, while 'Science is the Tree of Death.' Newton's atomism was 'to Educate a Fool how to build a Universe with Farthing Balls'. There was no doubt in Blake's mind as to whom to blame for the degradation inflicted on men by the Industrial Revolution:

> I turn my eyes to the Schools and Universities of Europe
> And there behold the Loom of Locke, whose Woof rages dire,
> Wash'd by the Water-wheels of Newton: black the cloth
> In heavy wreathes folds over every Nation: cruel Works
> Of many Wheels I view, wheel without wheel, with cogs tyrannic
> Moving by compulsion each other, not as those in Eden which,
> Wheel within wheel, in freedom revolve in harmony and peace.

And thus Blake prayed:

> May God us keep
> From single vision and Newton's sleep.

Just as Coleridge believed that his own living being interacted with a universe that was itself alive, so did Wordsworth. In many of his poems the impression is given that his image of nature was not unlike that of the Wintu Indian woman:

> To every natural form, rock, fruits or flower
> Even the loose stones that cover the highway,
> I gauge a moral life: I saw them feel,
> Or linked them to some feeling: The great mass
> Lay embedded in a quickening soul, and all
> That I beheld respired with inward meaning.

We are not surprised to read that Wordsworth expressed deep unease at the progress of science:

> Man now presides
> In power, where once he trembled in his weakness;
> Science advances with gigantic strides;
> But are we aught enriched in love and meekness?

Wordsworth could understand only too well why scientists were able to see only 'a universe of death':

> Our meddling intellect
> Mis-shapes the beauteous forms of things.
> We murder to dissect.

It is rather, advised Wordsworth,

> with an eye made quiet by the power
> Of harmony, and the deep power of joy,
> We see into the life of things.

In his rejection of Holbach's philosophy Goethe was equally emphatic. Writing that 'the word "freedom" sounds so beautiful that we cannot do without it, even though it should denote an error', he called on his contemporaries to reject the experimental method that had produced a universe 'so grey, so chimerical, so deathlike':

> Friends, avoid the darkened prisons
> Where they pinch and tweak the light.[44]*

The 20th-century revolt against the new science

If the Romantic protest against the scientific image of nature and all its implications produced no sympathetic response from the scientific community, nevertheless the Romantic protesters had produced a cry of anguish that was to reverberate in the hearts and minds of some 20th-century authors. A mechanized image of nature produces a mechanized culture that reinforces the mechanized image of (disappearing) nature—a feedback process that must ultimately turn image into reality. Thus Lewis Mumford:

> As the outer world of perception grew in importance, the inner world of feeling became more and more impotent. . . . What was left was the bare, depopulated world of matter and motion: a wasteland. . . . If science presented an ultimate reality, then the machine was . . . the true embodiment of everything that was excellent. Indeed in this empty, denuded world, the invention of machines became a duty. . . . Machines —and machines alone—completely met the requirements of the new scientific method and point of view: they fulfilled the definition of 'reality' far more perfectly than living organisms.[45]

Similarly in his influential work *Life against Death* Norman Brown sees modern science as 'one aspect of a total cultural situation which may be described as the dominion of death-in-life'. 'The mentality,' he writes, 'which was able to reduce nature to "a dull affair; soundless, scentless, colourless; merely the hurrying of material endlessly, meaninglessly" . . . is lethal. It is an awe-inspiring attack on the life of the universe.'[46]

---

* In 1853 Goethe was explicitly answered by the German scientist von Helmholtz: 'But Goethe . . . cannot rest satisfied until he has stamped reality itself with the image and superscription of poetry. This constitutes the peculiar beauty of his poetry, and at the same time fully accounts for his resolute hostility to the machinery that every moment threatens to disturb his poetic response, and for his determination to attack the enemy in his own camp.

'But we cannot triumph over the machinery of matter by ignoring it; we can triumph over it only by subordinating it to the aims of our moral intelligence. We must familiarize ourselves with its levers and pulleys, fatal though it be to poetic contemplation, in order to be able to govern them after our own will, and therein lies the complete justification of physical investigation, and its vast importance for the advance of human civilization.'[43n]

Bertrand Russell, too, expressed similar fears concerning the long-term implications of the scientific enterprise. Suggesting that we seek knowledge of an object either because we love it or because we wish to have power over it, his opinion was that 'in the development of science the power impulse has increasingly prevailed over the love impulse'.[47] 'As physics has developed,' he explained, 'it has deprived us step by step of what we thought we knew concerning the intimate nature of the physical world. Colour and sound, light and shade, form and texture . . . have been transferred from the beloved to the lover, and the beloved has become a skeleton of rattling bones, cold and dreadful.' Thus, reasoned Russell, 'disappointed as the lover of nature, the man of science is becoming its tyrant'. More and more it has 'substituted power-knowledge for love-knowledge and as this substitution becomes completed science tends more and more to become sadistic'.[48] Russell was therefore apprehensive concerning the desirability of a scientific society. Indeed he thought it probable 'that the sadistic impulses which the asceticism [of such a society] will generate will find their outlet in scientific experiment. The advancement of knowledge will be held to justify much torture of individuals by surgeons, biochemists, and experimental psychologists. As time goes on,' he warned, 'the amount of added knowledge required to justify a given amount of pain will diminish. . . .'[49] A society ruled by scientists according to a scientific image of nature and man was therefore definitely not desirable. In short, wrote Russell, 'the power conferred by science as a technique is only obtainable by something analogous to the worship of Satan, that is to say, by the renunciation of love'.[48]

Russell's incisive words bring to mind the tragedy of Doctor Faustus and a question that will be increasingly asked in the troubled decades to come: To what extent have scientists made a pact with the Devil in their quest (?) for power and prestige? Let us remind ourselves of a few lines from Marlowe's play. Faustus is meditating in his study:

> These necromantic books are heavenly,
> Lines, circles, scenes, letters and characters:
> Ay these are those that Faustus most desires.
> Oh, what a world of profit and delight
> Of power, of honour, of omnipotence,
> Is promised to the studious artizan!
> All things that move between the quiet poles
> Shall be at my command.

The evil angel eggs him on:

> Go forward, Faustus, in that famous art
> Wherein all nature's treasure is contained.
> Be thou on earth as Jove is in the sky,
> Lord and commander of these elements.

Lord and commander of these elements! Masters and possessors of nature! The opponents of science would undoubtedly claim there to be more than a streak of Faustian man in the modern scientist.

Since the author of one of the most recent and impassioned attacks on the world of science does not appear hesitant as to the answer he would give to the question just asked, let us turn to his impassioned attack on science which turns out to be, at the same time, an impassioned defence of that emerging culture of magic and mysticism that the new scientists of the 17th century had gone to so much trouble to try to demolish.

Theodore Roszak and 'The Making of a Counter-Culture'

Capitalist society will in time, Roszak simply asserts, solve its problems and lay golden eggs for everyone. It is a mistake to believe the danger to truly human values comes from capitalism, or for that matter from socialism; the danger comes from technocracy—that society in which 'the relentless quest for efficiency, for order, for ever more extensive rational control' takes priority over all else, a society 'in which those who govern justify themselves by appeal to technical experts who in turn justify themselves by appeal to scientific forms of knowledge'. 'And beyond the authority of science,' Roszak informs us, 'there is no appeal.'[50] The dissenting young are confronting a fatally diseased culture, a civilization 'sunk in an unshakeable commitment to genocide', the 'prime symptom of which is the shadow of thermonuclear annihilation'.[51] The scientific ethos is all-pervasive. Indeed the reason why so many of our men of science, our scholars, even would-be revolutionaries make their peace with the technocracy is not (Roszak argues) because they lack intellect or are ignorant of human values but because technocratic assumptions about the nature of man, society and nature form the basis on which all else is built.

The experts, the men who run the modern industrial state, have access to reliable knowledge, i.e. scientific knowledge. Such knowledge is 'true . . . real . . . dependable . . . It works.' In order to acquire such knowledge, and with it the right to run society, it is necessary 'to cultivate a state of consciousness cleansed of all subjective distortion, all personal involvement'. Thus scientific knowledge is guaranteed to demythologize, not to remythologize. It is, after all, that residue which is left when all myths have been filtered away. 'This is the bedrock on which the natural sciences have built, and under their spell all fields of knowledge strive to become scientific.'[52] Indeed, claims Roszak, the mentality of the ideal scientist has become the very soul of the society. For just as the ideal scientist unemotionally performs the most gruesome experiments on animals and writes them up in a similar vein, so, for example, Barbarella and James Bond 'keep their clinical cool while dealing out prodigious sex or sadistic violence'.

The scientific mentality, argues Roszak, necessarily divides reality into two spheres, the 'In-Here' and the 'Out-There', the aim being to push as much as possible of the In-Here into the Out-There in order to produce the widest possible reality for the completely detached and impersonal objective consciousness to tabulate, categorize and manipulate. Since on no account is the In-Here, on pain of violation of scientific objectivity, to be allowed to empathize with the Out-There, this is best accomplished by the In-Here's viewing Out-There as completely deprived of any natural beauty or dignity, '*as if* it were completely stupid, meaning without intention or wisdom or purposeful pattern'.[53]

Unfortunately, however, the ideal scientist does not exist, and despite all his good intentions human feelings are always threatening to penetrate the In-Here, the objective consciousness of the flesh and blood scientist. It is for this reason that the machine becomes the scientist's ideal. And since it, and it alone, achieves the perfect state of objective consciousness the machine necessarily becomes the standard by which all things are to be gauged. Life itself must be reduced to what is measurable, orderly, predictable. Ultimately, warns Roszak, 'we shall find ourselves moving through a world of perfected bureaucrats, managers, operations analysts and social engineers who will be indistinguishable from the cybernated systems they assist'.[54]*

If the scientific method is characterized by objective consciousness, what kinds of men, Roszak asks, are likely to become scientists. The answer is predictable. The protective strategies of the pursuit of truth, pure research, etc., are 'especially compatible with natures that are beset by timidity and fearfulness; but also with those that are characterized by plain insensitivity and whose habitual mode of contact with the world is a cool curiosity untouched by love, tenderness or passionate wonder'. Furthermore, Roszak warns us, 'behind both such timidity and insensitivity there can easily lurk the spitefulness of a personality which feels distressingly remote from the rewards of warm engagement with life'.[55] The creative and the joyous embarrass the scientific mind; indeed 'more and more the spirit of "nothing but" hovers over advanced scientific research: the effort to degrade, disenchant, level down'. Clearly one should not be deceived when the scientist speaks of the beauty of nature—what he means is 'pigeonholed order'! Roszak is in no doubt as to the direction technocratic society is taking: 'the progress of expertise, especially as it seeks to mechanize culture, is a waging of open warfare upon joy.'[56]

It is therefore not surprising that Roszak welcomes the anti-scientific spirit he sees in the counter-culture. For science has declared that objective consciousness—the alienated life *par excellence*—is the sole way of understanding reality. And yet 'the art and literature of our time tell us with ever more desperation that the disease from which our age is dying is that of alienation'.[57] No wonder then the young are rejecting science. For what we should want, declares Roszak, is to expand our personalities, to make them more

---

* Writes Jacques Ellul: 'Man is to be smoothed out, like a pair of pants under a steam iron.'[54n]

beautiful, more creative, more humane. The young, quite rightly, are increasingly realizing that the myth of objective consciousness is a poor mythology which diminishes life rather than expands it. In so far as the rebellion of the young is anti-scientific in spirit, there lies in this rebellion, believes Roszak, the only hope for mankind.

It is perhaps only to be expected that Roszak welcomes the magical vision of nature as the life-enhancing competitor to what he sees as the life-destroying scientific image. For magic, says Roszak, 'is a matter of communing with the forces of nature, as if they were mindful, intentional presences'— obviously the very opposite attitude to that of the scientist.[58] Built into the magical vision of reality, he argues, is a sense of the sacredness of nature, a sacredness conspicuous in the scientific image only by its absence. Although in a technocratic society, writes Roszak, it is heretical 'to believe that this magical vision is anything but a bad mistake', nevertheless 'from such a vision of the environment there flows a symbiotic relationship between man and not-man in which there is a dignity, a gracefulness, an intelligence that powerfully challenges our own strenuous project of conquering and counterfeiting nature'. Roszak is here, of course, talking of what he calls good magic, not of bad—of the magic that confers responsibility, not privilege, of the magical 'culture in which power, knowledge, achievement recede before the great purpose of life. Which is, as an old Pawnee shaman taught: to approach with song every object we meet.'[59] With this appeal for mystical union with nature Roszak concludes his indictment of the scientific mentality.

Nightmare visions of the future

It is obvious that belief in science as a liberating force has drastically declined; here and there a light flickers but the preponderant mood is one of foreboding, if not of despair. The promise of science, it is argued, rested on an illusion. If science does not destroy all life in the immediate future, then it will give rise to a 'life' devoid of warmth, tenderness, love and imaginative thinking, a life-in-death, the only kind of life, its critics argue, that science is willing to recognize. Certainly this is the end result of the scientific quest as envisaged by some 20th-century authors who have dared to look into the future. For while the inheritors of the Copernican world system dreamed of a society of men, freed by science from pain, toil and fear, a society of men living creatively and joyously together, the 20th-century inheritors of the Galilean image of nature have been able to experience only the nightmare vision of a world of robot 'men', of a world either of pain and terror or of meaningless 'happiness'. Thus it is that the torturer O'Brien, assisted, of course, by a man in a white coat, tells Winston that the world 'the Party' is creating is 'the exact opposite of the stupid hedonistic utopias that the old reformers imagined', and whose symbol will be 'a boot stamping on a human face—for ever'.[60] Paradoxically, so Orwell writes in *1984*, the idea of an earthly paradise in which men should live together in a state of brotherhood, without laws and without brute labour, which had haunted the human imagination for thousands of years, had been discredited at exactly the moment when it became realizable.[61] Marcuse pinpoints what he claims to be one of the reasons: 'Within the total mobilization of man and nature which marks the period, science is one of the most destructive instruments—destructive of that freedom from fear which it once promised. As this promise evaporated into utopia, "scientific" becomes almost identical with denouncing the notion of an earthly paradise.'[62] But Marcuse is here not severe enough: the fear is that science will bring into being the very opposite of an earthly paradise. Let us therefore look briefly at some central features of a few of these nightmare visions of the future, societies which contrast dramatically with the utopian visions with which chapter 4 opened.

Common to E. M. Forster's *The Machine Stops*, Zamiatin's *We* and Huxley's *Brave New World* is the rejection of love. Machines do not fall in love. Love is a dangerous emotion, irrational and unpredictable. In Forster's nightmare world, human beings rarely meet face to face, an event considered crudely physical, while in Brave New World, as Mustapha Mond tells the Savage, 'The greatest care is taken to prevent you from loving anyone too much.' Eggs are fertilized in test-tubes and the degradation of parenthood is unknown. Similarly in Zamiatin's *We*, the chief mathematician D-503 is appalled and amazed at the feelings he experiences after seduction by the rebels' most voluptuous agent E-330; his feelings cannot be mathematized, they are irrational.

In all three dystopias nature has been rejected. While the inhabitants of Forster's world live underground in luxurious, solitary cells, the citizens of Brave New World are conditioned from birth to fear and reject the non-mechanical. In Zamiatin's *We* the United State has sealed itself off from the outside world of nature where hairy human beings, a constant threat to the civilized world, are still to be found. Machines do not grow hair; the growth of too much hair obviously and legitimately attracts the attention of the secret police (although patches of hair on the hands of D-503 also raised the hopes of E-330 that her quest would be successful!).

Brave New World is a stable society; there is no possibility of change. For science has made possible the resolution of the conflict between freedom and happiness. Since people cannot enjoy both, the rulers of Brave New World take upon themselves the anguish of freedom in order that the ruled can be unfree but happy. Even the epsilons are happy for they have been carefully conditioned into joyfully accepting their most inferior social status; clearly their intellectual ability has also been carefully designed to be of epsilon standard. And science itself, which made Brave New World possible, is recognized to be 'potentially subversive'. Its further progress is neither necessary nor allowed.

The United State, however, is not stable for the dissenters within the city are in league with the hairy ones outside. At the moment of crisis, however, the scientists of the United State locate that part of the brain responsible for creative thinking, for fantasy. The Well-Doer, who endures freedom so that the ruled might be unfree but happy, orders all citizens to have the offending part of the brain removed. D-503 is forcibly taken to the operating theatre. Law and order will be preserved. E-330 is captured and tortured while D-503 watches unperturbed. Although his seductress refuses to talk, the chief mathematician remains optimistic: 'I hope we win. More than that; I am certain we shall win. For Reason must prevail.'

The underground world of E. M. Forster's short story has only one lone dissenter. But his impassioned outburst against the omnipotence and omnipresence of the Machine (which directs and controls all activities in the underground city) is unforgettable.

> Cannot you see, cannot all you lecturers see [he appeals to his mother] that it is we that are dying, and that down here the only thing that really lives is the Machine? We created the Machine to do our will, but we cannot make it do our will now. It has robbed us of the sense of space and of the sense of touch, it has blurred every human relation and narrowed down love to a carnal act; it has paralysed our bodies and our wills; and now it compels us to worship it. The Machine develops—but not on our lines. The Machine proceeds—but not to our goal. We only exist as the blood corpuscles that course through its arteries, and if it could work without us, it would let us die.[63]

Eventually the impossible happens, the Machine itself breaks down, all lights go out,* nobody can repair the Machine, and everybody dies in the ensuing panic.

Orwell's *1984* is a nightmare world of the immediate future. No longer are the rulers benign father-figures who suffer freedom in order to give happiness to their peoples. On the contrary, their motivation is the desire for power. Since men and women must not form loyalties the Party might not be able to control, love and eroticism are proscribed. It was therefore the Party's policy 'to remove all pleasure from the sexual act'. When, serenaded by a thrush and not far from a meandering stream, Winston *enjoys* his first love-making with Julia, he later feels, as Julia sleeps by his side on the grass, that 'it was a blow struck against the Party. It was a political act.' Although Winston is tortured into believing that two and two make five, O'Brien's victory is not complete until, in the infamous room 101, Winston is tortured

---

* On a night in November 1965 all electric power failed in the north-eastern United States and Canada over an 80 000 square-mile area. According to Commoner, in his essay *Science and Survival*, 'The breakdown was a complete surprise. For hours engineers and power officials were unable to turn the lights on again; for days no one could explain why they went out; even now, no one can promise that it won't happen again.'[63n] The citizens of New York did not panic, however; nine months later (as legend has it) the birth rate in New York showed a sudden and dramatic bump. But had, say, the power failed during the Cuban missile crisis the outcome might not have given birth to amusing stories.

ON THE ETHICAL NEUTRALITY OF SCIENCE

into betraying the only person he has ever loved and who loved him. The way is then clear for Winston to 'love', as in the end he does, Big Brother.

There is one all-pervasive fear common to these nightmare worlds, the fear that human beings will become dehumanized, depersonalized, mere objects to be manipulated by the Machine or the Ruling Élite, lacking even the awareness that any change is possible. By conditioning, by lobotomy, by the restricted use of language, the thing-beings of the future are made to believe that their society is the only possible, the only conceivable, reality. There will be paradigm articulation but never again paradigm change. The subversive universals are non-existent in *1984*. Since the word 'freedom' stands for all the freedom so far won as well as for that (undefined) freedom yet to be won, it is not to be found in the vocabulary of Newspeak. There will be no Goethes in the world of 1984.

And of course it is science that has helped bring the nightmare worlds into existence, that has at last created the thing-beings whose behaviour is at long last no different from that of the psychologists' rats they were modelled after! The scientists' victory is complete: unpredictable human behaviour, creative thinking, are no more. Of course, since it was clear to Huxley that, paradoxically enough, scientific thinking itself is creative and therefore potentially subversive, in Brave New World scientific research is put under the strictest control. Science therefore finds itself a victim of the society it has brought into being. But not before it has achieved the success that was inherent in its very structure from the 17th century onwards.

So runs this literary indictment of science.

5. Remarks Towards an Evaluation

Anti-physicalism as an alternative paradigm

The impression has been given perhaps that the physicalist research-action programme goes unopposed within the scientific community. This is decidedly not the case. Scientists themselves are to be found expressing increasing unease at the implications of the physicalist paradigm. According, for example, to Heitler, who is a physicist, 'the mechanistic superstition is ... dangerous. It leads to a general spiritual and moral drying-up which can easily lead to physical destruction. When once we have got to the stage of seeing in man merely a complex machine, what does it matter,' he asks, 'if we destroy him?'[64] Von Bertalanffy, a biologist, agrees: 'The acceptance of living beings as machines,' he warns, 'the domination of the modern world by technology, and the mechanization of mankind are but the extension and the practical application of the mechanistic conception of physics.'[65] Psychologists such as Rogers are explicitly opposed to the behavioural programme as advocated by Skinner. Rogers would rather choose 'to value man as a self-actualizing process of becoming; to value creativity, and the process by which knowledge becomes self-transcending', seeing in science the means for 'the discovery of new modes of richly rewarding living', and for the creation of a society in which 'individuals carry responsibility for personal decisions'.[66]

Rogers' remarks are particularly appropriate, serving to underline one of the themes running throughout this book—that values and goals inform all significant human activity. Without values and without goals, human enterprises, if they can be conceived as starting at all, must inevitably peter out into a wasteland of trivialities. Champions of 'value-free' enterprises are deceiving themselves. Nowhere, of course, is self-deception more evident than in the social sciences. Value-free social science does not and cannot exist (as we argued in chapter 6).

In addition, although perhaps less obviously so, we have seen that paradigms underlie and inform the natural sciences just as much as they do the social. Indeed, scientific research is conceived only within the framework of paradigms, each paradigm supplying not only criteria for the distinguishing of phenomena into problematic and unproblematic but also procedures for 'explaining' and resolving the problematic within the paradigm's value-orientated and conceptual framework. Within the physicalist programme, as we have seen, nearly all characteristic human properties and behaviour automatically become problematic; such behaviour becomes unproblematic only when the physicalist is able to account for it in terms of the known laws of physics and chemistry.

Now the consistent physicalist (if such a being really exists) is deceiving himself as badly as is the 'value-free' social scientist. His own consciousness, his values and his goals, his own actions are not explicable in any meaningful sense in terms of current physical theories. (One is curious to know how 'in principle' physicalists intend to explain on the basis of current physical theories the striking phenomenon that 'large numbers of atoms' are in certain cases able to develop a 'collective' consciousness and undertake creative research in molecular biology.) The physicalist does not himself behave like a billiard ball. Not only must he make tactical research decisions. (Which experiment shall I do next?) but he must also make ethical decisions (Do the probable results of this experiment warrant the mutilation of this animal?). Of course, the consistent physicalist, since he believes himself to be so much matter obeying the inexorable laws of physics will fail to understand the meaning of self-responsibility for his actions; he can do no other, he will claim, than what he does. Despite such claims, however, the physicalist will certainly not attempt the futile task of analysing his own actions in terms of the many-body Schrödinger equation. Undoubtedly, what will happen instead is that his own values, both aesthetic and ethical, will tend to become devalued. Less and less will he subject his own behaviour to critical and ethical self-examination. He will simply assume that his behaviour 'obeys the laws of physics'.

The implications of the physicalist programme are sobering. If the physicalist has really convinced himself that human beings are in principle no different

ON THE ETHICAL NEUTRALITY OF SCIENCE

from rats and rats no different from machines, then there is every likelihood that the physicalist will tend to treat human beings as he treats rats and rats as he treats machines. Ultimate success for the physicalist research-action programme would appear to be the construction of a world society of thing-beings, perfectly predictable and controllable in their behaviour, just so much object-matter for a ruling élite of emotionless and non-introspective physicalist scientists. Such is Roszak's nightmare. (No misunderstanding! Of course a physicalist can strive to help build a creative society of self-determining, self-responsible human beings. But he does so as an inconsistent physicalist. An exchange of views between Professors Blanshard and Skinner makes the point. After noting that 'it is hard to see why, in a behaviourist world, any consequences should be better than any others', Professor Blanshard comments, 'Fortunately, Professor Skinner is so unreasonable a behaviourist as to be a very kindly and considerate man.')[67]

There are, however, other paradigms in the sciences. Let us therefore turn to the implications of a paradigm that is anti-physicalist at its core, whose proponents view man as a 'pilot' rather than as a 'robot'. To be sure, such anti-physicalists no doubt admit that human actions, viewed as so much matter in motion, in all probability do not violate the laws of physics, at least where the latter can in any way be applied. But they maintain that human actions, and in particular the *meaning* of such actions, transcend the world of events described by physical theories—they point out that the world picture of physics does not even contain a niche in which the sentient and conscious *creator* of the picture can in principle be inserted* (a stronger argument can in fact be made—that in a non-trivial sense quantum mechanics actually presupposes the existence of consciousness).[68] The anti-physicalist position is in essence holistic, certainly anti-reductionist: the world of nature is pictured as a multiplicity of interacting levels, 'higher levels' being non-reducible to 'lower'. Essential to this view is the belief that the observed properties of higher levels cannot be deduced from the properties of lower levels, that the properties of a *whole* cannot necessarily be predicted from a knowledge of the properties of its component parts. In addition it is possible for the anti-physicalist to believe that the universe is still evolving in the sense that new 'orders' are continually coming into being, that 'higher' levels are continually being created, and that in this world process man is an active and creative participant. Unlike the physicalist, therefore, the anti-physicalist will have every reason to emphasize the importance of values and goals, to constantly subject his own behaviour to analysis and to encourage others to be similarly self-responsible. An unbeliever in the central tenet of the physicalist

---

* Note part of a conversation between the intellect and the senses (as given by Democritus):
INTELLECT Ostensibly there is colour, ostensibly sweetness, ostensibly bitterness, actually only atoms and void.
SENSES Poor intellect, do you hope to defeat us while from us you borrow your evidence? Your victory is your defeat.

paradigm that *all* human behaviour can in principle be shown to be 'lawful and orderly', he acknowledges, and believes it desirable to acknowledge, that the most interesting human behaviour (of which scientific activity is but one kind) is essentially unpredictable and *can in no sense be adequately described by physical theories*. Thus writes Polanyi:

> Recognition of the impossibility of understanding living things in terms of physics and chemistry, far from setting limits to our understanding of life, will guide it in the right direction. And even if the demonstration of this impossibility should prove of no great advantage in the pursuit of discovery, such a demonstration would help to draw a truer image of life and man than that given us by the present basic concepts of biology.[69]

Given that the anti-physicalist believes in the possibility and desirability of consciously planned creative action orientated towards a future that is at least to some extent open, what are the social goals that he is likely to choose? In the long run he will surely not continue to place his scientific techniques at the service of capitalism. Although we shall look more fully in the next chapter at the long-term divergence of interests between capitalism and science, it will nevertheless be convenient if we digress briefly at this point to glance at the 'courtship' stage between science and capitalism in the 17th and 18th centuries, the 'married state' in the 19th and 20th centuries and the divorce that must surely come.

Science and capitalism

In the courtship stage, the goal set by leading spokesmen of the new science was, as we have seen, domination and control over nature (and very reasonably so since men were living lives dominated by scarcity and subject to the vicissitudes of nature). Obviously this was a goal not only acceptable to capitalist society but very much welcomed. However, just as exploited races so often come to be regarded as inferior by their exploiters (with attempts made to 'prove' them so), so it became convenient for the new scientists to regard the nature they intended to dominate and exploit as so much mere matter in motion, lacking in all qualities that 'rational' men could come to love and revere. Obviously therefore the new science provided not only techniques but a quantifiable image of nature very appropriate for a society in which the monetization of life—the measuring of all human needs in monetary units—was to become an overriding goal, in which working class people were in many ways to be regarded by the 'superior class' as so much manipulable matter.*

---

* In giving advice to 'the Superintendents of Manufactures' on how to increase their profit rates, this is how even Robert Owen chose to talk of working men: 'From the commencement of my management I viewed the population, with the mechanism and every other part of the establishment, as a system of many parts, and which it was my duty and interest so to combine, as that every hand, as well as every spring, lever, and wheel, should effectually cooperate to produce the greatest pecuniary gain to the proprietors'. He explained

ON THE ETHICAL NEUTRALITY OF SCIENCE

Certainly the combination of capitalism and science has led to the development of the means whereby scarcity can now be overcome. However, the abolition of scarcity would undoubtedly bring with it the rejection of materialistic values and indeed a rejection of any way of life originally based on the pervasiveness of scarcity. Today's ruling élites, therefore, apparently unable or unwilling to move towards such radical 'paradigm changes' (both in advanced capitalist countries and in the Soviet Union), are thus faced with the task, in order to perpetuate themselves, of preserving an economic insufficiency that their own achievements, together with those of the new science, have made eliminable. It is not an easy task; the contrast between the reality and the society *that could be*, at least to the educated young, is becoming increasingly obvious. Yet it is a task that the marriage of science and capitalism seems once again successful in tackling. (The position of the Soviet Union and other 'socialist' countries is as discussed at the end of chapter 9.) The willing cooperation of the scientific partner is, however, crucial. It is also forthcoming. In weapons production and design (leading towards that final triumph (?) of physicalism—nuclear holocaust and the creation of a world of mere matter in motion); in the exploration of the lunar surface; in the construction of resource-consuming apparatus for 'pure' science (such as high-energy accelerators); in the development and production of new 'necessities' of life for the rich of the earth [70]; in all these new versions of pyramid-building in the interests of ruling élites* the scientists more than play their part. Yet while the achievement and acknowledgement of material sufficiency would be a

---

that 'after experience of the beneficial effects, from due care and attention to the mechanical implements, it became easy to a reflecting mind to conclude . . . that the more delicate, complex, living mechanism would be equally improved by being trained to strength and activity; and that it would also prove true economy to keep it neat and clean; to treat it with kindness, that its mental movements might not experience too much irritating friction; to endeavour by every means to make it more perfect; to supply it regularly with a sufficient quantity of wholesome food and other necessaries of life, that the body might be preserved in good working condition, and prevented from being out of repair, or falling prematurely to decay.' 'Time and money so applied,' Owen reassured his readers, 'would return you not five, ten or fifteen percent for your capital so expended, but often fifty and in many cases a hundred percent.' [69n]

* The reader is reminded of Keynes' bitter comment of the 1930s that 'in so far as millionaires find their satisfaction in building mighty mansions to contain their bodies when alive and pyramids to shelter them after death, or, repenting of their sins, erect cathedrals and endow monasteries or foreign missions the day when abundance of capital will interfere with abundance of output may be postponed'. In 1969 Sir Peter Medawar, in a Presidential address to the British Association for the Advancement of Science (an address significantly entitled 'On "The Effecting of All Things Possible" '), commented thus on the launching of a space rocket: 'It must have occurred to many who saw pictures of it that the great steel rampart or nave from which the Apollo rockets are launched had the size and shape and grandeur of a cathedral, with Apollo itself in the position of a spire.' [70n] He went on to say: '*Like a cathedral it is economically pointless*, a shocking waste of public money; but like a cathedral it is also a symbol of aspiration towards higher things' (emphasis added). Clearly some scientists know not what they do. Nevertheless, one cannot fail to be impressed. To provide the heavens with an adequate supply of the new cathedrals is going to create an awful lot of scarcity on earth.

mortal blow to capitalist society—a blow that capitalist ruling élites will try to continue to use science to ward off—it is far from obvious that it is in the long-term interests of science to help preserve social systems that depend for their existence on the very difficult and contradictory task of maintaining and increasing material need while at the same time profitably exploiting that need. Indeed, although capitalism has undoubtedly been a benevolent despot towards science, yet as scientists become increasingly aware of the limitations imposed by capitalism on the application of scientific techniques towards the solution of global problems now endangering the very survival of man on this planet, it is to be expected that they will attempt to mobilize themselves for the construction of a society radically different from the one that has put the existence of man—and science—in jeopardy.

Scientists and the future

But towards what kind of radically different society? Towards more repressive ones with scientists infiltrating or even replacing existing ruling élites? Or towards less repressive ones—even possibly towards a liberated world society?

Let us return to consider the attitude of proponents of the anti-physicalist paradigm—those scientists who *do* stress values and goals—to ask what is likely to be their attitude as capitalism proves itself manifestly unable to solve the problems its scientific and technological successes have generated (and for which problems scientists will, no doubt, increasingly believe themselves to be receiving too large a proportion of the blame). Novelists (as we have seen) have already pointed to some of the more unpleasant social goals that could be adopted. The anti-physicalist who welcomes power over nature and his fellow men could join forces with the less aware and emotionally stunted physicalist in the attempt to achieve the goal of rule by scientific élites. (In the pursuit of such an objective the power-seeking anti-physicalist, employing many of the same *techniques* used by his physicalist colleague, would undoubtedly be the more successful of the two; the physicalist would no doubt quickly find himself behaving at the rat end of the spectrum in any confrontation with his more aware and *consciously* ambitious colleague.) But for what reasons might the anti-physicalist seek such power? Certainly he could do so because he enjoys manipulating nature and other men, perhaps rejoicing in the humiliation and mutilation it would be in his power to inflict; such were the motives of Orwell's O'Brien. Somewhat more positively, however, he could seek power in order to take upon himself the anguish of freedom (as he sees it) and thereby to rule over an orderly society of men whom he has made unfree but contented; such were the motives of Zamiatin's Well-Doer and Huxley's Mustapha Mond (and also Dostoyevsky's Grand Inquisitor).

There is, however, the possibility that the anti-physicalist will not seek power at all. It is at least conceivable that despite the potential power his manipulatory knowledge gives him the anti-physicalist could nevertheless commit himself to the goal of helping to build a world society in which the manipulatory phase of science will, to a large extent, have been surpassed *as a result of its own achievements*. What kind of society would this be? Certainly it can only be a society in which all basic necessities of life have been satisfied for *all* people—and are freely acknowledged to have been satisfied— and that thanks to the manipulatory phase of science no major diseases remain and the vast majority of people live in excellent health. If such a society is a 'liberated' one then, by definition, the *problem* of consciousness will have been superseded by the gift of consciousness—the joy of consciousness. It would be a society in which people, in ever-growing self-awareness, would no longer tend to treat each other as hostile objects to be dominated and controlled but would become increasingly able to open themselves in trust and confidence to each other. It would therefore be a society in which the individual is increasingly and justifiably able to view the actions of other people as an *enhancement* of his own freedom, not as a *limitation* of it. The object of life would be to experience and enjoy it, to cooperate together in building a world that is increasingly beautiful (and yet perhaps that is increasingly exciting), a world that represents human care for each other and, wherever possible, for nature as well—indeed it would be a world in which men tend to see themselves more as nature's servants and custodians than as her masters and possessors.

The aesthetic and ethical spirit of such a 'liberated' society is reasonably clear. What is far from clear, however, is the extent to which the present ethos of science must itself be transcended if there is to be a good chance not merely of achieving a liberated society, but of ensuring its survival. In order to examine this problem—a problem that returns us directly to the indictment of science with which this chapter opened—let us investigate the extent to which each science as it is now practised would be compatible with the aesthetic and ethical values of a liberated society.

6. The Place of Science in a Liberated Society

As the reader will already suspect, my belief is that the ethos of physics need not be incompatible with the ethos of a non-repressive, non-robot-like society.

First of all, let us note, 20th-century physics lends considerable support to an anti-reductionist paradigm. That the properties of measuring apparatuses, describable by *classical* physics, are an essential feature of the *quantum mechanical* formalism means that not all macroproperties of matter can be explained in terms of microproperties.[71] If, then, the reductionist programme has come to grief even within physics itself, the commitment to reduce all sciences to physics (to quantum mechanics (?)) is scarcely likely to meet with difficulties any less serious.

Secondly, a study of the history of physics conspicuously fails to lend support to Roszak's indictment that the physical sciences have demonstrated that the path to progress lies in cultivating 'a state of consciousness cleansed of all subjective distortion, all personal involvement'. On the contrary, subjective and passionate commitment based to a large extent on aesthetic preferences is *the* major driving force of the physical sciences.

Thirdly, it is difficult to see how a 20th-century physicist could regard the progress of physics as a series of steps leading ever nearer towards an absolute ontology. At most the modern physicist knows only that quantum mechanics 'works', that it gives him power over nature (a power that in a capitalist society will *always* be exploited, if not by the physicists, then by their employers). In short, quantum mechanics does not tell the physicist how nature *is*.* That the behaviour of matter can be approximately predicted with the use of aesthetically pleasing mathematical theories in no way means that the redness of the sunset is not in part an intrinsic feature of nature. Indeed in a liberated society the very successes of physics would give people all the more reason to revere a nature that so generously enables them to experience beauty in so many different ways. Thus it is not strange that in the introduction to *Einstein on Peace* we find one of the editors writing the following words about a man whose life was devoted to the practice of physics and to the quest to build a socialist and liberated world: 'He was a deeply religious man—a religious unbeliever, as he once called himself—a man who revered nature with profound humility, who looked again and again in awe at the trees in the front of his study, reflecting upon the richness and beauty of their branches and leaves and upon the minuscule understanding which man had acquired of the laws of nature in thousands of years and observation.'[72]

Not in search of a definite ontology through physics, the men and women of a liberated society would regard the activity of physics as one designed to give pleasure and joy to its practitioners, above all, to help provide that beauty without which life becomes meaningless. Michelson, when asked why he measured the velocity of light so many times, humbly replied that he found it fun. Einstein added of Michelson: 'I always think of Michelson as the artist in science. His greatest joy seemed to come from the beauty of the experiment itself and the elegance of the method employed.'[73] For Schrödinger 'the chief and lofty aim of science' should be to enhance 'the general joy of living'. Unfortunately Roszak's tendency is to compare mutilating experiments on animals with the works of Blake. This is as unfair to physics (although possibly not to biology and experimental psychology) as a comparison between Einstein's Theory of Relativity and the products of the advertising industry would be to creative literature. But when Einstein's achievements are compared with those of Blake who will dare to say which is the greater in inspiration, magnificence and beauty? There can be little doubt, I think, that

---

* Note the very relevant remarks by Heisenberg: 'The atomic physicist has had to resign himself to the fact that his science is but a link in the infinite chain of man's argument with nature, *and that it cannot simply speak of nature "in itself"*.' Heisenberg tells us that 'science always presupposes the existence of man and ... we must become conscious of the fact that we are not merely observers but also actors on the stage of life.'[71n]

physics would continue as a joyful and liberating activity in a non-repressive society. To be sure, few people in a liberated society would devote their entire lives to the cultivation of physics as many physicists now do; there would be so much else to experience. The aim of life would be to develop and enjoy as many of one's abilities as possible, not just one of them. Nevertheless, physics would remain an acceptable activity in a world whose principal endeavour is the cultivation of beauty in all its forms.

Disease is ugly and in a liberated society medicine would not only be regarded as the noblest of the sciences but would have self-abolition as its ultimate (if unachievable) aim. A truly liberated world society, as we earlier stated, cannot exist unless disease has been virtually eliminated everywhere on the globe. (Value-judgements are, of course, being explicitly made in favour of man against such enemies as the tapeworm and the malaria-carrying mosquito; not all life can be regarded as sacred—a cause of malaria is held to be the malarial mosquito, not man, even though, 'objectively' speaking, the elimination of either could serve to eliminate the state of affairs called 'malaria'.)

Now were it not for the existence and constant threat of disease much biological research as it is currently practised would be considered unethical, even repulsive, by men and women whose goal is a beautiful world. Certainly to the people of a liberated society all experiments that degrade or mutilate animals in any way would be inconceivable unless such research was considered essential for the combating of still-existing disease. Unlike the physicist*, the biologist is inextricably faced with the dilemma that his research activity is not, and can never be, 'ethically neutral'. Charles Darwin, in recognizing not only that 'physiology cannot progress except by means of experiments on living animals' but also that disease afflicted a major part of mankind, consequently held 'the deepest conviction that he who retards the progress of physiology commits a crime against mankind'. Nevertheless, the practice of vivisection made him feel 'sick with horror' and he held it totally unacceptable that it should ever be practised for the satisfaction of what he called 'mere damnable and detestable curiosity'.[74] Darwin's attitude drives home the point that 'knowledge for its own sake' can never be an acceptable slogan for a biological scientist who holds a minimum respect for life, let alone one whose goal is a beautiful world. In a recent article entitled 'Scientific Counter Culture', such ideas are expressed quite forcefully by three biologists who, all sharing a basically anti-reductionist approach to biology, express unease at what they believe to be the arrogance of the scientific approach to the study of life.[75] Brian Goodwin disapproves strongly, as he puts it, of 'the attitude of mind that frees you from all moral obligation towards the

---

* To be sure, the physicist also disturbs the 'natural' order and must ask himself if he is justified in so doing. The answer depends on the probable consequences of his experiments. The point I wish to make here, however, is that the physicist sincerely believes he is not inflicting pain on nuclei when, let us say, he bombards them with protons.

proper respectful treatment of the object you are investigating'. In his opinion, 'that object is the subject with which we in science carry on a dialogue, whereby we ask a question and get a response. We ought to respect the subject of our experiments in the same way that we respect other people when we are carrying on a conversation with them.' He declared that he now finds his own experiments that result in the deformation of tadpoles unethical.

Goodwin's statements give grounds for *cautious* optimism. For since such statements are consistent with, although not demanded by, a holistic image of nature, it is at least conceivable that some anti-reductionist biologists could commit themselves to the goal of building a non-repressive and non-manipulative society. Certainly Goodwin's attitude of respect for nature is one that must necessarily underlie biological research in such a society. Indeed, biology, unlike physics, will continue to be practised in a *liberated* society only if it can be made beautiful, such that it gives joy and pleasure to its practitioners, and at the very least does not degrade or mutilate the organisms with which a dialogue is initiated. If this is found to be impossible then no great tragedy will ensue. The enjoyment of nature in a liberated society will prove a more enriching experience than any arising from the pursuit of a manipulative biology in a manipulative society. As Bertrand Russell so wisely wrote,

> The mystic, the lover, and the poet are also seekers after knowledge—not perhaps very successful seekers, but none the less worthy of respect on that account. In all forms of love we wish to have knowledge of what is loved, not for purposes of power, but for the ecstasy of contemplation. . . . Whenever there is ecstasy or joy or delight derived from an object there is the desire to know that object—to know it not in the manipulative fashion that consists in turning it into something else, but to know it in the fashion of the beatific vision, because in itself and for itself it sheds happiness upon the lover. . . . Love which has value contains an impulse towards that kind of knowledge out of which the mystic union springs.[76]

Perhaps similarly Charles Darwin towards the end of his life began to sense the necessity of not allowing the 'poetic' approach to life to be excluded by the analytic and manipulative. In his autobiography he lamented and warned:

> My mind seems to have become a kind of machine for grinding general laws out of a large collection of facts, but why this should have caused the atrophy of that part of the brain alone, on which the higher tastes depend, I cannot conceive. A man with a mind more highly organized or better constituted than mine, would not I suppose have thus suffered; and if I had to live my life again I would have made a rule to read some poetry and listen to some music at least once every week; for perhaps the parts of my brain now atrophied could thus have been kept active

through use. The loss of these tastes is a loss of happiness, and may possibly be injurious to the intellect, and more probably to the moral character, by enfeebling the emotional part of our nature.[77]

On a short vacation from science Darwin described thus an experience he had enjoyed: 'At last I fell asleep on the grass, and awoke with a chorus of birds singing around me, and squirrels running up the trees, and some woodpeckers laughing, and it was as pleasant and rural a scene as I ever saw, and I did not care one penny how any of the beasts or birds had been formed.'[78] Wordsworth, one of Darwin's favourite poets before he became, as he put it, 'a withered leaf for every subject except science' would have understood and approved of such a mood. In a liberated society—a society, to be sure, made possible through the application of techniques developed during the manipulatory phase of science—it will be the mystical experience of nature that is sought rather than the experience of knowing *in yet more detail* the 'mechanisms' underlying the nervous systems of squirrels and woodpeckers. The manipulatory phase of science will have been superseded.

The ethical problems just discussed are, of course, not peculiar to the biological sciences but are to be met in even more acute form in experimental psychology and the other human sciences. Since it is axiomatic that the goal of a non-repressive society is to make possible human relationships in which no human being need be treated by another as an object, social sciences that are liberating in their aim will necessarily be guided in their research-action programmes by such an ethic. To be sure, social sciences could and would continue in a liberated society but they would have as their goal the continuing development of an environment in which all human beings are increasingly able to live more creative, more joyful, more self-determining lives (as Rogers suggests). Indeed, all the sciences that are compatible with the ethos of a liberated society would share the aim of enhancing the joy of living, of helping to develop and preserve that natural and social environment in which humans—and animals—can live lives increasingly free from mental and physical degradation—all would share the aim of building an increasingly beautiful world. The advocacy of such a goal comes without embarrassment from the pens of anti-reductionist scientists. Writes David Bohm in an article advocating a holistic approach to life: 'A human being cannot do other than to seek to live in harmony and beauty, without which survival has no value.'[79]

Scientists and the 'Golden Age'

A harmonious and beautiful world! The very successes of capitalism and science have now made realizable a society qualitatively different from both capitalist society and one ruled by a scientific-technocratic élite. Mankind is approaching a doorway through which age-old dreams can become reality. The mentality that exploited and dominated—that of necessity exploited and dominated—has reached a historical turning point in which self-transcendence is now a possibility. The 'Faustian' endeavour to achieve the infinite, together with its view of men as locked in a series of endless struggles against each other and against nature, can now give way—and, if a human world society is to be built, must give way—to an 'Orphic' civilization in which the object of life will be no longer to dominate and control but to open oneself to life's liberated experiences—to aim at greeting, as Roszak advises and as Orpheus did, every object with a song. Clearly the elements of such a new mentality are now emerging among the young (although under the most adverse conditions of ridicule and repression): the openness of relationships, the tenderness involved, the lack of desire to work for its own sake, to work to produce waste and to perpetuate a repressive system, to work so that the undernourished of the earth can be violently repressed should they rise against their rulers. The growth of such a mentality would seem to present exploitative, repressive society with one of its gravest threats, overlapping as it does with the emergence of a 'new left' mentality and the rejection of manipulative science *for its own sake*. At long last young people are beginning the struggle to transform the scientific quest according to the advice of such radical thinkers as Norman Brown and Herbert Marcuse. Brown (for example) has advised them that what we now need is a 'science based on an erotic sense of reality, rather than an aggressive, dominating attitude towards reality'.[80] Marcuse gives the same advice: 'In order to become vehicles of freedom, science and technology would have to change their present direction and goals; they would have to be reconstructed in accord with a new sensibility—the demands of the life instincts.'[81] Marcuse recognizes the depth of the transformation he is asking of science: 'This development,' he writes, 'confronts science with the unpleasant task of becoming *political*—of recognising scientific consciousness as political consciousness, and the scientific enterprise as political enterprise.'[82]

But to what extent is there a desire among the scientific community to change the present direction and goals of the scientific enterprise? Are scientists at present a conservative force in society hostile to radical change of any kind (desiring only to exercise increasing influence in decision-making processes concerned with the allocation of resources to science?) Or is an appreciable number of the scientific community already discontented with capitalist society and, if so, for what reasons? Above all, to what extent are scientists able to conceive of the possibility of a 'non-Faustian' society and, if able to conceive of such a possibility, how willing to help in its construction? Let us note the following: in their 1969 and 1970 Presidential addresses respectively the two Presidents of the British and American Associations for the Advancement of Science chose to express concern at the possible disappearance of the will to power. Medawar, to be sure, compassionately pleads for continuation of the Baconian ideal of domination of nature in the service of mankind. But the tone of his article is somewhat alarming: there is no corresponding plea for radical political and social change; Hobbes is continually quoted with approval,* Medawar concluding his address with the following quotations from the *Leviathan*: 'There is no such thing as perpetual tranquillity of mind while we live here because life itself is but motion, and can never be without desire, or without fear, no more than without sense'; 'there can be no contentment but in proceeding.' 'I agree', says Medawar.[83] Bentley Glass, on the other hand, *is* able to conceive of the possibility of a 'Golden Age'. What we find in his address, however, is that (while the majority of mankind still lives in the most desperate poverty) he concentrates his attention not so much on how to achieve the 'Golden Age' (surely a formidable task) but on the desirability of transcending it should it arise:

> Man requires a challenge and a quest if he is to avoid boredom. The Golden Age towards which we move will soon look tawdry if we no longer see endless horizons. We must, then, seek a change within man himself. As he acquires more fully the power to control his own genotype and to direct the course of his own evolution, he must produce a Man who can transcend his present nature.... Perhaps the Golden Age of no progress will be but a passing phase and history may resume. We can only hope.[84]

I for one do not share Glass's hope. Should a Golden Age arise, I am content to let the fortunate inhabitants of such a world at peace decide to where, if anywhere, they wish to proceed. Our task, and it is one that will require the use of all our Faustian energies, is to join in the political struggle to ensure the coming of such a Golden Age.

The scientific community, facing an agonizing reappraisal of the nature of science and of its place in society, is clearly in crisis. In the next chapter we look more closely at the dilemma confronting the community as some of the more enlightened and/or frightened of its members in the United States and the Soviet Union debate the fact that although proper application of modern science could provide material well-being for all on this planet, science is not only failing to do this but is contributing very substantially to the increasing dangers threatening man's very existence on this planet.

* The 17th-century philosopher who wrote, 'So that in the first place, I put for a general inclination of all mankind, a perpetual and restless striving of power after power, that ceaseth only in death.'

Chapter 11

THE SCIENTIFIC COMMUNITY IN CRISIS — SCIENTISTS AS AGENTS OF CHANGE?

> What are you working for? I maintain that the only purpose of science is to ease the hardship of human existence. If scientists, intimidated by self-seeking people in power, are content to amass knowledge for the sake of knowledge, then science can become crippled, and your new machines will represent nothing but new means of oppression. With time you may discover all that is to be discovered, and your progress will only be a progression away from mankind. The gulf between you and them can one day become so great that your cry of jubilation over some new achievement may be answered by a universal cry of horror. . . . As things now stand, the best one can hope for is for a race of inventive dwarfs who can be hired for anything.
>
> BERTOLT BRECHT *The Life of Galileo*

THE paradoxical position in which scientists find themselves is illustrated very well in the remarks made by a physicist in 1967 with respect to the quantum mechanical revolution of the late 1920s and 1930s. This physicist's 'cry of jubilation' makes no reference to any subsequent 'universal cry of horror':

> Although the years from 1924 to 1935 with their grievous economic depression and the threat and finally the arrival of nazism were in many ways alarming, they will be remembered by theoretical physicists as a happy era. There was the feeling of a great spiritual breakthrough, followed by a surprisingly rich harvest, there was a feeling of belonging to a small and select inner circle headed by a few really outstanding men.[1]

Although this particular physicist did not include in his 'rich harvest' today's nuclear and thermonuclear weapons and those who died at Hiroshima and Nagasaki, and those who were maimed there and those who are still dying as a result of their exposure to radiation, nevertheless the same 'spiritual breakthrough' also gave rise to this very bitter harvest, a harvest with which the scientific community has yet to come to terms. In this chapter we shall not only look in detail at this particular bitter harvest but enquire more generally as to what extent scientists are becoming aware of the gravity of world problems, are attempting to identify their causes, and to what extent they believe they can and should play a major role in their solution.

We begin by describing briefly some past views on the role of scientists in society, after which we focus attention on the recent prediction by two economists that scientists will and must increasingly assume responsibility for the solution of present problems. We then investigate this issue by examining in some detail the political role of scientists in the Cold War, especially with respect to the disarmament negotiations.

1. Comments on Scientists and Society

As we emphasized in chapters 4 and 10, the 17th-century scientific revolution gave rise to the belief that mankind could achieve liberation through the development of science, a belief which did not significantly wane over the next two centuries. While Bacon, for example, had placed his utopia on an imaginary island, some two centuries later the French mathematician Condorcet was to confidently argue that the very next stage of the *existing* society would be the utopia of men's dreams.

Science, Condorcet argued, was flourishing and science was the enemy of superstition, religion and tradition. Freedom and science were inseparable. The new social orders, Condorcet believed, would set up national scientific societies out of which would develop a world scientific society, 'the universal republic of the sciences', which in turn would promote the cause of human freedom in a world at peace guided by reason.[2] Condorcet's optimism was without limit. Although in hiding from Robespierre's police and shortly to perish in prison, Condorcet declared that 'no limit had been set to the improvement of human faculties . . . that the progress of this perfectibility, henceforth independent of any power that would arrest it, has no other limit than the duration of the globe where nature has set us. No doubt,' concluded Condorcet, 'this progress may be more or less rapid, but there will never be any retrogression.'

Saint-Simon, writing after the French Revolution, was similarly optimistic. 'The Golden Age of the human species is not behind us,' he proudly proclaimed, 'it is before us.'[3] The scientists, he believed, were clearly the new and decisive force in society. 'I can assure you,' Saint-Simon told the owners of property, 'that these people will compel you, in the end, to sacrifice your pride and money to place their leaders in the highest position of esteem, and to provide them with the necessary means for the conclusive exploitation of their ideas. . . . Until, gentlemen,' he warned, 'you adopt the course I propose, you will be exposed, each in his own country, to misfortunes, of the kind recently suffered in France by those of your class who happened to be established there.'[4] By 1813, however, Saint-Simon had begun to criticize the scientists for failing to understand their mission. Indeed in that year Saint-Simon spoke harshly to the scientists of Europe: 'All Europe is slaughtering itself—what are you doing to stop the butchery? Nothing. What am I saying! It is you who perfect the means of destruction.'[5]

But scientists were not only perfecting the means of destruction but also the means of production. And it was their development of the latter, Marx and Engels believed, which would make possible the realization of a society in which could be given 'to each according to his needs'. 'Science was for Marx,' declared Engels at the graveside of his best friend, 'a historically dynamic, revolutionary force.'[6] But Marx and Engels stressed that only after the negation of capitalist society would scientists be able to put the Baconian ethic—science at the service of mankind—into practice.

In 1921 the American sociologist Veblen wrote prophetically that

> the progressive advance of this industrial system towards an all-inclusive mechanical balance of interlocking processes appears to be approaching a critical pass, beyond which it will be no longer practicable to leave its control in the hands of the businessmen working at cross purposes for private gain, or to entrust its continued administration to others than suitably trained technological experts, production engineers without commercial interest.[7]

And yet, as Veblen acknowledged, 'the technicians, engineers, and industrial experts who possess the requisite technical information and experience to operate properly the country's industry are a harmless and docile sort, well fed on the whole and somewhat placidly content with the "full dinner pail"' graciously allotted them by the owners of industry.[8] Therefore, argued Veblen, no radical social changes were to be expected 'just yet'. Neither the proletariat nor the engineers possessed, he believed, the necessary revolutionary consciousness.

Eight years later the American economy suffered a breakdown of the kind predicted by Marx. But there was to be no proletarian revolution. Neither, as Veblen had foreseen, did the scientists and engineers attempt to run industry so that all Americans would benefit. Instead the Second World War, Keynesian economic theory and the growth of corporate militarism served to restore full employment to the United States. However, as argued in the preceding chapters, the solution of the problem of prolonged unemployment within the framework of capitalist society has generated new problems of such gravity that radical political action is critically necessary. Although most social scientists seem to agree (or perhaps to hope) that Western proletariats no longer possess a *revolutionary* consciousness, some at least are of the opinion that scientists are beginning to develop, if not a revolutionary consciousness, an increasingly troubled one and that they are destined to play an important, perhaps decisive, role in the critical years ahead. We therefore turn our attention to the arguments of two social scientists that these new intellectual workers will and must assume responsibility for resolving the crushing problems now confronting mankind.

Scientists as agents for change: Galbraith's view

Although in what follows we shall not be able to do justice in any way to Galbraith's detailed treatment of the nature of American society, neverthe-

THE SCIENTIFIC COMMUNITY IN CRISIS

less we hope that some threads of his argument will be made clear and particularly his recommendations for action in view of the peril he perceives. For the picture he paints of the 'The New Industrial State' is a grim one. Significantly he opens his book with the warning that 'our present method of underwriting technology is exceedingly dangerous. It could cost us our existence.'[9]

Giant corporations dominate the American economy. Goods are what they and the 'industrial system' produce and massive advertising performs 'a relentless propaganda on behalf of goods in general' and in doing so makes the industrial system appear important. Yet all-pervasive advertising, while essential, is not by itself sufficient to ensure smooth functioning of the industrial system. It is also necessary for the state both to underwrite advanced technology and to maintain that aggregate demand at which the goods produced by the giant corporations can be sold at prices sufficient to give the desired profits. Arms expenditure achieves both goals. And therefore, argues Galbraith, it is necessary 'that there be an image of the world which justifies or rationalizes the military expenditures that the arrangement requires'.[10] The requisite image, Galbraith tell us, has been that of the Cold War. At first it appears that Galbraith conforms completely to the conventional wisdom when he writes: 'That this image owes its existence only to the needs of the industrial system is not suggested for a second. The revolutionary and national aspirations of the Soviets, and more recently of the Chinese, ... were the undoubted historical source.'[10] But later on we are told that 'it is extremely important to know that our imagery is, in part, derived from the needs of the industrial system' and that 'a drastic reduction in weapons competition following a general release from the commitment to the Cold War would be sharply in conflict with the needs of the industrial system'.[11] At the end of the book the message is unequivocally stated: the industrial system 'constructs images of public and foreign policy which, though they serve admirably the needs of the industrial system, could if unchallenged be mortal for civilization'.[12]

Whan then is to be done? The danger is great but Galbraith argues that there is 'a chance of salvation'. For although the industrial system will try to 'place a preclusive emphasis on education that most serves the needs but least questions the goals of that system', nevertheless the community of educators and scientists 'holds the critical cards' in so far as the industrial system, in committing itself to technology, 'has made itself deeply dependent on the highly trained manpower advanced technology requires'.[13] Thus, argues Galbraith,

> as the educational and scientific estate grows in numbers and self-confidence; and as it comes to realize that foreign policy is based on an imagery that derives in part from the needs of the industrial system; and as it realizes further that this tendency is organic; and as it sees that the

only corrective is its own scrutiny and involvement and that this involvement is not a matter of choice but an obligation imposed by its position in the economic and political structure, we can reasonably expect it to be more effective. Nothing in our time is more important.[14]

Indeed, 'the future of what is called modern society depends on how willingly and effectively the intellectual community in general, and the educational and scientific estate in particular, assume responsibilities for political action and leadership'.[15]

Already in 1967 Galbraith believed he saw evidence that the educational and scientific estate was beginning to assume such responsibilities. For on the whole, Galbraith felt himself able to argue, the mood of the intellectuals had been one of growing scepticism to the Cold War imagery. And in particular the scientists had been much alarmed by the escalating arms race. Significantly, writes Galbraith, 'it was they, not the university specialists on international relations or the professional diplomats, who instituted the steps leading to the partial test ban. They have similarly led on other discussions with the Soviets on weapons control and disarmament.'[16]

But what is it that Galbraith recommends the educational and scientific estate to actually do? For we are told that 'a simple increase in consumer spending resulting from tax reduction or in public spending for housing or pensions would be no substitute' for arms expenditure.[17] Clearly comparable expenditure on public welfare would not underwrite advanced technology. There is only one possible solution, argues Galbraith: the arms race can and must be replaced by competition in space exploration. Indeed, it is Galbraith's opinion that 'in relation to the needs of the industrial system, the space competition is nearly ideal. It requires very high spending on complex and sophisticated technology. It underwrites the same highly developed planning as does the weapons competition and, hence, is an admirable substitute for it.'[18] Promotion of the space race at the expense of the arms race is therefore the most important task facing the educational and scientific estate.

But it is by no means the only one. For it will also be the responsibility of the educational and scientific estate to ensure in particular that sufficient homes, schools, hospitals and parks are built and more generally to provide opposition to the belief promulgated by the industrial system that the sole objective in life is to achieve the ownership of the maximum possible number of commodity goods. 'For more than the test of production, which is far too easy, the test of aesthetic achievement is the one that, one day, the progressive community will apply.'[19]

Criticism of Galbraith's view

It is difficult not to believe that Galbraith knows more than he is willing to write. In a long book this former ambassador to India gives no indication that the economy of the United States strongly interacts with those of other nations. There is no mention of American manufacturing investments abroad, of America's dependence on minerals and other raw materials from underdeveloped countries. Not a single hint is given that America could possibly be imperialist with an economic empire to control. It is therefore easy for Galbraith to argue (ironically?) that space expenditure could be a substitute for arms expenditure. But what Galbraith overlooks is that arms expenditure by a major capitalist power, in forcing the so-called socialist countries to follow suit, no matter what their initial goals, impedes the development of the planned 'socialist' economies while helping to maintain, as Galbraith acknowledges, full employment in the capitalist countries. Clearly countries striving to become socialist would not feel similarly compelled to maintain parity with the United States in a proposed space race devoid of military applications. Neither might the American people accept the necessity of such a race and even less might they accept the wisdom of unilateral space exploration. But above all, of course, considerable arms expenditure is necessary in order to provide weapons for the privileged élites of many of the underdeveloped countries and to allow military intervention in these vitally important countries whenever this is considered necessary. For these reasons alone, Galbraith's proposed solution, while perhaps a notable contribution to the annals of irony, is of little practical significance.

For the same reasons Galbraith misconceives the nature of the political struggle. Pressure by the educational and scientific estate is necessary only to substitute the space race for the arms race and to beautify the environment. But unless imperialism is ended, and this would mean the transcendence of the capitalist system, considerable expenditure on arms is essential for the United States. The educational and scientific estate has therefore a more dangerous, more prolonged struggle facing it than promotion of the space race. Even so, that Galbraith believes that the only way of saving the American 'industrial system' is to allocate vast human and material resources for the exploration of space while a majority of the world's population lacks the basic necessities of life serves only to demonstrate the irrationality and inhumanity of the kind of industrial system with which Galbraith is concerned but chooses not to name. The conclusion is to me unavoidable. That American society cannot even put advanced technology at the service of the American people (and of course it is currently being used on a massive scale against one of the world's smallest and poorest nations) demonstrates that far more radical changes are necessary than the emergence of the educational and scientific estate as a political force, however necessary and desirable this change would be in itself.

Scientists as agents of change: Heilbroner's view

Although, unlike Galbraith, Heilbroner is very ready to admit that America is a capitalist society he too sees no possibility of immediate radical change.

For the rest of this century it is almost certain, Heilbroner states, that production for profit rather than production for use will characterize American society; it will likewise be impossible to nationalize the giant corporations, to end the private ownership of the mass media, or to end the concentration of wealth in private hands. 'Barring only some disaster that would throw open the gates to a radical reconstruction of society,' such changes, no matter how desirable, are not realizable.[20]

Also, unlike Galbraith, Heilbroner readily acknowledges that America is imperialist. For 'the fierce economic conflicts of capitalist nations prior to World War II, the long history of capitalist suppression of colonials continuing down to the present in some parts of Africa, the huge and jealously guarded interests of the United States and other capitalist nations in the oil regions of the Near East or Latin America—all make it impossible to dismiss such a picture of a belligerent capitalist imperialism'. Furthermore, 'American capitalism is now a semi-militarized economy and will very probably become even more so during the next decade'.[21]

Clearly Heilbroner paints a more realistic picture of American capitalism than does Galbraith and therefore does not suggest that competition in space exploration can or will replace the arms race. But the situation is not altogether without hope. Although the proposed agents of revolutionary change are different, Heilbroner believes, as Marx did, that capitalism is producing its own grave-diggers—not the proletariat but the scientists. For although science provides capitalism with a 'virtually inexhaustible source of invention and innovation to ensure its economic growth', nevertheless Heilbroner argues that these short-term advantages 'conceal long-term conflicts and incompatibilities' between science and capitalism.[22] In particular the side-effects of advanced technology tend to create problems requiring public control; automation especially will in the long run mean that those who are unable to obtain work will have to be given the right to share in society's output—a 'vital infringement of the market's function'.[23] And, even more important, science and technology will ultimately raise the general standard of living to such an extent that higher salaries no longer suffice to attract people into otherwise unpleasant jobs. Should this become the case, emphasizes Heilbroner, 'some authority other than the market must be entrusted with the allocation of men to the essential posts of society'.[24]

The essential point Heilbroner wishes to make is that although capitalism cannot exist without scientific and technological innovation yet since the profit motive allows little scope for the control of the results of such innovation capitalism finds itself 'essentially defenceless before the revolutionizing impact of its technical drive'.[25] There will be only one way out: market forces will have to give way to 'public' control.

It is therefore science that is undermining capitalism from within and it will be scientific élites who will rule in the future. So argues Heilbroner. Indeed whereas 'capitalism as an *idea* has never garnered much enthusiasm', science, on the other hand, has won 'from all groups, and especially from the young,* exactly that passionate interest and conviction that is so egregiously lacking to capitalism as a way of life'.[26] The hope is that in a world run by scientific élites, who exercise political, social and economic control, both local and global problems will succumb to 'rational' solution. Although Heilbroner finds it difficult to speculate with respect to the immediate future, let alone the nature of the new society, one thing is, however, certain: 'It is the profound incompatibility between the new idea of the active use of science within society and the idea of capitalism.'[27]

The issue is clear: whereas science as it is applied to society sees man as a being 'who shapes his collective destiny', capitalism sees man as a being 'who permits his common social destination to take care of itself'; whereas 'the essential idea of a society built on scientific engineering is to impose human will on the social universe', that of capitalism is 'to allow the social universe to unfold as if it were beyond human interference'. Thus beneath the surface of mutual interests there exists a deep and irreconcilable conflict between science and capitalism. For no system which makes a virtue out of the absence of a directing intelligence could in the long run gain the support of science. And therefore 'in the end', writes Heilbroner, 'capitalism is weighed in the scale of science and found wanting, not alone as a system but as a philosophy'.[27]

However, Heilbroner is far from unaware of the possible dangers of a scientifically organized society. 'There is,' he warns us, 'in the vista of a scientific quest grimly pursued for its own sake, a chilling reminder of a world where economic gains are relentlessly pursued for their own sake.' While he believes that 'Science is a majestic driving force from which to draw social energy and inspiration,' he expresses the fear that 'its very impersonality, its "value free" criteria, may make its tutelary élites as remote and unconcerned as the principles in whose name they govern.' Heilbroner therefore concludes with the following advice:

> Against these cold and depersonalizing possibilities of a scientifically organized world, humanity will have to struggle in the future, as it has had to contend against not dissimilar excesses of economic involvement in this painful—but also liberating—stage of human development. Thus if the dawn of a new age of science opens larger possibilities for mankind than it has enjoyed heretofore, it does not yet promise a society whose overriding aim will be the cultivation and enrichment of all human beings, in all their diversity, complexity and profundity. That is the struggle for the very distant future, which must be begun, nevertheless, today.[27]

* Heilbroner's essay was published in 1966; in 1968 Roszak felt able to give a public welcome to the increasing disaffection of the young with science.

Summary

Although both Galbraith and Heilbroner therefore single out scientists as an emerging political force, Heilbroner's analysis is I believe the more realistic. For Galbraith there is nothing inherently wrong with the 'industrial system' that concerned action by the educational and scientific estate cannot correct; the only problem, no matter how serious, is one of misplaced technological investment. In this he is surely wrong. Heilbroner, on the other hand, in acknowledging that America is both capitalist and imperialist, looks further ahead towards radical political change. Although he believes that this is all but impossible in the next few decades, nevertheless the near certainty that capitalism will be unable to solve the increasingly serious problems it constantly and unavoidably generates will eventually oblige scientists to attempt to make full use of their intelligence and therefore as a prerequisite to demand radical transformation of the institutions of capitalism. The prediction that Heilbroner makes is therefore *not* the Galbraithian one that scientists will in the long run begin to exert a restraining, moderating influence, helping to solve those problems generated by their own scientific successes in an essentially anarchic, exploitative society (although Galbraith would disagree that the 'industrial system' is anarchic) but rather that scientists will attempt to assume political, social and economic control—his warning that the rule of a scientific élite will not necessarily be liberating is one the dissenting young have already taken to heart.

However, the world's problems are already immense and none more immediately threatening than the escalating arms race with respect to which scientists are very intimately involved, not to say responsible. It is therefore of great relevance to sketch the growing involvement of scientists in government brought about by the Second World War and its consequences and, with the theses of Galbraith and Heilbroner in mind, to ask to what extent the scientific community is developing a political consciousness not simply as a result of its involvement with government but, more importantly, as a result of the political failure of American and Soviet governments to achieve an end to the arms race.

2. Nuclear Weapons and the Political Role of Scientists

The explosion of the first atomic weapon weighed heavily on the conscience of its makers. As Oppenheimer later wrote in 1948, 'In some sort of crude sense which no vulgarity, no humor, no overstatement can quite extinguish, the physicists have known sin; and this is a knowledge which they cannot lose.'[28] The scientists committed themselves wholeheartedly to political action: *they* had made such a terrible weapon, *they* would ensure that this weapon would never again be used in anger.

In the autumn of 1939 Einstein had sent a letter, drafted by Szilard, to President Roosevelt advising him of the possibility of constructing atomic weapons and of the danger of allowing Germany to obtain a lead over the United States in atomic weapons research. By July, 1945, however, when the first atomic bomb was tested, the war had been three months over in Europe. Subsequently, on August 6 and 9, atomic weapons were dropped on Hiroshima and Nagasaki and the war with Japan was brought to a dramatic end. But, as has been suggested in chapter 9, although Japan was the physical victim of the atomic attack, the intended psychological target was the Soviet Union. Although the expected response was Soviet withdrawal from Eastern Europe, the result, however, was the consolidation of Soviet power in that half of Europe and the precipitation of the nuclear arms race.

Several scientists had feared the consequences of such a use of atomic weapons. Bohr in particular had hoped to persuade both Roosevelt and Churchill of the wisdom of making a generous offer to Stalin concerning the sharing of atomic information. While his interview with Churchill in April, 1944, was a 'sad fiasco', Roosevelt was, Bohr felt, reasonably sympathetic to his views.[29] Nevertheless, at the Roosevelt-Churchill meeting of September, 1944, the two leaders agreed not to give Stalin any information about the atomic weapons project. Undaunted, Bohr was trying to arrange another meeting with Roosevelt when the President died on April 12. 'Looking back on these days,' Bohr wrote in 1950, 'I find it difficult to convey with sufficient vividness the fervent hopes that the progress of science might initiate a new era of harmonious cooperation between nations, and the anxieties lest any opportunity to promote such a development be forfeited.'[30]

The panel of four scientists appointed to the Interim Committee of the War Department (Compton, Fermi, Lawrence and Oppenheimer) reached the conclusion that atomic weapons should be dropped on Japanese cities. According to Stimson, 'On June 1, after its discussion with the Scientific Panel, the Interim Committee unanimously adopted the following recommendations: (1) The bomb should be used against Japan as soon as possible. (2) It should be used on a dual target—that is, a military installation or war plant surrounded by or adjacent to houses and other buildings most susceptible to damage; and (3) it should be used without prior warning.'[31] Alternative policies the Interim Committee rejected as 'impractical'. Considering some of the opinions of the military cited in chapter 9, such a political recommendation agreed to by the four scientists at first sight seems very surprising. But, as Oppenheimer later admitted, the scientists 'didn't know beans about the military situation. We didn't know whether they [the Japanese] could be caused to surrender by other means or whether invasion was really inevitable.'[32] The invasion of Kyushu had been planned for November 1, 1945. Atomic bombs were dropped on Hiroshima and Nagasaki *nearly three months before that date.*

The Franck Report

At the Metallurgical Laboratory in Chicago a group of atomic scientists had been for several months analysing the international implications of the new weapons they were helping to make. 'It is our opinion,' states one of the summaries of the June 4 (1945) meeting, 'that the manner in which this new weapon is introduced to the world will determine in large part the future course of events.'[33] In their final report, which was considered by the Interim Committee on June 21, the scientists warned that in Russia 'the basic facts and implications of nuclear power were well understood in 1940, and the experience of Russian scientists in nuclear research is entirely sufficient to enable them to retrace our steps within a few years, even if we should make every attempt to conceal them.'[34] They emphasized that, 'If no efficient international agreement is achieved, the race for nuclear armaments will be on in earnest not later than the morning after our first demonstration of the existence of nuclear weapons.' And if atomic weapons were used against Japan then it might be 'very difficult to persuade the world that a nation which was capable of secretly preparing and suddenly releasing a weapon as indiscriminate as the rocket bomb and a million times more destructive, is to be trusted in its proclaimed desire of having such weapons abolished by international agreement'.[35] It was their opinion that 'much more favourable conditions for the eventual achievement of such an agreement could be created if nuclear bombs were first revealed to the world by a demonstration in an appropriately selected uninhabited area'.[36] The Interim Committee, however, was not impressed and reaffirmed the recommendations made three weeks earlier. Further efforts by scientists to reverse the decision, such as Szilard's petition to the President, proved fruitless. As President Truman wrote in his *Memoirs*: 'The final decision of where and when to use the atomic bomb was up to me . . . I regarded the bomb as a military weapon and never had any doubt that it should be used.'[37]

It is significant that the scientists received a directive from Groves requiring the first atomic weapon to be ready for testing by mid-July and the second to be available for military purposes in August. One of the nuclear scientists, Philip Morrison, stated that 'I can personally testify that a date in the neighbourhood of the 10th of August was the mysterious deadline which we, who were working daily on the job of finishing the bomb, had to observe at all costs, irrespective of money troubles or the mass of development work still to come.'[38] August 8, was the date on which the Russians were to declare war against Japan. On August 9, the day the second atomic weapon was dropped on a Japanese city, President Truman declared that the East European countries were 'not to be spheres of influence of any one power'.[39] The scientists who had opposed the dropping of atomic weapons on Japan had attempted to warn their government of those very consequences which President Truman and his advisers intended their use of the atomic bomb to bring about: the intimidation of the Soviet Union and the end of the wartime alliance. *The nuclear scientists had completely failed to understand not only the intentions of the government employing them but also, and more importantly, the nature of the social system in which they now had such an important role.* In 1942 two American sociologists, H. H. Gerth and C. Wright Mills, had written: 'Precisely because of their specialization and knowledge, the scientist and technician are among the most easily used and coordinated of groups in modern society . . . the very rigour of their training typically makes them the easy dupes of men wise in political ways.'[40] The wartime behaviour of American scientists would appear to confirm this assessment.

Only too aware, however, that they had failed to achieve their limited, although important, objectives, the scientists began to organize themselves for piecemeal political action. On October 31, the Federation of American Scientists was formed in order 'to meet the increasingly apparent responsibility of scientists in promoting the welfare of mankind and the achievement of a stable peace'[41], and on December 10, 1945, the first issue of the *Bulletin of the Atomic Scientists* appeared. Many of these politically-conscious scientists strongly opposed the 'May-Johnson' bill (despite the support it enjoyed among such senior scientists as Oppenheimer and Fermi) since they held that such a bill would place the field of atomic energy under the direct control of the US War Department. Urey, a Nobel prize-winning physicist, passionately argued against the bill before a congressional committee on the grounds that its military emphasis 'will hamper all attempts to secure international control of atomic bombs'.[42] The War Department's bill was defeated and on August 1, 1946, President Truman signed the new Atomic Energy Act establishing civilian control over atomic energy combined, however, with extremely severe security measures. Unfortunately (Lapp comments), the Atomic Energy Commission's first chairman allowed the administration of the AEC to be 'marked by conservatism and deference to the military to the point where the AEC became a service agency fulfilling the requirements of the Pentagon'.[43] Thus the victory of those scientists who had opposed Oppenheimer and the military was to prove only of very limited extent.

The Baruch Plan

The Chicago scientists had predicted that if atomic weapons were used on Japan then America would forfeit the trust necessary to achieve post-war international control of atomic energy. Nevertheless, leading atomic scientists did their best to contribute towards what they believed to be government plans for such international control and enthusiastically worked on proposals that were substantially incorporated by the American government into the official 'Baruch Plan'. However, when on June 14, 1946, Baruch presented the American plan, it quickly became obvious that the Soviet leaders were highly suspicious of the American government's intentions. For not only would the Soviet government have to permit international inspection of the

Soviet Union, thereby depriving the Russians of one important asset (the secrecy maintained about military strength and resources), but only after 'an adequate system for control of atomic energy' had 'been agreed upon and put into effective operation', together with rules for punishing violators, would the American government stop the manufacture of atomic bombs and dispose of existing ones. 'My country,' declared Baruch, 'is ready to make its full contribution towards the end we seek, subject of course to our constitutional processes, and to an adequate system of control becoming fully effective, as we finally work it out. But,' he added, 'before a country is ready to relinquish any winning weapons, it must have . . . a guarantee of safety, not only against the offenders in the atomic era, but against the illegal users of other weapons—bacteriological, biological, gas—perhaps—and why not? against war itself.'[44] In effect, the Russians were being asked to permit United States control over their atomic resources in return for an American promise to stop the production of and to dispose of existing stocks of atomic weapons at such a time when the American government felt fully secure. But, as Baruch implied in his speech, there could be no guarantee that the American Senate would necessarily agree with the government's views on American security at any particular time. How could the Soviet government ever have accepted such proposals? Indeed, Blackett's assessment of the Baruch Plan appears inescapable: 'What was intended by the atomic scientists to bring co-operation with Russia became an instrument in the hands of the American Government to coerce her.'[45]

However small the probability of the Russian government's acceptance of the Baruch Plan, that probability must have shrunk to zero when the American government decided that the weeks immediately following Baruch's initial speech were the most appropriate time to conduct the first peacetime tests of atomic weapons. Despite much misgiving expressed by the Federation of Atomic Scientists, the United States exploded on July 1, its fourth atomic bomb. On July 3, *Pravda* declared that the Bikini test proved that the United States was interested only in perfecting atomic weapons, rather than in their control. The Soviet Army's *Red Star* also condemned what it called United States belligerence.[46] On July 25, a second atomic weapon was exploded. The Soviet Union did not accept the Baruch Plan. Partly because of the 'genuine idealism', however, of the atomic scientists' original proposals, it became relatively easy, Blackett argues, for the American government to present the Russian rejection as a conclusive demonstration that 'Soviet intransigence' was the major obstacle to world peace. 'In short, the atomic scientists were outmanoeuvred—for a second time.'[47] This is Blackett's concluding opinion and, once again, I find it inescapable.

Of considerable interest, consequently, is the following passage from Gilpin's analysis entitled *American Scientists and Nuclear Weapons Policy*:

> The United States could afford *such a discrepancy between its disarmament policy and its military policy* when there was little likelihood that

one would interfere with the other. For example, the Baruch Plan was no threat to the implementation of the recommendation of the Commission on Air Power that American defence be based on the strategic employment of atomic bombs, because in the years immediately following World War II *there was little likelihood that the Soviet Union would subscribe to the Baruch Plan* (emphases added).[48]

Indeed, the American government's policy was to attempt to ensure that no meaningful agreement of any kind would be reached between the United States and the Soviet Union. Although, of course, this was certainly not the way American policy appeared to the overwhelming majority of American scientists, nevertheless, with the failure of the Baruch Plan, the first serious division appeared within their ranks.

According to Gilpin, a minority of scientists, but including several in the leadership of the Federation of American Scientists, believed that 'the danger of uncontrolled atomic energy' was the cause, not the result, of the differences between the United States and Soviet Union, and that these differences would be only aggravated by further rearmament and the formation of a Western alliance.[49] Instead, Gilpin writes, they urged their colleagues to join them in devising new means of international control and to pledge abstention from military research, while urging the American government not to continue to promote the clearly unacceptable Baruch plan.

On the other hand (Gilpin writes), the majority of scientists, including such prominent physicists as Oppenheimer, Teller and Bethe, believed that the Cold War was primarily the result of the Soviet Union's ambition to control Western Europe and ultimately the whole world and that therefore the United States should retain its atomic lead over the Soviet Union. Thus the scientists of the General Advisory Committee (GAC) to the Atomic Energy Commission concluded, in Oppenheimer's words, 'that the principal job of the Commission was to provide atomic weapons and good atomic weapons and many atomic weapons.' Towards the end of 1947 Oppenheimer told the National War College that 'the very bases for international cooperation between the United States and the Soviet Union were being eradicated by a revelation of their deep conflicts of interest, the deep and apparently mutual repugnance of their ways of life, and the apparent conviction on the part of the Soviet Union of the inevitability of conflict—and not in ideas alone, but in force'.[50]

Nevertheless, this 'containment school' was itself divided as to whether the nuclear rearmament programme should be accompanied by attempts to achieve international control of atomic energy. Although Oppenheimer for a period of years believed that the latter effort should be temporarily dropped, both he and his associates on the GAC changed their minds when confronted in 1949 with the decision as to whether hydrogen bombs should be constructed with all possible speed. Other scientists disagreed strongly. The conflicting commitments are described by Gilpin as follows.

The decision to build the hydrogen bomb

The GAC scientists decided against escalation of the arms race. A crash programme to develop the hydrogen bomb was not, they believed, the most suitable response to the Soviet Union's first atomic test of August, 1949. After their October, 1949 meeting, the GAC scientists reported: 'We all hope that by one means or another, the development of these weapons can be avoided. We are all reluctant to see the United States take the initiative in precipitating this development. We are all agreed that it would be wrong at the present moment to commit ourselves to an all-out effort towards its development.'[51] An adequate and desirable response, they believed, would be to initiate a programme of expansion, improvement and diversification of atomic weapons combined with a general increase of conventional military strength. Fermi and Rabi thought it 'wrong on fundamental ethical principles to initiate the development of such a weapon' while Oppenheimer and Conant expressed the view that 'in determining not to proceed to develop the super bomb, we see a unique opportunity of providing by example some limitations on the totality of war and thus of eliminating the fear and arousing the hope of mankind'.[52] Bethe later expressed the opinion that he 'thought that the alternative might be or should be to try once more for an agreement with the Russians'.[53]

The scientists of the 'infinite containment' school thought differently. Led by Teller, they argued that the nuclear arms race would necessarily continue and that the United States should retain its lead over the Soviet Union by initiating a crash programme to develop the hydrogen bomb. The advice of the GAC scientists was rejected by President Truman who on March 10, 1950, authorized such a crash programme to prepare for quantity production of hydrogen weapons. However, he did accept their advice that the United States should develop more fully the capacity to wage limited war with conventional weapons, advice that was rendered clearly sound (from the government viewpoint) by the outbreak of the Korean War in June of that year.

Project Vista

During the years 1951–2 the scientists of the 'finite containment' school attempted to substantiate their case that the United States should not develop nuclear weapons with the sole objective of instant, massive retaliation. In the report of the so-called Project Vista, these scientists argued that the primary response to limited Communist aggression, as they saw it, should be with conventional and tactical nuclear weapons and that the latter weapons should be made available to the United States Army and Navy. Since the US Air Force had at that time a complete monopoly of nuclear weapons and wished this monopoly to continue, the leading scientists in Project Vista antagonized influential people in the US Air Force and in both political parties. And as Gilpin has commented, 'In part these feelings of hostility would account for the security trial of Robert Oppenheimer in 1954.'[54]

THE SCIENTIFIC COMMUNITY IN CRISIS 301

Meanwhile, due principally to important discoveries by Teller, the first hydrogen bomb had been completed and was scheduled for testing in early November, 1952. Many scientists were still alarmed at the prospect of yet another escalation of the nuclear arms race and in the summer of 1952 Vannevar Bush visited the Secretary of State to plead for at least postponement of the proposed test. As he declared in 1954 in his testimony in defence of Oppenheimer: '. . . I felt strongly that that test ended the possibility of the only type of agreement that I thought was possible with Russia at that time, namely, an agreement to make no more tests. . . . I still think we made a grave error in conducting that test at that time, and not attempting to make that type of simple agreement with Russia.'[55]

In the summer of 1953 Oppenheimer published an article in the *Bulletin of the Atomic Scientists* criticizing, although in 'the most guarded and cautious language', the policy of massive retaliation subscribed to by the US Air Force. 'The very least we can conclude is that our twenty-thousandth bomb, useful as it may be in filling the vast munitions pipelines of a great war, will not in any deep strategic sense offset their two-thousandth.'[56] Shortly after Oppenheimer's article was published, the Soviet Union exploded its first hydrogen bomb. Seven months later in March, 1954, the United States tested its first 'droppable' bomb. The new Secretary of State, John Foster Dulles, called for a policy of instant, massive retaliation while Vice President Nixon declared in March, 1954, that 'rather than let the Communists nibble us to death all over the world in little wars, we would rely in the future primarily on our massive mobile retaliatory power . . . against the major source of aggression . . .'.[57] Thus Gilpin argues that the Vice President was endorsing a military policy against which the majority of scientists had fought for almost ten years, and against which they would continue to fight. But they were to be deprived of their most influential spokesman. On June 29, 1954, the AEC deprived Oppenheimer of his security clearance, thus disbarring him from any future government service. At his 'trial', the Chief Scientist for the Air Force declared '. . . that there was a pattern of activities all of which involved Dr. Oppenheimer. One of these was the Vista project . . .'[58] When Teller was asked whether he believed Oppenheimer to be a security risk, he replied: 'In a great number of cases . . . [Oppenheimer has acted] in a way which for me was exceedingly hard to understand. I thoroughly disagreed with him in numerous issues and his actions frankly appeared to me confused and complicated. To this extent I feel that I would like to see the vital interests of this country in hands which I understand better, and therefore trust more.'[59]

Soviet acceptance of Western disarmament proposals, May 1955

In April, 1955, the now militarily stronger Soviet Union, two years after Stalin's death and after nine years of fruitless negotiation between the Soviet

Union and the West, agreed to withdraw Russian troops from Austria and to sign a 4-power pact guaranteeing Austrian independence. A month later on May 10, the Soviet Union put forward proposals accepting positions that had long been advocated by the Western powers. In particular, the Soviet government accepted the Western proposals of comprehensive disarmament in stages and showed a willingness to consider at least some international controls. On May 12, 1955, the US delegate, James Wadsworth, said he was 'gratified to find that the concepts which we have put forward over a considerable length of time ... have been accepted in a large measure by the Soviet Union'.[60] Six days later, however, the Western delegates decided to postpone further discussion, despite Soviet protests, until after the Heads of Government Conference scheduled to take place in July. At the summit conference, Moscow proposed 'that both NATO and the Warsaw Pact be replaced by an all-European security pact' within which a neutralized Germany could be reunited. The Western powers, however, rejected the proposals, insisting 'that Germany should be reunited on the basis of free elections and a free hand in foreign and military policy', conditions that the Soviet government immediately rejected.[61] That September, Washington placed a 'reservation' on its previous disarmament proposals, which had been 'in large measure' accepted by the Soviet Union, and stressed instead inspection measures and investigation of the technical problems of control. Although this dramatic reversal of Western policy was bitterly criticized by Soviet diplomats at the time, there was, as three MIT analysts have noted, 'a long silence where a propaganda line publicly "exposing" the US might normally have been expected'. Nevertheless, the same authors surmise 'that Western conduct in the months following the May 10, 1955, proposal resolved any doubt that Moscow might have entertained about the readiness of the "moderate" forces in Western governments to enter into meaningful disarmament measures with the Communist states.'[62]

The Bethe Panel and the Geneva System: April to August, 1958

American scientists then split into two distinct factions, one faction, the control school, arguing strongly for a ban on *all* nuclear tests as a first step to general disarmament (a petition to this effect, drafted by Linus Pauling and signed by 9 000 scientists, was delivered to the United Nations in January, 1958), and the other, the infinite containment school, arguing strongly against a control system (because they claimed it could not be made foolproof) and in favour of continued testing and ever-greater American nuclear strength. The leading scientists representing the two conflicting positions were Bethe and Teller. Bethe, Gilpin writes, believed that the political advantages to be gained by a test ban far outweighed the small risks involved and that, in any case, further testing would be much more to the advantage of the Soviet Union than to the United States. Teller, on the other hand, argued that evasion techniques would progress faster than detection techniques and that 'in the contest between the bootlegger and the police, the bootlegger has a great advantage'.[63] After the Soviet 'sputnik' success, President Eisenhower appointed as his Special Assistant for Science and Technology, James Killian, President of MIT, who was sympathetic to the position of the control school and to the Soviet proposal that there should be a ban on all nuclear tests. Aided by the AEC and the Defense Department, Killian appointed a committee under Bethe's chairmanship to examine the value of continued testing and the feasibility of a nuclear test ban. In April 1958 the Bethe Panel reported to the President that American nuclear weaponry was sufficient to permit a test ban without prejudice to American security and that, although no foolproof detection system could be devised, the risks involved could be made 'acceptable'. As a result, President Eisenhower wrote to the Soviet Premier, Khrushchev, at the end of April suggesting that technical talks be held on the feasibility of a detection system. The Soviet government agreed and the talks commenced on July 1, 1958. After seven weeks of discussion, the American and Soviet scientists concluded 'that the methods of detecting nuclear explosions available at the present time ... make it possible, within certain specific limits, to detect and identify nuclear explosions, and [we] recommend the use of these methods in a control system'.[64] 'Although the control system would have great difficulty in obtaining positive identification of a carefully concealed deep underground nuclear explosion', the experts believed that 'there would always be a possibility of detection of such a violation by inspection' and that 'whatever the precautionary measures adopted by a violator he could not be guaranteed against exposure, particularly if account is taken of the carrying out of inspection at the site of the suspected explosion.'[65] Subsequently, the Soviet and American governments agreed that negotiations to implement the Geneva system should begin on October 31, in Geneva.

Before this conference began, however, the United States government held a series of underground nuclear explosions, the analysis of which (the so-called Hardtack data) convinced AEC technical experts that the Geneva system would be far less efficient in distinguishing earthquakes from underground nuclear tests than had been stated in the Geneva report. According to the nuclear physicist Ralph Lapp, 'the faulting of the Geneva system,' ... a development in which Teller played a leading role, 'appeared to lead to the conclusion that an acceptable control system would require a "large expansion in the number of inspection stations"'.[66] The Eisenhower administration requested the Soviet government to agree to a technical reassessment of the Geneva system. The Soviet Union's response was to accuse the United States government of wanting to sabotage the Geneva discussions then taking place. In particular, there appeared an article in the March, 1959, issue of *Fortune Magazine* describing in detail the controversy within the United States government over the proposed test ban treaty, an issue that 'split the Administration, and set the scientists, including the members of the President's own

Science Advisory Committee, in opposing camps'.[67] The names of those who opposed the decision to suspend testing were actually stated. Bernard Bechhoefer, who resigned from the State Department in 1958 (where he had been a senior officer on the armaments control question since 1946), has written in his *Postwar Negotiations for Arms Control* that the detailed material for the ideas expressed in the *Fortune* article

> could not have been assembled without the active cooperation of certain sectors of the United States Government. The motivation clearly was to build popular sentiment to reverse the previous presidential decision to support the ending of the tests. The Soviet leaders, with some reason, concluded that the true ground for the reversal of United States policy would be not that a test cessation ban was unenforceable but that the United States in the interest of its security should resume tests ... *The Soviet actions in the light of their justified suspicions of United States motives seem quite moderate* [emphasis added].[68]

Indeed by December, 1959, the Soviet technical experts had already begun to consider the new data on the underground tests.

The research-action programme of Teller and Latter

Convinced that the Soviet Union would devise secret methods of testing after a successful test ban treaty, Teller asked scientists at the Lawrence Radiation Laboratory and RAND Corporation to consider ways in which secret violation of the Geneva control system would be possible. The scientists explored several possibilities such as exploding a nuclear device in a mountain, in either a porous or expandable container or in a large cavity several thousand feet beneath the earth's surface. Latter's estimate was that detonation in such a large cavity would allow nuclear explosions to be 'decoupled' by a factor of several hundred and he therefore chose to present this theory to the President's Science Advisory Committee (PSAC) in January, 1959. The members of PSAC, including Bethe, agreed that such 'decoupling' was theoretically possible. Thus it appeared that the Geneva Control System was obsolete for another reason quite distinct from the conclusions drawn from an analysis of the Hardtack data.

However, in March, 1959, the Berkner Panel on Seismic Improvement, set up by Killian at the end of December, reported that the original sensitivity of the Geneva Control System could be partly restored by increasing the number of seismometers at each control station from 10 to 100; furthermore, the Panel pointed out, there would almost certainly be continual technical advance. The decoupling theory was played down: although Latter had reported to the Panel that the decoupling factor could be as much as 300, while Bethe estimated it at 700, the Panel concluded 'that decoupling techniques existed which could reduce the seismic signal by a factor of ten or more'.[69]

Although many scientists did not dispute Latter's theoretical calculations, they held that the construction of a cavity sufficiently large and deep to decouple a 10-kiloton explosion by a factor of 300—a cavity 360 feet in diameter and 3 000 feet underground—would be 'a formidable engineering feat, and even if feasible, an incredibly expensive task'. The Panel emphasized the need for further research to try to improve the Geneva Control System, a project eventually allocated by the Secretary of Defense to the Defense Department. However, since (as Jacobson and Stein point out in a masterful understatement) the Defense Department 'was less than enthusiastic about the wisdom of a test ban',[70] it is scarcely surprising that although by the end of 1959 Latter's decoupling theory was being tested with underground chemical explosions the research recommended by the Berkner Panel had hardly begun. Indeed, the following comments by Lall, a disarmament expert at Cornell University, are worth noting:

> Throughout most of the period from 1958 to 1963, test-ban negotiators experienced considerable difficulty in obtaining data from defence scientists: annual expenditures of several dozen million dollars appeared to have borne few tangible research results in nuclear test detection. Perhaps if scientists working on such matters as the identification of underground tests had had greater motivation or encouragement from their superiors to solve the technical problems involved in disarmament and arms control, results might have been more fruitful.[71]

The Sino-Soviet break

During this time Khrushchev had been pursuing his policy of peaceful coexistence with the Western powers. Indeed, according to the Chinese, '... the Soviet Government unilaterally tore up the agreement on new technology for national defence concluded between China and the Soviet Union on October 15, 1957, and refused to provide China with a sample of an atomic bomb and technical data concerning its manufacture'.[72] In September, 1959, Khrushchev visited the United States and agreed to veto-free, on-site inspections by mixed national-foreign teams on a quota basis. Thus it was in this more 'hopeful' atmosphere that the Geneva negotiations recommenced on October 27, 1959.

Soviet reaction to the decoupling theory

Agreement, however, was not achieved. The Soviet scientists continued to object to the American analysis of the Hardtack and more recent data. But it was the decoupling theory which caused the final rift. After Bethe and Latter had presented the theory, the Soviet scientist Fedorov commented bitterly that 'the contribution to our work which Dr. Latter tried to make in his report is quite similar to the contribution that he made previously to this problem; I mean the book that he published on the subject along with Dr. Teller' (*Our Nuclear Future*).[73] The Soviet scientists admitted the theoretical validity of Latter's calculations but declared that since there was no empirical confirmation of the theory further discussion of it would be 'premature'. On December, 18, the day the Working Group finally recessed, Fedorov stated, 'I will be very frank. I have absolutely nothing to do with your [decoupling] report. It is no longer of any interest to me.'[74] In their final report appeared the sentence: 'The Soviet experts submit that here their United States colleagues are on the brink of absurdity.'[75] Because of these and similar remarks made by Russian scientists, Gilpin concludes that 'the Soviet treatment of the decoupling theory defended by Bethe and Latter was most condescending and even insulting'.[76]

Political evaluations and research programmes

What, however, was Bethe's opinion? In August, 1960, he summed up thus his attitude towards the Soviet Union and the decoupling theory. 'I do not think the Russians intend to violate a treaty banning weapons tests,' he wrote. 'I do not think that the Russians could risk cheating, even if there is only a small likelihood of being detected.... I believe that the Soviet Union, which is posing as a peace-loving nation, whether rightly or wrongly, simply cannot afford to be caught in a violation, and therefore I think that it will not try to cheat.'[77] And thus with respect to the decoupling theory Bethe wrote: 'I had the doubtful honour of presenting the theory of the big hole to the Russians in Geneva in November, 1959. I felt deeply embarrassed in so doing, because it implied that we considered the Russians capable of cheating on such a massive scale. I think that they would have been quite justified if they had considered this an insult and had walked out of the negotiations in disgust.'[78] Bethe's opinion in his *Scientific American* article was expressed no less frankly than Fedorov's: 'We are all behaving like a bunch of lunatics to take any such thing as the big hole seriously.'[79] As Bethe told the US Joint Committee on Atomic Energy: 'The next round ought to go to the detection rather than to the concealment [of nuclear explosions].'[80]

But Teller and Latter shared a position diametrically opposed to Bethe's. In his January 1958 *Foreign Affairs* article, Teller had asked: 'Is it wise to make agreements which honesty will respect, but dishonesty can circumvent? Shall we put a free, democratic government at a disadvantage compared to the absolute power of a dictatorship.'[81] In *Our Nuclear Future*, published in 1958, Teller and Latter had concluded that 'According to past experience, an agreement to stop tests may well be followed by secret and successful tests behind the iron curtain.'[82] And thus they undertook a highly successful research programme to show that it is always theoretically possible to secretly violate a test ban treaty.

Indeed, because of their different political judgements, Bethe and Teller not only committed themselves to different research programmes but also

THE SCIENTIFIC COMMUNITY IN CRISIS

evaluated the significance of the existing data on testing in radically different ways:

Bethe: 'I believe, therefore, that it is technically feasible to devise a system of detection stations and inspections which give reasonable assurance against clandestine testing with the possible exception of very small, decoupled tests.'[83]

Teller: 'This is the impasse at which we find ourselves today. We can say simply, surely, and clearly that if we agree on test cessation today, we have no way of knowing whether the Russians are testing or not. There are no technical methods to police a test ban.'[84]

In July, 1960, the Federation of American Scientists summed up the feelings of the pro-ban scientists: '... we do not know that our efforts have failed because of Soviet intransigence. Our efforts have not been complete enough for us to be confident that we would have found a basis for agreement if such existed. Our Government has not undertaken on a sufficient scale the hard work and intensive research that are necessary for an informed political judgment on specific arms control proposals.'[85]

The experimental testing of the decoupling theory

Given the importance of the decoupling theory in contributing to the breakdown of the Geneva talks, it is interesting to assess the attempts made to test the theory experimentally. In early 1960 the so-called Dribble programme was established by the AEC but had to be suspended in early 1962 for lack of funds after only exploratory drilling had been performed. Supported by the Defense Department, the programme was resumed in September, with the estimate that construction for the first cavity to accommodate a 100 ton detonation would cost $3 200 000 and would require almost a year. By mid-1964 construction had not yet started on the cavity although more than the estimated cost had been spent. Jacobson and Stein report that as of September, 1965, no nuclear decoupled explosion had been made.[86]

The Kennedy administration and the partial test ban treaty

The U2 flights and satellite reconnaissance had demonstrated to American planners the relative military inferiority of the Soviet Union. Nevertheless, President Kennedy, after campaigning on a supposed 'missile gap' (the existence of which was publicly repudiated by his own Secretary of Defense), requested on March 28, May 25 and July 25, 1961, additional appropriations for military purposes. Subsequently in the two years from the end of 1961 to the end of 1963 the United States doubled its number of nuclear warheads and increased the number of operational long-range missiles from 45 to 500. In September, 1961, the Soviet Union resumed testing, to be followed by atmospheric testing by the United States in April, 1962. Thus the test ban negotiations resumed in March, 1962, during a period of unprecedented military build-up. Noting that even in early 1960 the US Strategic Air Command delivery capability 'far outweighed the total Soviet capacity to strike at North America', several commentators have argued that the subsequent rapid American mobilization 'probably helped precipitate the Cuban missile adventure', for Khrushchev had been provoked in attempting to achieve parity with the United States in one decisive move.[87] Since Kennedy, however, had no intention of losing 'overnight' the overwhelming American military superiority there was therefore no choice for Khrushchev but to admit defeat. The confrontation, however, apparently convinced both administrations of the desirability of pursuing the policy of 'peaceful coexistence' initiated by Khrushchev. In a major speech delivered to the American University on June 10, 1963, Kennedy called for a re-examination of American attitudes towards the Cold War and for increased understanding with the Soviet Union. Khrushchev was impressed with the speech, Soviet jamming of the Voice of America stopped completely, and on July 2 Khrushchev proposed a partial test ban treaty omitting underground testing completely.

On July 15 the Soviet Communist Party made public its indictment of the Chinese Communist Party. Seventeen days later Kennedy declared in a press conference that he saw the combination of '700 million [Chinese] people, a Stalinist internal régime, and nuclear powers, and a government determined on war as a means of bringing about its ultimate success, as potentially a more dangerous situation than any we faced since the end of the Second World War, because the Russians pursued in most cases their ambitions with some caution'.[88]

Having established a common enemy, the Soviet Union and the United States rapidly reached agreement on a Partial Test Ban Treaty which Kennedy presented to the Senate for ratification at the beginning of August. Bethe spoke in favour of the Treaty, Teller against, the latter arguing that the United States needed to know more about the effects of nuclear explosions in the atmosphere and that the Treaty would hamper the development of an anti-ballistic missile system. However, in the Senate hearings on the Treaty McNamara declared that 'by limiting Soviet testing to the underground environment, where testing is more difficult and more expensive and where the United States has substantially more experience, we can at least retard Soviet progress and thereby prolong the duration of our technological superiority'. Furthermore, according to the Chairman of the Joint Chiefs of Staff Kennedy had promised them 'the conduct of comprehensive, aggressive and continuing underground test programs designed to add to our knowledge and improve our weapons in all areas of significance to our military posture for the future'.[89] On September 24, 1963, the US Senate ratified the Treaty by a vote of 80 to 19, the Soviet Presidium unanimously ratifying the Treaty on the next day. A first and very tentative step to control the arms race had been taken.

3. Scientists and Engineers for Social and Political Action?

Clearly the test-ban treaty was not to prove the international breakthrough the scientists (in their political misreading of Western society) had hoped it would be. While human beings were spared from more radioactive fall-out (apart from French and Chinese nuclear testing), Western imperialism continued and with it the arms race. Following Kennedy's assassination, scientists and engineers mobilized support in 1964 for Johnson and Humphrey, the 'peace' candidates. Of prominent scientists, only Teller and Libby supported Goldwater. However, despite Johnson's overwhelming victory, the scientists' hopes were not to be realized. American intervention in the Vietnamese civil war, initiated by Kennedy, rapidly escalated until, by 1967, more bombs were being dropped each year on that small peasant country than the Americans had dropped on the entire Pacific throughout the Second World War. Some 3 000 university faculty members, in sympathy with the growing youth movement against the war, protested against their government's policy in Vietnam using such terms as brutalizing and degrading, reckless and barbaric, disastrous, illegal, senseless, inhumane and immoral. The breakdown of the faculty protesters into their respective disciplines, shown in Table 11·1, is particularly interesting (a discipline's index

Table 11.1

Physics	2·53
Social sciences	2·51
Medicine and Health	1·37
Humanities	1·32
Mathematics	1·27
Biology	0·85
Engineering	0·60
Fine arts	0·60
Chemistry	0·54
Education	0·18
Business	0·10

is given by the percentage of academics in the discipline who protested divided by the percentage of academics in general who protested).[90] Of the natural scientists, physicists were clearly the most radical in their appraisal of American policy in Vietnam and indeed share the same average index as their colleagues in the social sciences. Mathematicians were also reasonably radical but biologists, engineers and chemists were under-represented among the disciplines.

The MIT 'strike'

As the escalation in Vietnam continued, and as the American government prepared to construct anti-ballistic missile systems around American cities and to test 'multiple independently targeted reentry vehicles', so the protest movement of scientists gathered momentum. In early 1969 a statement was made public, signed by 46 faculty members of the Massachusetts Institute of Technology, containing the declarations that 'misuse of scientific and technical knowledge presents a major threat to the existence of mankind', and that 'through its actions in Vietnam our government has shaken our confidence in its ability to make wise and humane decisions'.[91] The 46 faculty members acknowledged that 'a handful of eminent men' had tried but had largely failed 'to stem the tide from within the government', while the 'concerned majority' had been on the sidelines and ineffective. There followed the paragraphs:

> We therefore call on scientists and engineers at MIT, and throughout the country, to unite for concerted action and leadership: action against dangers already unleashed and leadership towards a more responsible exploitation of scientific knowledge. With these ends in mind we propose: . . .
>
> To convey to our students the hope that they will devote themselves to bringing the benefits of science and technology to mankind, and to ask them to scrutinize the issues raised here before participating in construction of destructive weapons systems. . . .
>
> To explore the feasibility of organizing scientists and engineers so that their desire for a more humane and civilized world can be translated into effective political action.

They ended their statement by calling on their colleagues—faculty and students—to stop their research activity on March 4, in order to join with them in discussion of the issues raised by their statement. The idea of a one-day 'strike' was taken up at some 30 other American universities including Cornell, Columbia, Yale, Stanford, Minnesota and Berkeley. At MIT the Nobel prize-winning biologist George Wald (of Harvard), as if in answer to Jules Henry's bitter accusation that 'together with the engineers and technicians, they [the scientists] constitute the well-fed, comfortably housed culture of death',[92] passionately advised his audience of scientists not to accept what he called 'the facts of death'. 'Our government,' he said, 'has become preoccupied with death and the preparation for death.'[93] But the business of scientists 'is with life not death'. Subsequently 182 faculty and graduate students wrote to the Presidential Science Adviser, DuBridge, urging him

> as the principal spokesman for the scientific community within the Nixon Administration to call for closer university ties not with the Department of Defense but with the Department of Health, Education and Welfare, of Housing and Urban Development, and of Transportation . . . to promote scientific endeavour not for its military potential, but for its

potential to alleviate human suffering; not 'from the point of view of international prestige', but from the point of view of science as a cultural achievement worthy in its own right of public support.[94]

At the Spring Meeting of the American Physical Society, the recently formed Scientists for Social and Political Action, aided by the Federation of American Scientists, helped organize opposition to the proposed ABM system. About 250 physicists demonstrated outside the White House, delivering to DuBridge a petition for Nixon signed by 1 100 physicists (between a third and a quarter of the total attendance at the meeting). The petition read:

> We as scientists and citizens urgently seek the withdrawal of plans to build and deploy the Safeguard ABM system. Our concern springs from two basic sources:
> 1. As scientists we are wholly unconvinced that any presently proposed ABM system can defend against a determined missile attack.
> 2. As citizens we deplore the beginning of a particularly dangerous, yet ultimately futile round of nuclear arms escalation when our expanding domestic crisis demands a re-allocation of the national resources.[95]

After an impassioned debate, the Senate accepted by one vote Nixon's proposal to proceed to the first stage in the construction of the ABM system. Despite the opposition of many scientists, the nuclear arms race had been given yet another escalation.

Perhaps symbolic of the radical thinking of *some* American scientists is the letter sent to *Nature* by three Harvard molecular geneticists concerning the publicity given to an article in which they had announced an important technical achievement. Dramatically the scientists wrote:

> We did not publicize our work in order to add to our own or Harvard's prestige or to make a plea for more money for basic research. In a country which makes a prodigious use of science and technology to murder Vietnamese and poison the environment, such an enterprise would be at best terribly irrelevant, at worst criminal. On the contrary, we tried to make the following political statement. In and of itself, our work is morally neutral—it can lead to benefits or to dangers for mankind. But we are working in the United States in the year 1969. The basic control over scientific work and its further development is in the hands of a few people at the head of large private institutions and at the top of government bureaucracies. These people have consistently exploited science for harmful purposes in order to increase their own power.[96]

The three American scientists drew the logical conclusion: 'What we are advocating is that scientists, together with other people, should actually work for radical political change in this country.'

Sakharov's manifesto

It is not only in the United States but also in the Soviet Union that scientists are demonstrating increasing concern over their failure to ensure that science be used solely for the benefit of man. In his courageous 1968 essay entitled *Progress, Coexistence and Intellectual Freedom*, Sakharov, a leading Soviet physicist who had contributed very greatly to the development of thermonuclear weapons for the Soviet Union, explained that his views 'were formed in the milieu of the scientific-technological intelligentsia, which manifests much anxiety over the principles and specific aspects of foreign and domestic policy and over the future of mankind'.[97] In calling for an end to the Cold War, for the removal of censorship, for support to Czechoslovakia and for greatly increased aid to the underdeveloped world, Sakharov declared that the 'all-encompassing scientific and technological revolution . . . will require the greatest possible scientific foresight and care and concern for human values of a moral, ethical and personal character' and 'will be possible and safe only under highly intelligent worldwide guidance'.[98] Sakharov left his readers in no doubt that in his opinion the Soviet Communist Party is not at present contributing to the solution of the world's problems—indeed, is acting so as to aggravate them further. Sakharov's indictment has not been published in the Soviet Union. The fact, however, remains: both Soviet and American scientists are expressing increasing concern at the reluctance or inability of their respective governments to act in what the scientists conceive to be a responsible and humane manner.

Scientists in the service of life?

Since Hellbroner published his essay in 1966 scientists have indeed become more socially concerned. The immensity of the problems confronting humanity is becoming increasingly apparent. In a well-known article the American biophysicist John Platt, after considering the more immediately pressing social problems, made some alarming predictions. 'Whether we have 10 years or more like 20 or 30, unless we systematically find new large-scale solutions, we are in the gravest danger of destroying our society, our world, and ourselves in any number of different ways well before the end of the century.'[99]

It is interesting to read Platt's prescriptions in the light of Heilbroner's speculations. 'What can we do? I think that nothing less than the full application of the full intelligence of our society is likely to be adequate. These problems will require the humane and constructive efforts of everyone involved. But I think they will also require something very similar to the mobilization of scientists for solving crisis problems in wartime.' Yet Platt offers no easy cure. 'Even a great mobilization of scientists may not be enough. There is no guarantee that these problems can be solved, or solved in time, no matter what we do. But for problems of this scale and urgency, this kind of focusing of our brains and knowledge may be the only chance we have.' The point is that while 'it is the new science and technology that have made our problems so immense and intractable . . . it may be only more scientific understanding and better technology that can carry us past them'.[100] Indeed, says Platt, 'the whole human experiment may hang on the question of how fast we now press the development of science for survival'.[101]

Similarly George Wald in his famous talk at the MIT 'strike' of 1969 recognized that the problems were global in nature, requiring therefore global solutions. 'The thought that we are in competition with Russians or Chinese is all a mistake and trivial. We are one species with a world to win . . . Our challenge is . . . what becomes of men—all men of all nations, colours and creeds. It has become one world, a world for all men. It is only such a world that can now offer us life and the chance to go on.'[102]

Such comments by scientists are now easy to find. As a third and final example, we note that King-Hele, deputy chief scientific officer at Farnborough, asserts on the first page of his 1970 essay that 'man has many possible futures, but the most likely ones are disastrous'.[103] Like Wald, King-Hele writes that if we are to survive, 'we shall have to develop world loyalty, world-wide sympathies and international government, in place of the narrow nationalism and the racial and ideological hatreds so rife today'. Again we find a scientist condemning the lack of a directing intelligence in world affairs. 'The car of human life seems to be careering down the road to the future without proper brakes and steering.' 'The world,' says King-Hele, 'has always been governed, and never more so than today, by the men of energy rather than the men of intellect . . . they mouth optimistic platitudes; they lack the scientific knowledge needed to grasp the real world.'[104]

There is, therefore, as these several examples suggest, growing social concern on the part of natural scientists, a *concern* that is clearly to be welcomed. It is, however, less clear, as we argued in chapter 10 and as these last three examples themselves suggest, that this concern will give rise to a commitment by scientists to help in the building of a *liberated* world society. For while Platt puts great stress on the application of 'intelligence' to solve social problems, one searches in vain through his paper for an ethical commitment to use science in the service of the oppressed of the earth against their Western exploiters. Of course, such an undertaking would be very difficult. Platt would find that Western governments and corporations are singularly reluctant to employ scientists with such aims—although not scientists whose aim is to put science at the service of exploitation. The most essential component of 'the solution', therefore, is not more science, as Platt suggests, but radical social change towards a *liberated* world society. Wald, too, in his talk failed to analyse the nature of the society he lives in. While it is very comforting to be told that 'the thought' that the United States is in competition with the Russians and Chinese 'is a mistake', unfortunately this 'competition' is part of the *reality* of the world situation; and the 'mistake' can only be remedied in a cooperative and liberating way if and when there is revolutionary social change in the United States. Basically the same remarks apply to King-Hele's essay: it is not more 'intelligence' that the world primarily needs (which is King-Hele's thesis) but the defeat of Western imperialism; the unfortunate fact of the matter is that 'intelligence' only too readily finds itself profitable employment in what it fails to see as the defence and, indeed, promotion of an intrinsically exploitative social system.

Should one wish to feel even more doubtful of the liberating potential of science at this historical juncture, the following remarks by Stafford Beer, President of the (British) Operational Research Society, might be borne in mind. (If he had intentionally wished to incense Theodore Roszak, he could not have expressed himself to better effect.) 'At present,' Beer writes in his article entitled 'Managing Modern Complexity',

> the most obtrusive outcome of the system we have is a gross instability—of institutional relationships and of the economy. This cannot last. The society we have known will either collapse, or it will be overthrown. In either case a new kind of society will emerge with new modes of control; and the risk is that it will be a society which no one actually chooses and which we probably will not like.... And let us so engineer our systems that their latent outcomes suit our social purpose. It is true that the outcomes cannot be fully determined, because there is a noise (or shall we call it free-will?) in the system. But a systemic design taking due account of cybernetic laws may be expected to produce behaviour which is predictable in terms of the overriding need for stability.[105]

And in a later article: 'How can cyberneticians say that we know best and everyone else is wrong? It is that we are *responsible*. We are not responsible because we have been elected to govern affairs; we are responsible because cybernetics, that science of effective organization, is our profession.' Beer writes that the socio-political sciences 'have forfeited their claim to respect by inventing or embracing ideological frameworks for discussion and research which are antipathetic to scientific advance. As a result mankind does not at all understand, and cannot therefore control, the complex forces which have pushed him to the edge of the evolutionary cliff.' Thus it is 'that the moment has come for us [cyberneticians] to start work, and we must do the best we can'.[106]

Peccei, founder of The Club of Rome and member of the multinational corporate élite, shares Beer's view that the world is reaching 'a critical moment'. In a discussion with Shonfield (Director of the Royal Institute of International Affairs and author of *Modern Capitalism*), Peccei argues that the disastrous results of uncontrolled material growth on a world scale necessitate a detailed study of 'the global system' in order to see 'how we can best rationalize the world's productive system'. Shonfield agrees with Peccei that there are problems. In his view: 'I do not see how the poor two-

THE SCIENTIFIC COMMUNITY IN CRISIS

thirds of mankind can achieve what I regard as a tolerable standard of living. ... If the Third World succeeds in developing rapidly along traditional lines, this will exacerbate the problem of the shortage of resources of the planet to an intolerable extent in the twenty-first century.' He then spells out what he believes to be the political implications of Peccei's programme: 'What we are really talking about is the need to exert far more central authority over masses of people. ... The point I want to make is that implicit in the Peccei system of global management is the reassertion of imperial power on a world scale.' Peccei comments: 'No, I wouldn't use those words. I think there would simply be a rethinking among *the decision makers* towards taking responsibility for the long-term development of the economy or society on a global basis' (emphasis added). Peccei believes that what is first necessary is to see 'if this huge, complex equation can be put through our minds and then our computers'.[107] MIT systems analysts have offered The Club of Rome their services.

Yet it would be misleading to conclude the chapter on this pessimistic note. Many groups of scientists are well aware of the ideological service rendered by modern science to ruling élites. Just one example: in a manifesto for a conference organized by the British Society for Social Responsibility in Science, a group of scientists avowed an anti-élitist programme of action. Declaring that 'the mystique and exercise of expertise are central to the anti-democratic structure of our society', and that 'science underpins the authority of "experts",' they expressed solidarity with 'the struggle for workers' self-management in science', a struggle which can 'contribute to breaking down deference to specialized knowledge'. In declaring opposition, however, to 'a movement for scientists only', they expressed their desire to 'fight alongside other working people' in order 'to achieve workers' self-management in science and throughout society'. They stated their aim was to 'place control of the sources of funding for research and the objectives of our work in the hands of the people'. For 'we must socialize knowledge just as we must socialize property. Neither process alone can achieve an egalitarian society.'[108]

Liberation or no, these examples lend considerable support to Heilbroner's thesis that there is a basic incompatibility between capitalism and science. Of course, many of the scientists who stress so explicitly the need to apply man's intelligence to global problems do not at present recognize explicitly that this is precisely what capitalism is basically unable to permit. But, if Heilbroner is right, such scientists will be led to this conclusion since (as he argues) 'each small step taken to correct [capitalism's] deficiencies only advertises the inhibitions placed on the potential exercise of purposeful thought and action by its remaining barriers of ideology and privilege'.[109] In the end, believes Heilbroner, the world of capitalism will be condemned by the world of science. Certainly, as the problems intensify, we can reasonably expect both natural and social scientists to become increasingly concerned and politically conscious. We can reasonably expect them to realize that to remain 'apolitical' and 'value-free' is to support the *status quo* and thereby to invite disaster. What, however, cannot be taken for granted is that scientists will commit themselves to the building of a liberated society and not to the building of a society directed by 'intelligence' in the service of that 'intelligence'.

The struggle will be long and hard. Perhaps the chapter is best summarized and ended by noting the following insights by Marcuse. 'Within the total mobilization of man and nature which marks the period, science is one of the most destructive instruments—destructive of that freedom from fear which it once promised.'[110] It should not be forgotten, Marcuse tells us, agreeing with Veblen, that the scientists, researchers, technicians, engineers and 'even psychologists' are 'today the pet beneficiaries of the established system'. Yet there is a chance. For, declares Marcuse, the scientists 'are also at the very source of the glaring contradictions between the liberating capacity of science and its repressive and enslaving use'. The task of the revolutionary is therefore clear: it is 'to activate the repressed and manipulated contradiction, to make it operate as a catalyst of change'.[111] But the chance is only a slim one. Marcuse quotes an engineer as saying: 'My real love is minimum weight structures . . . but I'm willing to work on minimum cost structures or how to kill the Russians better because the organization survives by doing research that's saleable.' 'Potential revolutionaries?' asks Marcuse.[112] We return to this question in the concluding chapter.

Chapter 12

TOWARDS A BEAUTIFUL WORLD
SOME CONCLUDING THOUGHTS

> For the Western World already has the resources and the technique, if we could create the organization to use them, capable of reducing the Economic Problem, which now absorbs our moral and material energies, to a position of secondary importance. Thus the author of these essays, for all his croakings, still hopes and believes that the day is not far off when the Economic Problem will take the back seat where it belongs and that the arena of the heart and head will be occupied, or reoccupied, by our real problems—the problems of life and human relations, of creation and behaviour and religion.
> JOHN MAYNARD KEYNES *Essays in Persuasion*

ONE of the principal arguments underlying this series of sketches has been that current global problems cannot be resolved within the institutional framework and class structure of Western societies but require the construction of a radically different social order for their successful resolution. In this concluding chapter I wish to elaborate on this theme, first of all by summarizing the argument of the essay so far, and then by sketching some features of the new paradigm whose adoption and articulation by the now capitalist countries is a prerequisite, I believe, for the successful continuation of man's quest to build a beautiful world.

1. Résumé

Chapters 1–3: Objectivity and commitment in the physical sciences

We began this essay by recognizing that although the world is in a crisis state, it is by no means clear how one should act, how one can justify one's actions to oneself, to others. Yet action cannot be avoided. Does one commit suicide . . . , continue to 'keep politics out of science' . . . , commit oneself to radical political action? No matter how one decides to act, how can one be sure that one's actions are not the very opposite of those one would undertake if one had only more relevant information or even simply more time to think? Yet of course it is far from apparent that one can *ever* have enough information to warrant certainty, ever enough time to digest even that information which is available. We noted that prominent thinkers of reasonably similar values had come to diametrically opposed evaluations of the world situation and some had even changed their minds from one 'extreme' to another. We asked how this was possible and it was in order to throw some light on these questions that we looked first at the undoubted progress of the physical sciences and at attempts made to understand the nature of that activity which appears so readily to result in consensus among its practitioners.

Now whether or not scientific activity is characterized by long periods of consensus, it was immediately apparent that no consensus is shared among contemporary philosophers of science concerning either the nature of science or the nature of that scientific practice most likely to achieve progress in science. Whereas Popper argues that the true scientist as opposed to the pseudo-scientist strives constantly to falsify his theories and immediately abandons them as soon as there is disagreement between theoretical predictions and experiment, Kuhn takes issue with this point of view in stating that all theories confront problems at all times and that the flesh and blood scientist quite rightly does not reject the scientific community's 'paradigm' because of disagreement with experiment but rather commits himself to resolving the ever-emerging 'puzzles' within the established framework of concepts, hypotheses and standards. There is no point (Kuhn argued in 1962) at which it becomes illogical or unscientific to continue to believe that a difficult puzzle cannot be so resolved. A crisis state occurs, however, when a minority of scientists not only convince themselves that a radically new paradigm is needed in order to resolve some outstanding puzzles (on which perhaps the attention of the scientific community has long been focused) but also conceive of the framework of a new paradigm whose articulation (they believe) will succeed in solving outstanding puzzles. A period of revolutionary science then follows in which scientists articulating different paradigms no longer necessarily agree on the nature of science and in particular on criteria for identifying problems or for evaluating proposed solutions to common problems. Eventually, however, one paradigm is declared the most fruitful guide to future research problems, a success in which Kuhn argues that charisma and rhetorical persuasion play no inconsiderable part. Textbooks are then rewritten and a new generation of scientists is inculcated with the belief that the community's new paradigm can be articulated so as to resolve all the many problems with which it is and will be confronted. Without such conviction, Kuhn argues, science cannot progress to the next revolutionary period.

We noted that Kuhn's radical thesis has been attacked by several prominent philosophers of science on the grounds that it appears to allow scientists to act in whatever way they want to act, to believe whatever they want to believe, no matter what the evidence against the paradigm—that it can be used to justify commitment to any form of belief and action. In short, these philosophers argue that Kuhn not only gives no objective criteria to distinguish truth from falsity, rational action from irrational action, but implies that none are possible. Nevertheless, despite the acerbity of their polemic against Kuhn not one has succeeded in establishing objective criteria which would allow scientists to choose unambiguously between competing theories. Lakatos' attack, although of great importance, is clearly not decisive. Like Kuhn, he agrees that since all theories are 'refuted' at all times, 'refutation' must not mean automatic rejection; also like Kuhn he puts emphasis not on competition between theories but between series of theories or research programmes. However, unlike Kuhn, by assuming that all research programmes eventually share problems in common together with criteria for acceptability of solutions, Lakatos is able to argue that that research programme able to solve problems its competitors 'after sustained effort' cannot solve is the research programme to which the rational scientist should commit himself. Commitment to the hard core of the losing paradigm is rational only in that period in which the network of protective surrounding theories is continually being theoretically enriched although without the empirical corroborations that would constitute acceptable solutions of outstanding problems. Prolonged failure to achieve solutions must compel transfer of commitment, on pain of irrationality, to the successful research programme. We noted three difficulties with Lakatos' attempt to make scientific revolutions 'rationally reconstructible'. In the first place, his crucial assumption mentioned above is questionable. In the second place, he gives no objective criteria for deciding on what constitutes 'sustained effort'. If there is no point at which it becomes irrational to make one last ingenious effort to solve the problem, why not just keep on trying? In the third place, Lakatos gives no indication as to why scientists decide to commit themselves to a particular research programme and suggests no criteria for deciding when commitment to a new and undeveloped research programme must be considered irrational. Is it rational to try to develop any research programme whatsoever? Are there no criteria guiding scientists in their choice and construction of research programmes? It was in order to throw light on these questions, although not necessarily to answer them, that in the next two chapters we examined two major revolutions in physics, the Copernican revolution of the 16th and 17th centuries and the Einsteinian revolution at the beginning of the 20th century.

Although both established research programmes were confronting many puzzles when Copernicus and Einstein made their revolutionary proposals, this had been a continual feature of the Aristotelian-Ptolemaic and Newtonian paradigms. It is therefore of great significance that both Copernicus and Einstein expressed aesthetic dissatisfaction with the existing paradigm and committed themselves to the articulation of a paradigm each believed to be more simple and more harmonious than the existing one. Furthermore, for at least several years during the two revolutionary periods, there existed no experimental or observational evidence of such a nature that it compelled all scientists to accept, on pain of charges or irrational behaviour, only one of the competing paradigms as worthy of further attention. Moreover, the revolutionary Copernican and Einsteinian paradigms faced serious problems which did not exist for the established paradigms. (What force moves the earth? How are waves propagated through empty space?) Despite such difficulties, however, the revolutionary scientists retained faith in the eventual fruitfulness of their 'utopian' paradigms, a commitment rewarded by a series of progressive problem shifts that appeared to establish beyond all 'reasonable' doubt the superiority of the revolutionary paradigms. The Tychonic and Ritzian compromise paradigms received little attention; not only did they lack the simplicity, mathematical beauty and unifying power of their revolutionary rivals but also they were sufficiently different from the established paradigms so as not to win over conservative opposition. As to Einstein's later refusal to help articulate the successful quantum mechanical paradigm, we decided that this could not be considered irrational action since Einstein had both acknowledged the power of the new paradigm and had demonstrated his understanding of its basic principles. Although Einstein's critics saw only justification for their own paradigm where Einstein saw problems, nevertheless could Einstein have predicted the behaviour of individual atomic systems then another revolution would have occurred. At least until the next revolution in physics does occur it is perhaps only prudent not to pronounce too definite a verdict on Einstein's apostasy.

To what extent did our excursion into physics throw light on the basic questions asked earlier? In this very important way: we established a strong case that even this most mathematical of all sciences is far from being the 'objective', 'value-free' science it is so often made out to be. On the contrary, aesthetic criteria and personal commitment play a crucial role, particularly in periods of crisis. Although the goal of physicists is to explain and predict from a single set of principles the behaviour of the inanimate world—this world, our world, not a world that exists only in our imagination—nevertheless in order to achieve this objective not only is great imaginative ability necessary but personal decisions must continually be made as to when problems confronting a research programme challenge only the ingenuity of its supporters or, on the other hand, demand revolutionary changes that imply rejection of one or more of the hard-core principles underlying the entire programme. We have seen that the making of such decisions is guided in very large part by subjective criteria as to what is beautiful and what is not, to what unifies and what does not. Aesthetic criteria to a physicist are as winds to a sailing ship—without them he would be becalmed. That physicists have discovered a value-free method by which progress in science may be ensured is not merely a myth but, as we argued in later chapters, a dangerous myth.

Chapters 4–6: Objectivity and commitment in the social sciences

We noted that the early 17th century gave rise to several Hermetic utopias in which it was confidently argued that the wise use of science would ensure the construction of societies free from disease, poverty, back-breaking toil and illiteracy, societies in which men would help each other develop their individual abilities. Both Campanella and Andreae stressed that for such societies to be possible private ownership of wealth would have to be abolished, just as More before them had stressed that, no matter how many reforms were made, a society in which a few owned most of the wealth would remain inherently diseased. Science, however, was to prosper in those emerging societies founded on private ownership of the means of production. Although Adam Smith argued in the 18th century that the 'general good' would nevertheless result from the pursuit of private interests, Karl Marx in the 19th century argued a very different thesis: that the problems confronting capitalist societies would intensify to such an extent that 'a universal crisis' would inevitably occur, so giving a grateful proletariat the opportunity of creating a classless society in which each man, in cooperation with all other men, would be free to develop and improve on the kind of society of which the earlier utopian thinkers had but dreamed. When, however, the Western world did suffer a universal crisis in the 1930s, capitalist society proved itself able to overcome this particular crisis *with its class structure and basic institutional framework unchanged*. We were, of course, particularly interested to see how social scientists, and especially economists, had reacted to the alarming (or hopeful!) social situation in which they found themselves in the 1930s.

Apart from the occasional heretic, economists in the forty or so years before 1929 had been practising 'normal science', firm in their belief that their paradigm could be articulated to resolve all problems with which it might be confronted; indeed, according to their paradigm there would be no *serious* problems to resolve since a self-adjusting mechanism automatically ensured maximum economic growth by conveniently preventing anything more serious than temporary departures from full employment. Prolonged large-scale unemployment was declared to be an impossibility in a perfectly competitive capitalist society. The moderately large-scale unemployment following the Great Crash was first treated as a puzzle (or even as a normal and welcome event signifying the inevitable downturn of the trade cycle) but four years later as a very serious problem when in 1933 one quarter of the American labour force became unemployed. Of those social scientists committed to preserving the 'hard-core principles' of capitalist society, the majority held that actions such as trade union activity and government interference in the economy were responsible for preventing the economy from functioning 'properly'. However, a minority lost faith—not in capitalism—but in the paradigm with which they had been brought up and sought to show that even a perfectly frictionless, perfectly competitive capitalist society could generate prolonged large-scale unemployment. For our purposes, we took Robbins and Keynes to be representative of, respectively, conservative and revolutionary theorists committed to the preservation of capitalist society. Both wished to eliminate the existing large-scale unemployment while preserving the fundamental features of capitalism; both saw as a threat not only fascism but also the possibility of proletarian revolution and the construction of a planned economy in which no private ownership of the means of production would be legally possible. We briefly discussed basic differences between the neo-classical and Keynesian theories and the possibility of deciding 'objectively' between prescriptions for action based on the two competing theories. The results of Roosevelt's New Deal Policy appeared inconclusive. It was not until some time after the commencement of the Second World War—which ended the unemployment crisis of capitalism—that it became evident that Keynes' prescriptions were the more valid and that even under the totally unachievable conditions of perfect competition capitalism would almost certainly not be the self-adjusting socio-economic system the neo-classical economists had theorized it to be. We recognized, however, that for several years prior to this all but conclusive test of his theory, Keynes had been trying to persuade governments to test his theory in a way which, if he were right, would help preserve the kind of society in which he, Keynes, most wished to live. Like Copernicus and Einstein before him Keynes wholeheartedly believed in the validity of his theory before convincing 'empirical' evidence in its favour had been produced by the actions of the capitalist nations.

We then discussed similarities and differences between the physical and social sciences noting the major and obvious difference that, whereas the subject matter of the physical scientists manifests a stoic indifference to whatever paradigm happens to be in vogue, the 'subject matter' of the social scientists not only generates its own ideas and acts on them but is scarcely indifferent to the paradigms, and changes in paradigms, of the community of social scientists. There can be no pure social science in the sense that pure physics exist; it is useless to look for social laws that human beings obey as the planets obey (or approximately obey) Newton's laws. What must guide the social scientist in choosing between different research-action programmes is to a very considerable extent his choice of future. And choosing between different futures is not likely to be value-free! Just as in the physical sciences aesthetic criteria play an important role in both problem identification and problem solving, so in the social sciences aesthetic criteria in problem identifying and solving are reinforced by moral, ethical and political criteria. Whereas in the physical sciences it is simply wrong to claim that there exist 'value-free', 'objective' criteria which *at all times* enable the physical scientist to decide to which research programme he should turn in the hope of solving outstanding problems, in the social sciences it is not only wrong but dangerously so. For the ideology of 'value-free' social science promotes a belief that in so far as it is scientific to attempt to solve social problems at all, it is so only if solutions are sought within the social framework whose values the social scientist has internalized. Thus in tending to deprive contemporary society of speculation with respect to 'alternative futures', the ideology of 'value-free' social science is not only intellectually stultifying but also, given the direction of existing trends, exceedingly dangerous.

Chapters 7–9: *A world in crisis*

Guided, above all, by the objective of building a non-exploitative world society we attempted in the next three chapters to analyse the development of American capitalism, the interrelationship between the capitalist and underdeveloped countries and the attempt to construct a socialist society in the Soviet Union. The results of this analysis, summarized below, were sobering.

In the 19th century, capitalism, while a harsh and exploitative social system, was nevertheless paving the way through its development of the productive forces towards the possible elimination of material scarcity. At the same time its internal dynamic gave birth first to large corporations which have in the 20th century dominated the economies of both the advanced capitalist and, through investment, the underdeveloped countries, and then in the 1930s to widespread economic crisis—a possible turning-point for mankind. However, the Great Depression in the United States was terminated not by the construction of a socialist society but by the Second World War and the application of Keynesian economic theory in the maintenance of effective demand through the growth of a permanent war economy. The threat of material sufficiency, which would mean great difficulties for the capitalist system, is thus contained by expenditure on space exploration and, more importantly, by massive expenditure on armaments (beyond what is necessary for controlling and defending the overseas empire). The arms race, we noted, tends to be self-perpetuating and self-fulfilling; constant technological innovation maintains its momentum and thereby permanently underwrites the capital goods industries. Income inequality together with advertising, built-in obsolescence and constant innovation also contribute to the perpetuation of material desire in the minds of income earners, and thus help to maintain near full-employment and a profit rate that serves to keep the capitalist economy ever-expanding. Since constant expansion is a functional necessity of capitalist society, it is not surprising that although before the Second World War the United States was a net exporter of minerals, it is now a net importer and each year becomes increasingly dependent on minerals from underdeveloped countries. Furthermore, since competition from Europe and Japan for these minerals is increasing, it is therefore also not surprising that attempts by peoples of underdeveloped countries to achieve meaningful economic independence are met with repression by the United States and the minor capitalist powers. (For were the underdeveloped countries to begin to industrialize on a scale anywhere approaching that of the United States, a critical shortage of minerals and other raw materials would first of all bring the economies of the capitalist countries to a halt and then the economies of the industrializing countries.) In the meantime irreversible ecological damage continues to occur. Yet despite the fact that mineral sources are being depleted, that industrial pollution is increasing and is already becoming a serious health hazard, that one-half of the world's population lacks the benefits of a limited amount of industrialization, the capitalist countries continue to produce and consume an ever-increasing volume of goods—and waste—each year. Unless therefore there is radical change in the capitalist countries, the future will see increased pollution, decreasing sources of minerals, an increasing and ever-more undernourished world population and ever-more extensive repression by the capitalist countries of the non-white races as the latter attempt to achieve a redistribution of the world's wealth.

It was of particular significance, therefore, to ask whether the past brutalities and the present extremely disagreeable features of the Soviet Union must mean rejection of the commitment to build a socialist alternative to capitalist society, i.e. an *egalitarian* society in which men *cooperate* together for the wellbeing of all. Our answer was in the negative. An understanding of the appalling conditions under which the Soviet experiment was tried, while in no way excusing Stalin's brutal excesses, nevertheless more than suggests that the Soviet Union cannot be regarded as a test case of socialist practice. Indeed, in so far as the so-called socialist countries, emerging out of the most desperate poverty, must arm themselves against aggressive acts by the capitalist nations there can be no adequate approximation to the socialist ideals of equality, cooperation and individual self-fulfilment. It is only realistic to observe that the necessity of defence against capitalist imperialism tends to force 'socialist' countries to assume a way of life that is far from socialist. Although much has been and can be achieved by impoverished countries which have undergone 'socialist' revolutions, nevertheless the crucial test for socialism will not come until the most scientifically and technologically advanced of all nations, the United States, attempts the socialist experiment and thereby gives the world the opportunity to avoid barbarism by creating a world socialist society.

Chapters 10 and 11: Science in crisis

In the next two chapters we looked at the role of scientists—at the role of those men who once proclaimed the faith that the progress of science means the progress and welfare of all mankind but who are now faced with the task of convincing a growing and increasingly sceptical proportion of the peoples of advanced industrial countries that science is not the root cause of the ills with which these countries are afflicted. Their task is no easy one.

For, it is argued, sustaining and perpetuating the arms race with atomic, biological and chemical weapons, sustaining and perpetuating a system of waste embedded in an ocean of deprivation and suffering, are the scientists and technologists. The B-52s that patrol the skies, the minuteman missiles that lie in the bowels of the earth, the polaris submarines that thread the seas—all these instruments of power and death are beginning to be seen not so much as the result of international rivalry in a scientific era but rather as the very symbols of a scientific culture. Emphasis on the traditional belief that scientific knowledge is neutral, that the scientific enterprise ought not to be blamed for the misuse man has made of the knowledge and techniques it has provided, is now being challenged by the belief that scientific goals are far from ethically neutral, that the most important goal of the scientific enterprise has never been 'the relief of man's estate' but rather the attainment of power over the non-living and the living, so that 'the effecting of all things possible' might be achieved. Scientific reason has not been betrayed (it is claimed); on the contrary, since the goal of scientific reason has never been to build an egalitarian, beautiful world (science is inherently an élitist enterprise) but to attain power, it is all too obvious that scientific reason has been loyal to its primary purpose. To be sure, critics of science readily agree that beneficial side effects have resulted and will continue to result from the scientific enterprise, but such triumphs serve no longer to illustrate the desirability of science in the eyes of these critics but rather the failure of science to bring into being a world at peace, a liberated world society. The progress of science, the critics claim, serves only to take mankind ever further from the achievement of such a goal. It is argued that the slogans of 'ethical neutrality', 'disinterested curiosity' and 'knowledge for its own sake' serve as ideological masks to allow scientists to explore the limits of the possible and thereby to achieve the power over nature and other men that they intend their science to bring. Is it really surprising, it is asked, that scientists should have served so faithfully in the prosecution of the arms race? Not only has the continuation of the arms race meant the allocation of vast annual sums for science and technology and the guarantee of employment for scientists and technologists (an important requirement in a society not ruled by a scientific élite in the interests of science) but also it has provided the opportunity and the social encouragement to push back the limits of the possible in all possible directions. The fear is expressed, moreover, that scientists will not submit indefinitely to being ruled and that, therefore, in the long run, the probability is great that scientists will become the new ruling élite, replacing the capitalist ruling élite as that élite replaced aristocratic rulers. And that event, in the eyes of these critics of the scientific enterprise, will spell ultimate disaster for humanity. It follows, therefore, that as educated, sensitive young people all over the world reject the goal of power for the enhancement of power, it is with wisdom that they turn their gaze long and hard at power incarnate—the scientific enterprise.

Since the discussion on the ethos of science and the role of scientists will be reasonably fresh in the minds of readers I will not attempt to précis the discussion of chapters 10 and 11 but will give a further example of a scientific attitude to life that is so feared by critics of the scientific enterprise and then briefly summarize a few of the conclusions resulting from the main discussion.

Of particular relevance to an understanding of the inherent goals of the scientific enterprise and of the current reaction against science is a little work by a major scientist, J. D. Bernal, first published in 1929, and reprinted in 1970 with a short introduction by the author in which none of his original ideas is explicitly withdrawn. Certainly the year 1970 was an inauspicious choice in which to have such a book reprinted if the author's intention was to reassure a young and potentially hostile audience that the long-term goal of science is the service of humanity. The following are some of the ideas that Bernal expresses. The people of the future, he writes, 'will have anything from sixty to a hundred and twenty years of larval, unspecialized existence' during which to exhaust the possibilities of the body 'in dancing, poetry and love-making'[1] When the body eventually begins to wither, the brain will be removed from it and carefully rehoused in a short, cylindrical container that not only keeps it alive but whose movements the brain can direct. (As to the possibility of keeping a brain alive removed from its body, Bernal has this to say: 'After all, it is brain that counts, and to have a brain suffused by fresh and correctly prescribed blood is to be alive—to think. The experiment is not impossible; it has already been performed on a dog, and that is three-quarters of the way towards achieving it with a human subject.') The future man, therefore, that undergoes the process described by Bernal 'would emerge as a completely effective, mentally directed mechanism', 'a person mechanized for science rather than for aesthetic purposes'. Such a revitalized man is to be welcomed. For 'normal man is an evolutionary dead end; mechanical man, apparently a break in organic evolution, is actually more in the true tradition of a further evolution'.[2] Bernal, however, sees the danger that with the coming of the universal peace he desires 'the human race may well find itself statically employed in leading an idyllic Melanesian existence of eating, drinking, friendliness, love-making, dancing and singing, and the golden age may settle permanently on the world.' Yet there is more than a good chance that 'we may find the capacity to live at the same time more fully human and fully intellectual lives', as described above, 'for a sound intellectual humanity . . . will need a real externalization in the transforming of the universe and itself'. The point is that scientific and technological changes 'give more and more power to those groups of men which are involved in them and are bringing them about, so that, up to the present, in the war of the machines, the mechanists have always been the victors'. Nevertheless, Bernal sees that 'if the emotional reaction of the mass increased more rapidly than the power of the mechanists, the reverse would be the case. A severe crisis in mechanical civilization brought about by its inherent technical weakness or, as is much more likely, by its failure to arrange secondary social adjustments, is likely to be seized upon by the emotional factors hostile to all mechanism, and we may be closer to such a reversion than we suppose.'[3] That was written in 1929; the reprinting comes at a time when Roszak's essay *The Making of a Counter-Culture* is both a symptom and a contributory factor to the spreading of an emotional revulsion against science. Scientists are alarmed. The scientific adventure (equals the human adventure) must not end. The Faustian (anti-physicalist) scientist, seeking the infinite and welcoming perpetual discontentment, is appalled at the possible curtailment of the scientific quest. The risks are to be welcomed, not dreaded—they are the essence of life. Bernal describes this Faustian attitude admirably: 'The immediate future which is our own desire, we seek; in achieving it we become different; becoming different we desire something new, so there is no staleness except when development itself has stopped. Moreover, development, even in the most refined stages, will always be a very critical process; the dangers to the whole structure of humanity and its successors will not decrease as their wisdom increases, because, knowing more and wanting more, they will dare more, and in daring more they will risk their own destruction. But this daring, this experimentation, is really the essential quality of life.' To think of a non-developing situation is anathema to the Faustian mind. Writes Bernal: 'Even a scientific state could maintain itself only by perpetually increasing its power over the non-living and living environment. If it failed to do so, it would relapse into pedantry and become a perfectly ordinary aristocracy.'[4]

In this minor work—the work to be sure of a young man *whose ideas would change with advancing years*—is to be found an aspect of the scientific mentality against which the current attack against science is directed.

Now whether or not scientists consciously wish to become members of a ruling élite, a few predictions can be made with some plausibility. If it *is* the primary aim of the scientific enterprise to achieve 'the effecting of all things possible', then eventually scientists will have to realize that the social structure of capitalism must be transcended if they are to have the opportunity to continue their quest. For capitalism is not 'merely' threatening in the long run the continued existence of life on this planet. Of more immediate significance to scientists, capitalism is generating (through its manifestly exploitative use of scientific techniques) a growing revulsion against the scientific enterprise itself before that enterprise has made all the discoveries that would give a would-be ruling élite of scientists 'the means of directing the masses in harmless occupations and of maintaining a perfect docility under the appearance of perfect freedom' (as Bernal writes).[5] It therefore seems reasonable to predict that, as problems intensify, those scientists committed to 'the effecting of all things possible' will be forced as an act of self-preservation to seek more power and effective control over nature and other men than that allowed them within the social structure of capitalism.

Until recently, of course, as Bernal admits, scientists have never been overtly political, seeking always to accommodate themselves to the powers-that-be rather than risk destruction of the scientific enterprise. Furthermore, it is Bernal's opinion that scientists have in the past always managed to convince themselves that whether 'inventing submarines or depth charges' they were acting in the interests of humanity. The interests of science *were* the interests of humanity. Such an age of 'innocence' for science is, however, drawing to a close. Although it is true that during the 1940s and 1950s scientists swallowed easily the ideology of fund-awarding governmental, industrial or military organizations, convinced (perhaps) that in exploring the limits of the possible they were defending freedom against tyranny, yet during the decade of the 1960s it became evident that the world was sliding inexorably towards total disaster. American scientists responded to the crisis in different ways. Weisskopf, a leading father-figure in the physics community, called for more basic research as a means for eventually solving the problems; others, such as Platt, called for mobilization of scientists to solve the problems directly by scientific and technological means; others, such as Shapiro, began to look more closely at the social system housing the scientific enterprise, reaching the conclusion that it was in effect criminally irresponsible to continue to practise science in an exploitative, repressive society and that the scientific community should commit itself towards achieving radical political and social change.

In the coming years, therefore, it is probable that scientists will be forced to examine very carefully their values and goals. What is it they truly desire? What is it they truly seek? Is it possible, as one Professor of Law has predicted, that the United States will, in the near future, be ruled by 'a coalition of military men, scientists, technocrats, politicians and "realist" intellectuals who would combine a virulent anti-communist ideology with an unrestrained primacy of military and strategic needs', a rule that 'would lead to the gradual suppression of dissent and move the United States closer towards the society of *1984*'.[6] Or is it possible that at least some scientists will reject the goal of ever-increasing power and commit themselves to building a society in which life can flourish. We argued in chapter 10 that such a hope is not entirely unrealistic. To a considerable extent physics is practised not because of the power it gives but because of the aesthetic pleasure it gives. Although, to be sure, the physicalist image of nature is one that is directly hostile to life, the holistic image of nature and a scientific practice based on such an image need not be.

It is, of course, still true that many contemporary scientists appear to have little social conscience, that they talk only too easily about megadeaths, inflict pain and degradation only too readily on laboratory animals, and appear only too willing to manipulate their fellow human beings in the service of whoever pays them. But if it is borne in mind that many of these scientists live in societies that have institutionalized an aggressive, competitive, individualistic mode of behaviour as a way of life, then it is not too surprising that so many scientists (though far from all) have come to adopt an image of nature and to practise a science that reflect the harshness and brutality of the historical period in which they live, a period which has consequently resulted in (I believe) the erroneous belief that science is not and can never be a liberating force. Erroneous (in my opinion) because nature has oppressed man and is still oppressing man in most parts of the world and that a necessary (although not sufficient) condition for the liberation of man from the domination of nature is through the development of scientific techniques. Erroneous because (from my point of view) it is not science in the abstract that has given rise to the existence of intercontinental missiles and to the threat of thermonuclear warfare, so rightly regarded by Roszak as the symptom of a diseased civilization, but science performed in a world of nations and ideologies in conflict, dominated by the life-and-death struggle between the *reality* of capitalism and the *idea* of socialism. Although it is perhaps only to be expected that physicalism and power-seeking holism should be dominant scientific ideologies in our historical epoch, it is at least reasonable to hope that a qualitatively different epoch will see life-enhancing ideologies dominating the sciences. Encouraging, therefore, is the observation that despite the typical 'education' given to science students by their seniors (an education in which very specialized knowledge is usually presented in a complete historical, social and cultural vacuum), there are indications that scientists are increasingly coming to realize that *if* science is to serve mankind—and, wherever possible, nature as well—then science must be practised in a society in which, at the very least, the allocation of resources is planned according to both human needs and the needs of nature—in other words, in what I have called a socialist society.

Since socialism and the Soviet Union are so often equated, it is perhaps advisable to repeat that the Soviet Union is not a socialist society and cannot become one, I have argued, until capitalism and its ethos have been transcended in Western countries. And certainly it needs to be stressed that even after the socialist transcendence of capitalist society *serious problems arising from an over-bureaucratized world society will remain*. Nevertheless, order, in itself, is not to be feared: just as it is the very orderliness of the solar system that allows life to flourish on this planet, so only an appropriately ordered world society will allow human life to flourish in all its potential richness and diversity. But it must be a society ordered from 'below', not from 'above'. The task of socialist scientists is surely to help their colleagues understand that the scientific quest is embedded in values, to encourage them to make their own values explicit, and, hopefully, to bring about commitment to the goal of *helping* to build a liberated world society. To be sure certain kinds of science would continue to exist in such a liberated society but they would be sciences guided by the desire not to cause pain, not to approach every organism with a scalpel, sciences guided instead by the objective of helping to create that natural and social environment in which all human beings can best develop to the full those of their creative potentialities that serve to enhance the joy and beauty of living—a task that would deservedly win back the faltering allegiance of the young. The American biologist Barry Commoner is so right when he says that 'if we would know life, we must cherish it—in our laboratories and in the world'.[7]

2. Towards a Beautiful World

Crisis and commitment

The main argument of this book, let us repeat, has been that the continued existence of capitalism and its ethos is the *principal* cause of the more serious problems confronting mankind. The 'articulation' of capitalist society has led to a set of problems than can no longer be solved within its institutional framework. If this conclusion was not sufficiently appreciated in the 1930s, post-war consequences resulting from the preservation of capitalism are making such a conclusion more compelling. A professor of political science, H. J. Morgenthau, for example, writes that 'it should by now have become obvious that the great issues of our day—the militarization of American life, the Vietnam war, race conflicts, poverty, the decay of the cities, the destruction of the natural environment—are not susceptible to rational solution within the existing system of power relations'.[8] By rational solution, of course, he means, among other things, the rebuilding of cities so that people want to live in them, cities in which all are adequately housed, in which the air is clean. He emphasizes that 'any approach towards reform that leaves the relative distribution of power intact will at best mitigate the social ills or at worst convey to the victims the soothing appearance of remedial actions while confirming the *status quo*'. Morgenthau is right. It should by now have become obvious that capitalism is totally unable to solve the problems it has generated and continues to generate. Certainly a new John Stuart Mill would no longer be able to plead that 'the principle of private property has never yet had a fair trial in any country'. Indeed, history has surely given its verdict: if plastic pellets and napalm are not to continue to mutilate the bodies of men, women and children of underdeveloped countries, then it is necessary to seek—and as quickly as possible—the goal John Stuart Mill unwisely rejected a century ago, 'the subversion of the system of individual property'.

Above all, for many of the new radicals, it has been the Vietnam war that has led step by step to a radical reappraisal of the nature of the capitalist world. Although at first it seemed possible to consider the war to be only a 'puzzle'—an unfortunate phenomenon that could easily be resolved within the basic capitalist Third World pattern of relationships—as the war intensified and as the massive military might of the United States was used against this peasant country, with its policy of free-fire zones and saturation bombing, the 'puzzle' turned into a counter-example that called into question not just the relationship of the United States to the Third World but the fundamental structure of American society itself. Thus we find the American historian Gabriel Kolko writing that 'for a growing number of Americans the war in Vietnam has become the turning point in their perception of the nature of American foreign policy, the traumatizing event that requires them to look again at the very roots, assumptions, and structure of a policy that is profoundly destructive and dangerous'. His conclusion is that 'Vietnam is the logical outcome of a consistent reality we should have understood long before the United States applied much of its energies to ravaging one small nation' and that 'the elimination of that American hegemony is the essential precondition for the emergence of a world in which mass hunger, suppression, and war are no longer the inevitable and continuous characteristics of modern civilization'.[9] Other problems will remain. Socialists have faith that it will be possible to solve them within the framework of a world socialist society but not otherwise.

Some necessary changes

Although I do not in this essay intend to discuss the radical changes that I would like to see occur in Western societies the basic changes necessary can be readily sketched and follow from the arguments already given in this book. Clearly it is to be hoped that as the life-destroying nature of capitalist society becomes increasingly manifest, so mass social and political movements will develop in each capitalist country whose aim will be to make possible, to achieve and to consolidate socialist government. If the mass socialist struggle is successful, then it will be very difficult for it to be succeeded by 'control from above'. For governmental measures will be totally unenforceable unless they accord with the aims of the majority of people—people who will have achieved total or almost total control of their places of work. The initial programme of each socialist government would characteristically involve the nationalization of all banks and major industries, the return to appropriate governments of all property owned abroad, equitable redistributions of income, the abolition of advertising in public places and the mass media, the immediate cessation of all technological 'prestige' projects, unilateral and major cutbacks in all arms expenditure, aid to the now exploited underdeveloped countries, and greatly increased expenditure on public welfare to include the redesigning and rebuilding of cities, the development of public transport, and the building of all necessary schools, hospitals, nurseries and of course, homes. Of special importance would be architectural design that gives people the opportunity to transcend, if they wish, the typical family structure and thus to develop ways of living together as 'extended family' communities.

Only when such changes are successfully underway in the United States and other capitalist countries will mankind have crossed the most dangerous of all chasms on the path to the unity it must achieve if it is to survive. The transcendence of capitalism in the Western world would mean the disappearance of the world's major source of aggression, an event that would certainly facilitate the emergence of régimes in the Soviet Union and in China less authoritarian than the present ones. Indeed, genuine aid to China from a Western socialist world, although it would in no way adequately compensate for the many decades of Western exploitation of China, would undoubtedly lead the world an important step forward towards unification. In addition, the withdrawal of Western support from exploiting élites in underdeveloped countries would rather quickly lead towards radical social, political and cultural changes in these countries. Such changes would take mankind another decisive step along the path towards world unification and liberation.

Furthermore, in a united world society it would become possible for education to become an enriching and life-long experience, oriented not only towards the acquisition of necessary technical skills but also and primarily towards the development of ways whereby people would be able to make creative use of the free time that they would intend their working hours to make increasingly available to all people. In such a united, self-educated and humane world society the probability would be very great that a large majority of people would quickly see the wisdom of bringing the population growth rate to zero, and indeed to less than zero, for several decades or even centuries in order to achieve a permanent solution to the ecological problems that now threaten the very existence of man and many other inhabitants of this planet. An overall aim would have to be the achievement of a stationary *material* state in which as many materials as possible are recycled at a level of consumption the entire world population could enjoy without destruction of the environment or of the wildlife with which the human species shares this planet.

Remarks on the nature of man

To be sure, there is no lack of men who tell us that, because of man's supposedly innate aggressive nature, such a world—a creative, cooperative world—cannot be built. 'Hatred,' wrote Freud, surveying the carnage of the First World War, 'is at the bottom of all the relations of affection and love between human beings; hatred in relation to objects is older than love.'[10] He did not change his mind. During the early years of the Great Depression he told the world that the commandment to love one's neighbour as oneself 'is really justified by the fact that nothing else runs so strongly counter to the original nature of man'.[11] Choosing a recent work (1968) we see that Storr informs his readers that man is by nature a 'competitive, aggressive, territorial animal', and that 'ideally men should live in communities which are sufficiently small for them to maintain their identities and to encourage individual productivity, and which are in perpetual rivalry with other neighbouring small communities'.[12] Basically, of course, this is none other than a view of man that tends to see capitalist society as the norm. The argument of such authors is clear. Since the predicament we are in is a result of man's basic nature, aggressive, violent behaviour cannot be eliminated through qualitative social changes, even should such changes be feasible. Apparently the best one can hope for is to achieve a redirecting of man's aggressive and violent nature into activities *slightly* less suicidal than his present ones.

Freud, for example, believes it necessary to attack what he sees as the communist dream of eliminating aggressive behaviour through the abolition of private property. Although sympathetic to the communist aim of ensuring a just distribution of society's wealth, Freud argues 'that the psychological premises on which the system is based are an untenable illusion'. For 'aggressiveness was not created by property', and property, although a strong instrument for aggression, is 'certainly not the strongest'. The abolition of private property, could it be achieved, would not alter human nature and aggressiveness would certainly find an outlet in competition for 'sexual property', giving rise to 'the most violent hostility among men who in other respects are on an equal footing'. Were the family also abolished and complete freedom allowed in sexual life then, Freud continues, it would be difficult to foresee 'what new paths the development of civilization could take'. But since he believes aggressiveness to be innate to man, Freud's pessimism remains unabated. Were both private property and the family to be abolished, then with certainty at least one prediction could be made, namely that 'this indestructible feature of human nature' would follow the development of civilization along its new paths.[13]

Storr similarly wishes to dispel any 'illusion' that totally cooperative societies are possible. Nevertheless, there are, as he admits, difficulties in explaining why 'the uncritical acceptance' of such an idea should have been and still be so common. For it cannot be wholly explained, writes Storr, 'as a wish-fulfilling day-dream compensating for some present state of misery', since, as he points out, it has 'appealed to intelligent people who are neither poor nor suffering under tyranny. It is,' explains Storr, ' . . . an archetypal phantasy; in other words, a mental content which lies dormant in the minds of all men, and which is therefore easily activated by enthusiasts.' But the 'idealists' are doomed to failure. For 'unless some biological mutation alters the whole character of man as a species, it is impossible', says Storr, 'to believe that there could ever be a society without strife and competition. Man's perennial capacity to imagine Utopia is exceeded only by his recurrent failure to achieve it.'[14] Storr's paradigm commitment could scarcely be more clearly expressed.

It is, of course, not disputed that men have been violent to men, that men are indeed capable of mutilating each other. But a proof that men, as they are now genetically structured, are inherently incapable of building a peaceful and cooperative and emotionally rewarding world is not possible. Those 'scientists' who claim to so *know* human nature have not as yet understood the nature of science. Kaufmann surely comes to the only possible conclusion when in his psychological study of aggression and altruism he writes, 'It is unrealistic, cowardly, and in all probability, incorrect to assign our present desperate dilemmas to "human nature". We do not know what human nature is, and therefore cannot plead this alibi for a failure to examine the problems with which we must in the course of our lives come to grips.'[15]

To be sure, however, a view of human nature more optimistic than that espoused by Freud and Storr does not lack adherents. For example, according to Robert Lynd, 'the pattern of the [American] culture stresses individual competitive aggressiveness against one's fellows as the basis for personal and collective security.' Yet Lynd believes it 'safe to say that most human personalities do not crave as pervasive and continuously threatening competition as they tend to be subjected to in our culture'.[16] Erich Fromm goes further and argues that 'the necessity to unite with other human beings, to be related to them, is an imperative need on the fulfilment of which man's sanity depends',[17] and that 'any society which excludes, relatively, the development of love, must in the long run perish of its own contradiction with the basic necessities of human nature'. Pointing out that capitalist society is exactly such a society, Fromm concludes that 'those who are seriously concerned with love as the only rational answer to the problems of human existence must, then, arrive at the conclusion that important and radical changes in our social structure are necessary'.[18]

Nevertheless, just as it is impossible to prove theoretically that mankind is *not* capable of building a peaceful and cooperative world, so it is equally impossible to prove theoretically that he *is* capable. Although it is certainly reassuring to be told that small-scale peaceful and cooperative societies have existed, the only conclusive proof that a united, peaceful *world* society is possible will be the actual existence of such a world. Theory cannot prove; it can only guide the way. It is practice that must demonstrate the validity of the theory's claims, if valid they are.

One's conception of man's nature implies a conception of what one *believes* man to be capable of. Clearly this whole essay has been based on, and has argued in favour of, the socialist conception of man: that 'man makes himself', that man's nature is not fixed, immutable for all time, but that man possesses the ability to change himself as, at the same time, he changes his natural and social environment; 'For us,' writes Sartre, 'man is characterized above all by his going beyond a situation and by what he succeeds in making of what he has been made.'[19] Just as physicists have faith in the harmony of nature, and base their research commitment on that faith, so the commitment of socialists is based on their *faith* that man is a being who *is* capable of building an harmonious world. Whereas the capitalist belief that man is innately aggressive and competitive gives rise to a research-action programme to resolve all problems within the existing capitalist system, the socialist conception of man gives rise to a *praxis* which is revolutionary in nature, based on the conviction that only a radically different society will enable man to take another step on his long journey to world civilization and that failure to take this step (which is a possibility) will mean either continued barbarism or the end of the road altogether.

Marxism and scarcity

Certainly it was not Marx's view that man is innately greedy and aggressive. Marx's point of view is demonstrated quite dramatically by a mistranslation by one of his critics who claimed Marx to have written: 'The only wheels that set political economy in motion are *greed* and the war between the greedy—competition.'[20] However, as Fromm was quick to point out,[21] the critic is in error. What Marx wrote was that 'the only wheels which political economy sets in motion are *greed* and the war *amongst the greedy—competition*'. A rather different thought!

The point is that if throughout his existence man has felt basically insecure, threatened in particular by scarcity and the fear of scarcity (indeed, Sartre has defined man as a 'practical organism living with a multitude of fellow men in a field of scarcity'),[22] then it is very reasonable that man should have exploited his fellow man and should have striven to conquer and subdue the natural world. And it is to be welcomed that a social system has originated as a result of this striving that, in institutionalizing competition and struggle as a way of life, has developed science and technology to the point where scarcity —and the threat of scarcity—could now be abolished and a more humane social system constructed. The ultimate absurdity, however, indeed the ultimate obscenity, lies in the fact that this capitalist social system, a system that has produced the means whereby scarcity can be eliminated, has been prolonged to the point where its *normal* functioning results in the creation of scarcity, not its elimination. The most important task now facing mankind is the transcendence of this now dangerously obsolete social system.

The importance of the 'new' consciousness

But socialism not only implies the transcendence of capitalist relations of production but also the transcendence of the entire capitalist ethos. For without the predominance of a basically non-exploitative, joy-seeking and freedom-emphasizing ethos, 'socialist' expropriation and control of the productive forces would run the very great danger of giving rise to a social order of mechanical uniformity, one in which love, beauty, joyful spontaneity and a sense of humanness would come to be remembered—if remembered at all—as pre-scientific ideals of pre-scientifically organized society. At this crucial time, however, educated and sensitive young people are becoming increasingly aware that a historical juncture has been reached in which, provided all human beings are content to live at a materially simple level of existence, then *all* human beings will be able to enjoy a sensually and spiritually rich way of life. And it is exactly this way of life that an increasing minority of socially privileged young people appear to be struggling to achieve.

Writing over forty years ago, Keynes remarked very aptly that 'we have been trained too long to strive and not to enjoy'. He felt sure 'that with a little more experience we shall use the new-found bounty of nature quite differently from the way in which the rich use it today; and will map out for ourselves a plan of life quite otherwise than theirs'.[23] Hesitant and stumbling though the 'dissenting young' may be, nevertheless they are attempting just this—to map out a plan of life that would give the Western world, and with it the entire world, the opportunity to progress towards a saner and more humane stage of civilization. Indeed, the attainment of public ownership of the productive forces combined with the triumph of a consciousness at least in some ways epitomized by the ethos of the 'counter-culture' would constitute an important stride forward towards the realization of 'communist' society. For, as Marx and Engels wrote in *The German Ideology*,

> In communist society, where nobody has one exclusive sphere of activity but each one can become accomplished in any branch he wishes, society regulates the general production and thus makes it possible for me to do one thing today and another tomorrow, to hunt in the morning, fish in the afternoon, rear cattle in the evening, criticize after dinner, just as I have a mind, without ever becoming hunter, fisherman, shepherd or critic.[24]

There will be many more activities than these in a cooperatively organized society characterized by a non-power-seeking, non-materialistic ethos. But the principle will be the same. Men and women will be free to develop their individual abilities to whatever extent *they themselves choose*.

Thus between nuclear (or ecological) holocaust and scientific enslavement there exists the possibility of the triumph of a liberated consciousness—the triumph of all those people who are content to live at a low level of material sufficiency, who believe that the richness of life consists in diversity of people and 'life-styles', in experimentation in life-styles, in being able to live, if one so chooses, in *communities* of people who have helped each other to become open, sincere, who are no longer afraid to touch each other, people (as Marcuse writes[25]) who are intelligent and healthy enough to dispense with all heroes and heroic virtues, who have no impulse to live dangerously, to meet the challenge, people who, above all, do not regard life as necessarily being, or wish life to be, a perpetual struggle of man against man, and of man against nature.

The 'Faustian' mentality

We have already referred in chapter 10, if far too briefly, to the development of a necessarily different mentality during the Age of the Scientific and Capitalist Revolutions. The men of the new science and of the new Age convinced themselves that they could and should become 'masters and possessors of nature' and that it was their duty to strive for the infinite. The fact that the infinite could never be reached was a source of pleasure to them —it ensured what they most wanted—that the journey would never end. Perhaps this mentality tended to characterize scientist and non-scientist alike during the next three centuries.* In 1778, for example, the German dramatist and critic Lessing would describe the Faustian mentality's desire for struggle in the following way: 'If God held enclosed in His right hand all truth and in His left hand the ever striving for truth, although with the qualification that I must for ever err, and said to me "Choose", I should humbly choose the left hand and say "Father, give! Pure truth is for Thee alone!"'[26] (One wonders

---

* At the beginning of chapter 5 we noted how economists in defining the economic problem usually refer quite unproblematically to what they conceive to be man's 'insatiable' and 'unlimited' material wants.

what Popper would say if God presented him with such a choice.) Early in the 19th century the Faustian mentality—the mentality that had grown to welcome eternal struggle and conflict—received its apotheosis in the immortal work of Goethe. The bargain struck between Faust and the Devil penetrates to the very core of that Western mentality which, *because of its very great achievements*, can now be—and must be—rejected by its present heirs:

> FAUST If I be quieted with a bed of ease,
> Then let that moment be the end of me!
> If ever flattering lies of yours can please
> And soothe my soul to self-sufficiency,
> And make me one of pleasure's devotees
> Then take my soul, for I desire to die:
> And that's a wager!
>
> MEPHISTOPHELES Done!
>
> FAUST
> And done again!
> If to the fleeting hour I say
> 'Remain, so fair thou art, remain!'
> Then bind me with your fatal chain,
> For I will perish in that day.[27]

The mysticism of the 'counter-culture'

Today's dissenting young have indeed realized that unless Western man is prepared to say—and say very soon—to the fleeting hour, 'Remain, so fair thou art, remain!', then the human race is damned. Rightly or wrongly, many of the dissenting young have proceeded to reject, as they see it, the 'rationality' that has produced such a violent, dehumanized and dying world as is this one. Theirs is the mentality that is now responding to the 'irrationality' of the cry of anguish with which Gilbert greeted the 17th century and the dawn of the 'mechanical philosophy': 'Pitiable is the state of the stars, abject the lot of the earth, if this high dignity of soul is denied them.'*[28] The dissenting young of the 'counter-culture' are opposed to the 'mechanical philosophy'. At their best they wish to build a world in which they can care for each other and for the earth—their home—because they love it. Since, as they see it, all scientific images of man and nature declare such attitudes to be irrational, the dissenting

---

* The Hermetic texts had foretold of a time in which 'The soul and all the beliefs attached to it—this will be laughed at and thought nonsense.' *The Lament* predicted: 'In that hour, weary of life, men will no longer regard the world as the worthy object of their admiration and reverence. This All . . . will be in danger of perishing; men will esteem it a burden; and thenceforward they will despise and no longer cherish this whole of the universe . . . made up of an infinite diversity of forms . . . assembled in one whole, in a harmonious diversity, all that can be seen that is worthy of reverence, praise and love. . . . Then the earth will lose its equilibrium, the sea will no longer be navigable . . . the fruits of the earth will moulder, the soil will no longer be fertile, the air itself will grow thick with a lugubrious torpor. . . .'[28n]

young therefore turn to 'irrational' images of nature in which the ideals of love, beauty and communion with nature are not excluded at the outset. Such romanticism has deep roots in the counter-culture and indeed in the ideologies of many of its 'intellectual' leaders. Marcuse, for example, expresses almost poetically this romantic attitude towards nature and what he sees as its liberating potential. Contrasting Orpheus against Prometheus, the latter being 'the culture-hero of toil, productivity and progress through repression', Marcuse tells us that Orpheus offers an 'image of joy and fulfilment', that the voice of Orpheus 'does not command but sings'. He explains that in this Orphic experience of the world 'being is experienced as gratification, which unites man and nature so that the fulfilment of man is at the same time the fulfilment without violence of nature. In being spoken to, loved, and cared for, flowers and springs and animals appear as what they are—beautiful, not only for those who address and regard them, but for themselves "objectively"'. Marcuse tells us that:

> The song of Orpheus pacifies the animal world, reconciles the lion with the lamb and the lion with man. The world of nature is a world of oppression, cruelty and pain, as is the human world; like the latter it awaits its liberation. This liberation is the work of Eros. The song of Orpheus breaks the petrification, moves the forests and the rocks—but moves them to partake in joy.[29]

It is, however, in Theodore Roszak's plea for the adoption of a magical image of nature that we most strongly sense an affinity with the Hermetic tradition—that tradition which the 'mechanical philosophers' had regarded as one of their principal opponents and which they had attacked so strongly in the early part of the 17th century. The relation of the Magus to nature could not have been more different from the relation of the men of the new science to the world of mere matter in purposeless motion. Indeed, according to the Hermetic texts, the Magus Man, a half-divine being, came to the earth because he fell in love with beautiful nature and 'then Nature having received her loved one, embraced him, and they were united, for they burned with love'.[30] It is this erotic attitude to nature that is so alien, so irrational to the scientific mind—but it was characteristic of many of the Hermetic philosophers. 'Why is Love called a Magus?' asked Ficino. His answer:

> Because all the force of Magic consists in Love. The work of Magic is a certain drawing of one thing to another by natural similitude. The parts of this world, like members of one animal, depend all on one Love, and are connected together by natural communion. From this community of relationship is born the communal Love: from which Love is born the common drawing together: and this is the true Magic.[31]

For these Hermetic philosophers the figure of Orpheus represented their fundamental approach to nature. For Ficino the secret of spiritual transformation was for the magician first to put himself into a receptive state by playing Orphic hymns on his lyre and then, aided by the appropriate magical symbols, to attract the celestial virtues into himself. To better achieve this goal Ficino's lyre was decorated with the figure of Orpheus taming the wild animals, the mythical Orpheus who was able to bring even the stones to life with his beautiful singing. The idea of compulsion was alien to these Hermetic philosophers * (note Ficino's comments on the sun, given at the beginning of chapter 4). The secrets of nature, when understood by the Hermetic philosophers, would be used by them only in the service of Love. Ficino in particular emphasized how a man, to deserve the name human being, must love all other men as his equals.

The 'dissenting young' of the 'counter culture' clearly subscribe to beliefs resembling this idealized mystical-Hermetic attitude to nature and society, deliberately rejecting the 'rational' consciousness that promotes and is promoted by (as they see it) capitalism on the one hand and science on the other. However, while I share their conviction that capitalist society and its materialistic, competitive and power-seeking ethos must be transcended if a less violent and more humane world is to be achieved, the dissenting young are perhaps deceived to the extent they believe that no possibility of overlap exists between scientific and mystical modes of consciousness, that scientists will never be able to commit themselves to the goal of building a beautiful world. For if scientists in the course of their current soul-searching come to realize how very little they understand,† that although science certainly provides the possibility of power over nature it does not provide knowledge of *nature-in-itself*, that although science can remove mystery and awe from life it must eliminate life itself in order to do so, that science, if it is to be life-sustaining, ought to be practised only within a framework of love for man and respect, even reverence, for nature, then—if scientists can be so persuaded —a constructive dialogue would surely become possible between the 'scientific' and 'irrational' modes of consciousness and perhaps out of such dialogue a mutually beneficial collaboration might emerge between the 'new' scientists and the dissenting young.

---

* So much so that Gilbert, who had been strongly influenced by the Hermetic tradition, used the term 'magnetic coition' rather than 'attraction' to describe the 'interaction' between the loadstone and the iron: 'For where attraction exists, there, force seems to be brought in and a tyrannical violence rules.' He describes approvingly how 'Orpheus, in his hymns, tells that iron is drawn to the loadstone as the bride to the embraces of her spouse,' and how Thales 'not without reason . . . declares the loadstone to be animate, a part of the animate mother earth and her beloved offspring'.[31n]

† Perhaps it is not inappropriate to recall the words of one of the 17th century's most successful explorers, Isaac Newton, as towards the end of his life he looked back on his scientific discoveries: 'I do not know what I may appear to the world; but to myself I seem to have been only like a boy playing on the sea-shore, and diverting myself in now and then finding a smoother pebble or prettier shell than ordinary, while the great ocean of truth lay all undiscovered before me.' The power and predictive ability of science may well continue to increase but 'the great ocean of truth' is unknowable.

Liberation and the aims of science

What, however, *are* the present aims of the scientific community? How, we must ask, do contemporary scientists and technologists conceive of the next stage in mankind's evolution? Do they tend to look upon a stage of physical rest, of world peace, of beauty, of sensuality and of love as a betrayal of the scientific enterprise? Bernal, for example, appalled at so much unnecessary suffering during the 1930s and which he recognized as primarily due to the existence of capitalist societies, nevertheless (in 1939) was anxious to stress that even after the transcendence of capitalist society major tasks confronting humanity will continue to exist and will in fact never become exhausted. In his well-known work *The Social Function of Science* he wrote: 'With a fully organised world society such tasks ['the conquest of the Arctic'] could be pushed far further. It will no longer be a question of adapting man to the world but the world to man.... Other necessary tasks which we cannot imagine must lie before a developing humanity, and science will have its part to play in carrying them out.' Above all, he was anxious to reassure his readers that 'a Utopia is not a happy ecstatic state but the basis for further struggles and further conquests'.[32] Will technological projects continue to appear so 'technically sweet', as Oppenheimer remarked of the hydrogen bomb, that scientists will push ahead with their development regardless of the consequences, intent only on 'the effecting of all things possible'? Both Medawar and Glass have, as we noted in chapter 10, expressed alarm in important addresses at the thought that the human quest for power might cease, and with it, the scientific enterprise—or that the scientific enterprise might come to a halt through either the discovery of all 'basic' laws of nature or through the reaching of impassable barriers. Let us therefore recall that in so far as the 'irrational' young are rejecting science they are rejecting that search for power for its own sake that they have good reason to believe lies at the heart of the scientific enterprise. They have found out, perhaps, as Bertrand Russell wrote in *The Scientific Outlook*, that 'the lover, the poet and the mystic find a fuller satisfaction than the seeker after power can ever know, since they can rest in the object of their love, whereas the seeker after power must be perpetually engaged in some fresh manipulation if he is not to suffer from a fresh sense of emptiness'.[33] 'Faustian' men would, of course, contemptuously reject such advice. Nevertheless, if the new consciousness is able to take root and spread in countries that are now predominantly power-seeking (the Soviet Union, forced into economic, military and scientific competition with the West, certainly falls into this category together with all capitalist countries) then the following decades will surely see a most welcome transformation in the human condition. To be sure, the young people who desire a basically 'Orphic' existence must nevertheless be prepared to make good political use of whatever 'Faustian' components they still have to their personalities. For not until mass political movements have established a social and political structure that promotes the enhancement of life will it be possible to cease to strive for power and to begin to enjoy to the full the erotic, the spiritual and the intellectual richness of life that human diversity and human solidarity will offer. The seeds of the scientific revolution will then at last give rise to fruit that is sweet to the taste. Such striving as does remain will be with the goal of creating a world that is increasingly beautiful, of exploring, not of conquering, a universe that is awe-inspiring.

'When I come to die,' wrote Bertrand Russell, 'I shall not feel that I have lived in vain. I have seen the earth turn red at evening, the dew sparkling in the morning, and the snow shining under a frosty sun; I have smelt the rain after drought, and have heard the stormy Atlantic beat upon the granite shores of Cornwall. Science,' he continued, 'may bestow these and other joys upon more people than could otherwise enjoy them. If so, its power will be wisely used. But when it takes out of life the moments to which life owes its value, science will not deserve admiration however cleverly and however elaborately it may lead men along the road to despair.'[34] The world, with the aid of scientific techniques, could be made such a beautiful place, as Bertrand Russell emphasized. And in a beautiful world people would be able to experience many, many moments in which they would surely tell the fleeting hour to 'remain, so fair thou art, remain'; they would know the joys of opening themselves in trust and confidence to other people, the experience of communing with nature, the delights of a creative art and of a science in which knowledge is wanted for the aesthetic pleasure it gives, the joys of experiencing life with children (not necessarily their own); and as death approaches they might even be prepared to choose the moment in which, with dignity and without regret, they end a fulfilled life, content to be entrusting to the children they have cared for the privilege and responsibility of nourishing life.

Where will the scientists stand in the coming critical decades? There are indications, although no more than this, that at least some scientists in the advanced industrial countries of the world, including both the United States and the Soviet Union, are realizing the necessity of mobilizing themselves in support of the oppressed and exploited peoples of this earth, in solidarity with all groups of people who are struggling to build a non-exploitative world society. Should such commitment by scientists gather momentum—and it is by no means certain—the balance might just be tipped in favour of life and liberation in the difficult years to come.

Appendix

EINSTEIN'S POLITICAL STRUGGLE

In this appendix a few extracts are given from Einstein's writings in the crucial period from the end of the Second World War to his death in 1955. They reveal and summarize the thoughts and anguish of an outstanding scientist and revolutionary thinker as he watches the physics he has loved and to which he had contributed so greatly being used not only in a way over which he had no control but which represented the negation of all he had been striving to achieve in life. As the reader will see, it was not only with respect to quantum mechanics that Einstein pursued a lonely path. Time and again he tried to warn his fellow scientists and American citizens * that the path they were following would lead to catastrophic consequences. At the end of his life Einstein felt himself forced to acknowledge failure. Although he had been able to keep his independence of judgement as incessant propaganda took its toll of so many other men whose avowed intention it was to retain the ability to criticize, to judge and to choose, nevertheless Einstein had been unable to change the programme to disaster which he saw so clearly American policy would invite. Given in chronological order without comment the following extracts from Einstein's letters, articles and talks tell their own story. (We begin the series of extracts with a letter written *to* Einstein; all remaining extracts are from Einstein's writings or talks.)

October 21, 1945 · From a letter written to Einstein by a physicist from the Radiation Laboratory of the Massachusetts Institute of Technology:

The scientists of Cambridge, as well as those throughout the world, need help urgently in these days of turmoil and unprecedented tension. What makes the present atomic power situation so full of anguish for all of us is the cruel irony wherein one of the greatest and most joyful triumphs of scientific intellect may bring frustration and death rather than spiritual uplifting and more audacious life. The final total confirmation of your principle $E = mc^2$ should mark the beginning of an era of light; but we stand perturbed and seem to see ahead an impenetrable night...[1]

December 10, 1945 · From an address to a Nobel Anniversary dinner in New York:

Alfred Nobel invented an explosive more powerful than any then known—an exceedingly effective means of destruction. To atone for this 'accomplishment' and to relieve his conscience, he instituted his awards for the promotion of peace. Today, the physicists who participated in producing the most formidable weapon of all time are harassed by a similar feeling of responsibility, not to say guilt. As scientists, we must never cease to warn against the danger created by these weapons; we dare not slacken in our efforts to make the peoples of the world, and especially their governments, aware of the unspeakable disaster they are certain to provoke unless they change their attitude toward one another and recognize their responsibility in shaping a safe future. . . . The war is won, but the peace is not. The great powers, united in war, have become divided over the peace settlements. The peoples of the world were promised freedom from fear; but the fact is that fear among nations has increased enormously since the end of war. The world was promised freedom from

---

* A refugee from Hitler's Germany, Einstein was granted American citizenship in 1940.

want; but vast areas of the world face starvation, while elsewhere people live in abundance. The nations of the world were promised liberty and justice; but even now we are witnessing the sad spectacle of armies of 'liberation' firing on peoples who demand political independence and social equality, and supporting by force of arms, those individuals and political parties which they consider best suited to represent their own vested interests. Territorial conflicts and power politics, obsolete as these purposes of national policy may be, still prevail over the essential requirements of human welfare and justice. . . .[2]

May 29, 1946 · From a radio address to a Chicago rally of the Students for Federal World Government:

Many people say that, in the present circumstances, fundamental agreement between the United States and the Soviet Union is impossible. Such an assertion might be justified had America made a really serious effort in that direction since the end of the war. It seems to me that America has done just the opposite: It was not wise to admit Fascist Argentina into the United Nations over Russia's protest. Further, there was no need to keep on producing more and more atomic bombs and to spend twelve billion dollars in a single year on armaments, when there was no military threat in sight. Nor was there any sense in denying Trieste to Yugoslavia, a former ally who was in real need of this port that has, in fact, little economic significance to Italy, a former enemy country. There is no further point in further enumerating all the details which indicate that nothing was done by the United Nations to mollify Russia's distrust. In fact, we have done much toward fostering this distrust which the events of the last several decades make only too understandable.[3]

January 20, 1947 · In reply to a question from the Overseas News Agency with respect to Professor Wiener's withdrawal* from a symposium on computers organized by Harvard University and the United States Navy:

I greatly admire and approve the attitude of Professor Wiener; I believe that a similar attitude on the part of all the prominent scientists in this country would contribute much toward solving the urgent problem of international security.

Non-cooperation in military matters should be an essential moral principle for all true scientists, i.e., for all who are engaged in basic research. It is true that it is more difficult for scientists in non-democratic countries to adopt such an attitude; but the fact is that, at present, the non-democratic countries constitute less of a threat to healthy international developments than the democratic nations which, enjoying economic and military superiority, have subjected scholars to military mobilization.[4]

February 21, 1947 · In a message broadcast over a New York station in support of the nomination of Lilienthal as head of the United States Atomic Energy Commission:

You all know that the policies of the United States since the end of the war have caused anxiety and distrust throughout the world. The destruction of large Japanese cities without adequate previous warning, the unceasing production of atomic bombs, the Bikini tests, the expenditures of many billions of dollars for military purposes despite the absence of any external threat, the attempt to militarize science —all this has impeded the development of mutual trust among nations which is indispensable to the establishment of a secure peace.[5]

---

* In November, 1948, Wiener wrote in the *Bulletin of the Atomic Scientists*: 'The degradation of the position of the scientist as an independent worker and thinker to that of a morally irresponsible stooge in a science-factory has proceeded even more rapidly and devastatingly than I had expected.'

July 17, 1947 · In a broadcast on 'The Immediate Need for World Law':

The Soviet experiences with the outer world have never been too good. We must remember the support of the West by anti-Soviet generals during the [Russian] Civil War, the long political and economic boycott against the Soviet Union, the constant propaganda campaign of the foreign press against Soviet Russia. Later on the Russians joined the League of Nations, but then they saw how the Fascist aggressions in Manchuria, in Spain, in Abyssinia, in Austria, were accepted and condoned, how agreements were made with the aggressors. And eventually, when they found themselves excluded from the most crucial European settlements in the first period of the Hitler regime, then, understandably enough, they changed their attitude.[6]

November 1947 · In an article entitled 'Atomic War or Peace' published in the *Atlantic Monthly*:

At present, the Russians have no reason to believe that the American people are not actually supporting a policy of military preparedness, which they regard as a policy of deliberate intimidation. . . . Not until a genuine, convincing offer is made by the United States to the Soviet Union, supported by an aroused American public, will one be entitled to be hopeful about a possible response by Russia.[7]

January 28, 1948 · From a letter to an army officer who had challenged Einstein's opposition to conscription:

I believe the danger lies at present in the possibility that America may totally succumb to that fearful militarization which engulfed Germany half a century ago. . . . We should never forget that it is totally unlikely that any country will attack America in the near future, least of all Russia, which is devastated, impoverished and politically isolated. . . .[8]

April 27, 1948 · From a message to the One World Award Committee:

The proposed militarization of the nation not only immediately threatens us with war; it will also slowly but surely undermine the democratic spirit and the dignity of the individual in our land. The assertion that events abroad are forcing us to arm is incorrect; we must combat this false assumption with all our strength. Actually, our own rearmament, because of its effect upon other nations, will bring about the very state of affairs upon which the advocates of armaments seek to base their proposals.[9]

June 17, 1948 · From a message to a public meeting organized by Professor Shapley:

The United States emerged from the war as the strongest military and economic power and, temporarily, is the only country to possess the powerful atomic bomb. Such power imposes a heavy obligation. To a large extent, the United States is responsible for the ominous competitive arms race which has taken place since the end of the war and which has virtually destroyed the postwar prospects for an effective supranational solution of the security problem.[10]

May, 1949 · From an article by Einstein entitled 'Why Socialism' which appeared in the first issue of *Monthly Review*:

The economic anarchy of capitalist society as it exists today is, in my opinion, the real source of the evil. We see before us a huge community of producers the members of which are unceasingly striving to deprive each other of the fruits of their collective labour—not by force, but on the whole in faithful compliance with legally established rules. . . . The result of these developments is an oligarchy of private capital the enormous power of which cannot be effectively checked even by a democratically organized political society. This is true since the members of legislative bodies are selected by political parties, largely financed or otherwise influenced by private capitalists who, for all practical purposes, separate the electorate from the legislature. ... Moreover, under existing conditions, private capitalists inevitably control, directly or indirectly, the main sources of information (press, radio, education). It is thus extremely difficult, and indeed in most cases quite impossible, for the individual citizen to come to objective conclusions and to make intelligent use of his political rights. ... This crippling of individuals I consider the worst evil of capitalism. Our whole educational system suffers from this evil. An exaggerated competitive attitude is inculcated into the student, who is trained to worship acquisitive success as a preparation for his future career.

I am convinced there is only *one* way to eliminate these grave evils, namely through the establishment of a socialist economy, accompanied by an educational system which would be oriented towards social goals. In such an economy, the means of production are owned by society itself and are utilized in a planned fashion. A planned economy, which adjusts production to the needs of the community, would distribute the work to be done among all those able to work and would guarantee a livelihood to every man, woman, and child. The education of the individual, in addition to promoting his own innate abilities, would attempt to develop in him a sense of responsibility for his fellow men in place of the glorification of power and success in our present society.

Nevertheless, it is necessary to remember that a planned economy is not yet socialism. A planned economy as such may be accompanied by the complete enslavement of the individual. The achievement of socialism requires the solution of some extremely difficult socio-political problems: how is it possible, in view of the far-reaching centralization of political and economic power, to prevent bureaucracy from becoming all-powerful and overweening? How can the rights of the individual be protected and therewith a democratic counterweight to the power of bureaucracy be assured?

February 13, 1950 · From a television programme conducted by Mrs. Eleanor Roosevelt, which also included Lilienthal and Oppenheimer:

Outside the United States we ... establish military bases at every possible, strategically important point of the globe as well as arm and strengthen economically our potential allies. And inside the United States, tremendous financial power is being concentrated in the hands of the military; youth is being militarized; and the loyalty of citizens, particularly civil servants, is carefully supervised by a police force growing more powerful every day. People of independent political thought are harassed. The public is subtly indoctrinated by the radio, the press, the schools. Under the pressure of military secrecy, the range of public information is increasingly restricted.[11]

March 16, 1950 · From a letter to Professor Sidney Hook:

I have endeavoured to understand why the Russian Revolution became a necessity. Under the circumstances prevailing in Russia at the time, I believe that the revolution could have been successfully undertaken only by a resolute minority. A Russian who had the welfare of the people at heart would, under the then existing conditions, naturally cooperate with, and submit to, this minority since the immediate goals of the revolution could otherwise not have been achieved. To an independent person, this must surely have entailed *temporary*, painful renunciation of his personal liberty. But I believe that I myself would have deemed it my duty, and would have considered it the lesser evil, to make this temporary sacrifice. This, however, should not be taken to mean that I approve of the Soviet Government's policy of intervention, both direct and indirect, in intellectual and artistic matters. I view such interference as objectionable, harmful and even ridiculous. I also believe that centralization of political power and limitation of individual freedoms should not exceed the limits determined by such considerations as external security, domestic stability and the requirements of a planned economy. An outsider is hardly in a position to appraise adequately the existing conditions and needs of another country. In any case, there is no doubt that the achievements of the Soviet regime in the fields of education, public health, social welfare and economics are considerable and that the people as a whole have greatly benefitted from those achievements.[12]

October, 1950 · From a message 'On the Moral Obligation of the Scientist' sent to the forty-third meeting of the Italian Society for the Progress of Science:

Thus the man of science, as we can observe with our own eyes, suffers a truly tragic fate. In his sincere attempt to achieve clarity and inner independence, he has succeeded, by his sheer super-human efforts, in fashioning the tools which will not only enslave him but also destroy him from within. He cannot escape being silenced by those who wield political power. . . . He also realizes that mankind can be saved only if a supranational system, based on law, is created to eliminate the methods of brute force. However, the man of science has retrogressed to such an extent that he accepts as inevitable the slavery inflicted upon him by national states. He even degrades himself to such an extent that he obediently lends his talents to help perfect the means destined for the general destruction of mankind. . . . If today's man of science could find the time and the courage to reflect honestly and critically about himself and the tasks before him, and if he would then act accordingly, the possibilities for a sane and satisfactory solution of the present dangerous international situation would be considerably improved.[13]

October 8, 1950 · From a letter to a professor of English literature who had asked Einstein to sign a statement opposing German rearmament:

I have not signed the statement you have sent me although I am in full agreement with its content. The reason is simply that I wish to avoid any impression of disagreeing only with the rearmament of Germany. In reality such rearmament is but a link in the chain of measures our government has followed since Roosevelt's death; such measures may, in my opinion, lead to disastrous consequences in the future.[14]

November 5, 1950 · From a letter to a physician whose son had been inducted into the army and who had written to Einstein urging him to redouble his efforts on behalf of peace:

You are very right in assuming that I am badly in need of encouragement. I have indeed the impression that our nation has gone mad and is no longer receptive to reasonable suggestions. Its whole development reminds me of the events in Germany since the time of Emperor William II: through many victories to final disaster.[15]

January 5, 1951 · From a letter to the editor of the *Bulletin of the Atomic Scientists:*

I will say, however, that in my opinion the present policy of the United States constitutes a more serious obstacle to peace in the world than that of Russia. The current fighting is in Korea, not Alaska. Russia is exposed to a vastly greater threat than the United States, and everyone knows it. I find it hard to understand why people here accept the fable that we are in peril. I can only assume that it is because of their lack of political experience. While the government policy is apparently directed toward preventive war, there is, at the same time, a concerted attempt to make it appear as though the Soviet Union is the aggressor.[16]

January 6, 1951 · From a letter to the Queen Mother of Belgium:

While it proved eventually possible, at an exceedingly heavy cost, to defeat the Germans, the dear Americans have vigorously assumed their place. Who shall bring them back to their senses? The German calamity of years ago repeats itself: people acquiesce without resistance and align themselves with the forces for evil. And one stands by, powerless.[17]

October 9, 1952 · From a letter to a correspondent in England:

I am a lonely man, as you are, much older than you but probably not much wiser. What we have in common is a deep skepticism about everything we are told by the press, radio, etc., which are considerably worse in the United States than in Old England. . . .

The reports we receive about Russia are, of course, one-sided and too black. Yet, it seems certain that in spite of her social and economic achievements, her political organization is still considerably more brutal and barbarous than ours. However, it is clear to me that the postwar change in power relations among the nations of the world has resulted in the West's being much more aggressive than the Communist world.[18]

November 18, 1954 · In a statement published in *The Reporter:*

You have asked me what I thought about your articles concerning the situation of scientists in America. Instead of trying to analyze the problem, I should like to express my feeling in a short remark: If I were a young man again and had to decide how to make a living, I would not try to become a scientist or scholar or teacher. I would rather choose to be a plumber or a peddler, in the hope of finding that modest degree of independence still available under present circumstances.[19]

January 17, 1955 · From a letter to Max Born:

The hired hacks of an accommodating press have tried to tone down the impact of this statement, either by making it appear as if I regretted having been engaged in scientific endeavour, or by trying to give the impression that I attached little value to the practical occupations I mentioned.

What I wanted to say was just this: In the present circumstances, the only profession I would choose would be one where earning a living had nothing to do with the search for knowledge.[20]

January 2, 1955 · From a letter to the Queen Mother of Belgium:

Yesterday the Nuremberg trials, today the all-out effort to rearm Germany. In seeking for some kind of explanation, I cannot rid myself of the thought that this, the last of my fatherlands, has invented for its own use a new kind of colonialism, one that is less conspicuous than the colonialism of old Europe. It achieves domination of other countries by investing American capital abroad, which makes those countries firmly dependent on the United States. Anyone who opposes this policy or its implications is treated as an enemy of the United States. It is within this general context that I try to understand the present-day policies of Europe, including England.[21]

March 19, 1955 · From a letter to the German physicist von Laue:

My action concerning the atomic bomb and Roosevelt consisted merely in the fact that, because of the danger that Hitler might be the first to have the bomb, I signed a letter to the President which had been drafted by Szilard. Had I known that that fear was not justified, I, no more than Szilard, would have participated in opening this Pandora's box. For my distrust of governments was not limited to Germany.

Unfortunately, I had no share in the warning made against using the bomb against Japan. Credit for this must go to James Franck. If they had only listened to him![22]

March 2, 1955 · From a letter to Niels Bohr seeking his participation in Bertrand Russell's plan to produce a statement calling on governments to renounce the use of war:

Dear Niels Bohr,

Don't frown like that! This has nothing to do with our old controversy on physics, but rather concerns a matter on which we are in complete agreement. . . .

My own participation may exert some favourable influence abroad, but not here at home, where I am known as a black sheep (and not merely in scientific matters). . . .[23]

On April 17, 1955, Einstein died. In his essay *Knowledge for What?* Robert Lynd has quoted the following advice offered by Einstein to his fellow scientists:

Concern for man himself and his fate must always form the chief interest of all technical endeavours. . . . Never forget this in the midst of your diagrams and equations.

NOTES

The following abbreviations have been used in the notes for the Introduction and Chapter 1:

CGK   I. Lakatos and A. Musgrave, *Criticism and the Growth of Knowledge* (Cambridge, 1970).
CMSRP   I. Lakatos, 'Criticism and the Methodology of Scientific Research Programmes', *Proceedings of the Aristotelian Society* 69 (1968), 149–86.
CR   K. R. Popper, *Conjectures and Refutations* (Routledge, 1963, 2nd edition, 1965).
LSD   K. R. Popper, *The Logic of Scientific Discovery* (1934; Harper, 1965).
OSE   K. R. Popper, *The Open Society and its Enemies* (1945; Routledge, 5th edition, 1966), Vol. 2: *Hegel and Marx*.
SS   I. Scheffler, *Science and Subjectivity* (Bobbs-Merrill, 1967).
SSR   T. S. Kuhn, *The Structure of Scientific Revolutions* (Chicago, 1962).

Introduction

1. 'The History of our Time: An Optimist's View' (a lecture given in 1956), CR, p. 375.
2. *One Dimensional Man* (Routledge, 1964), p. 257.
3. Lamont C. Cole, 'Playing Russian Roulette with Biogeochemical Cycles', in H. W. Helfrich, Jr (editor), *The Environmental Crisis* (Yale, 1970), p. 14.

Chapter 1   Objectivity and Commitment in the Physical Sciences

1. SS, p. 1.
2. Ibid., p. 4.
3. Ibid., pp. 4 and 5.
4. OSE, pp. 225 and 224.
5. Ibid., p. 224.
6. Ibid., pp. 225, 233, 234 and 236.
7. Ibid., p. 168.
8. M. D. King, 'Science and the Professional Dilemma', in J. Gould (editor), *Penguin Social Sciences Survey 1968*, p. 72.
9. CR, p. 93.
10. J. L. E. Dreyer, *A History of Astronomy from Thales to Kepler* (Dover, 1953), p. 424.
11. P. A. Schilpp (editor), *Albert Einstein: Philosopher-Scientist* (Harper, Torchbook, 1959), p. 45.
12. LSD, p. 40.
13. CR, p. 36.
14. LSD, pp. 280 and 279.
15. SS, p. 8.
16. E. Nagel, *The Structure of Science* (Routledge, 1961), p. 86; quoted in SS, p. 48.
17. SS, pp. 9 and 8.
18. LSD, p. 31.
19. Ibid., p. 276.
20. CR, p. 240.
21. Ibid., p. 247.
22. Ibid., pp. 114–15.
23. Ibid., p. 113.
24. Ibid., p. 114.
25. Ibid., p. 119.
26. LSD, p. 97.
27. Ibid., p. 80.
28. Ibid., p. 107.
29. Ibid., p. 50, footnote.
30. CR, p. 248.
31. Ibid., p. 235.
32. Ibid., p. 238.
33. SSR, p. 77.
34. Ibid., p. 145.
35. Ibid., p. 91.
36. Ibid., pp. 43–4.
37. Ibid., p. 11.
38. Ibid., pp. 42, 17 and 102.
39. Ibid., p. 5.
40. Ibid., p. 37.
41. Ibid., pp. 80 and 38.
42. Ibid., p. 80.
43. Ibid., p. 52.

44. Laplace, *Exposition du Systeme du Monde*, Livre IV, Chapter II, 1796 (quoted in CMSRP, p. 169).
45. R. P. Feynman, *The Character of Physical Law* (MIT paperback edition, 1967), p. 33.
46. CR, p. 312, footnote; quoted in CMSRP, p. 167. See also note 55 in CMSRP.
47. LSD, p. 50; quoted in CMSRP, p. 167.
48. SSR, p. 99.
49. Ibid., p. 78.
50. Ibid., p. 68.
51. Ibid., pp. 84 and 90.
52. Ibid., p. 84.
53. Ibid., pp. 77 and 79.
54. Ibid., pp. 84–5.
55. Ibid., p. 101.
56. Ibid., pp. 101–2.
57. Ibid., p. 147.
58. Ibid., p. 149.
59. Ibid., p. 143.
60. Ibid., pp. 109 and 108.
61. Ibid., pp. 147 and 150.
62. Ibid., pp. 152, 153, 156 and 157.
63. Ibid., p. 158.
64. Ibid., p. 165.
65. Ibid., p. 166.
66. Ibid., p. 137.
67. Ibid., p. 168.
68. Ibid., pp. 91, 92 and 93.
69. M. Polanyi, *Personal Knowledge: Towards a Post-Critical Philosophy* (1958: Harper Torchbook, 1964), pp. 138 and 154.
70. Ibid., p. 312.
71. Ibid., p. 64.
72. CMSRP, p. 151n.
73. Ibid., p. 150.
74. Ibid., p. 155.
75. Ibid., p. 162.
76. Quoted in ibid., p. 160.
77. Ibid., p. 165.
78. SSR, pp. 23–4.
79. CMSRP, p. 176.
80. Ibid.
80n. M. Born, in P. A. Schilpp (editor) op. cit., p. 163. D. ter Haar, in A. C. Crombie (editor), *Turning Points in Physics* (Harper Torchbook, 1961), p. 44.
81. CMSRP, p. 183.
82. Ibid., p. 170.
83. Ibid., p. 176.
84. Ibid., p. 176n.
85. CGK, p. 215.
86. Ibid., p. 53.
87. Ibid., p. 264.
88. Ibid., p. 235.
89. Ibid., p. 261.
90. Ibid., p. 248.
91. Ibid., p. 204.
92. Ibid., p. 262.
93. *Tract on Monetary Reform* (Macmillan, 1923), p. 80.
94. *What is Sociology?* (Prentice-Hall, 1964), p. 103.
95. In discussion of R. P. Feynman's paper 'What Is and What Should be the Role of Scientific Culture in Modern Society', *Nuovo Cimento*, Supplemento IV (1966) 492–524, p. 518.

Chapter 2   The Copernican Revolution: A Study of Paradigms and Commitments

1. Quoted by J. Needham, *Science and Civilization in China* (Cambridge, 1956), Vol. 2, p. 38.
2. Ibid., Vol. 4, Part 1, p. 56.
3. Quoted by S. Toulmin and J. Goodfield, *The Fabric of the Heavens* (Penguin, 1963), p. 80. See also Aristotle, *de Caelo*, in Sir David Ross (editor), *The Works of Aristotle*, Vol. II (Oxford, 1930), 293ª.
4. Quoted by B. Farrington, *Greek Science* (Penguin, 1961), p. 96.
5. Plato, *The Republic*, Part 8, Section 4; quoted by Toulmin and Goodfield, op. cit., p. 89.
6. *The Almagest*, Book 1, Section 7, in *Great Books of the Western World* (Chicago, 1952), Vol. 16.
7. Quoted by Toulmin and Goodfield, op. cit., p. 147.
8. Quoted by T. S. Kuhn, *The Copernican Revolution* (Modern Library, 1959), p. 110.
9. R. Wittkower, *Architectural Principles in the Age of Humanism* (London: Tiranti, 1962), p. 29.
10. See M. Kline's analysis of the 'Last Supper' in *Mathematics in Western Culture* (Allen and Unwin, 1954), pp. 141–2.

NOTES

11. *Ten Books on Architecture*, trans. J. Leoni (Tiranti, 1955); Book 6, Chapter 2; Book 9, Chapter 5; Book 6, Chapter 2.
12. Copernicus, *On the Revolutions of the Heavenly Spheres*, Preface, in *Occasional Notes of the Royal Astronomical Society*, No. 10, 1947.
13. *Narratio Prima*, in E. Rosen (editor), *Three Copernican Treatises* (Dover 1959), p. 138.
14. Copernicus, op. cit., Book 1, Section 10.
15. Ibid., Book 1, Section 5.
16. Ibid., Book 1, Section 8.
17. E. A. Burtt, *The Metaphysical Foundations of Modern Physical Sciences* (Doubleday Anchor, 1954), p. 38.
18. *A Description of the Intellectual Globe*, Works, J. Spedding and R. L. Ellis (editors), (London, 1861), Vol. V, p. 517.
19. Quoted by P. Duhem, *To Save the Phenomena* (Chicago, 1969), p. 95. See also R. M. Blake, C. J. Ducasse and E. H. Madden (editor), *Theories of Scientific Method* (Washington, 1966), pp. 31–7.
20. Quoted by R. Dugas, *A History of Mechanics* (Routledge, 1957), p. 109.
21. Quoted by Kuhn, op. cit., p. 190.
22. Quoted by Burtt, op. cit., p. 58.
23. Quoted by A. Koestler, *The Sleepwalkers* (Penguin, 1964), pp. 361 and 364.
24. *A Dialogue Concerning the Two Chief World Systems* (California, 1953), p. 133.
25. Ibid., p. 117.
26. *The Assayer*, in S. Drake (editor), *Discoveries and Opinions of Galileo* (Doubleday Anchor, 1957), pp. 237–8.
27. *Dialogue*, p. 328.
28. Ibid., p. 334.
29. *The Assayer*, p. 265.
30. *Dialogue*, p. 32.
31. Ibid., p. 19.
32. Ibid., p. 59.
33. S. Drake (editor), p. 45.
34. Quoted in M. Caspar, *Kepler* (Abelard-Schuman, 1959), p. 102.
35. Quoted by Koestler, pp. 486–7; see also Galileo, *Dialogue*, p. 455.
36. Quoted by Koestler, op. cit., p. 264.
37. Quoted by R. Hooykaas in G. S. Metraux and F. Crouzet (editors), *The Evolution of Science* (Mentor, 1963), p. 276.
38. Quoted by Koestler, op. cit., pp. 339 and 326.
39. See esp. E. M. Rogers, *Physics for the Enquiring Mind* (Princeton), pp. 265–7, from which the text description is taken.
40. Koestler, op. cit., p. 352.
41. Ibid., p. 345.
42. Quoted by A. Koyré, *From the Closed World to the Infinite Universe* (Johns Hopkins, 1969), p. 61.
43. Cotes' Preface to the Second Edition of *Principia* 1713 (California, 1966), p. xxxii.
44. Needham, op. cit., Vol. II, p. 220.
45. In H. S. Thayer (editor), *Newton's Philosophy of Nature* (Hafner, 1953), p. 54.
46. Quoted by H. Stein, 'Newtonian Space-Time', *The Texas Quarterly* (Autumn, 1967), p. 198.
47. Quoted by S. F. Mason, *A History of the Sciences* (Collier Macmillan, 1962), p. 318.

Chapter 3 The Breakdown of the Newtonian Paradigm: A Study of Crisis

1. *Mathematical Principles of Natural Philosophy* (California, 1966), p. 6.
2. Ibid., pp. 8, 6 and 8.
3. Quoted by A. O'Rahilly, *Electromagnetic Theory* (Dover, 1965), p. 623.
4. Quoted by E. N. da C. Andrade, *An Approach to Modern Physics* (Doubleday Anchor, 1957), p. 242.
5. W. D. Niven (editor), *The Scientific Papers of James Clark Maxwell* (Dover, 1965), p. 775.
6. 'On the Relative Motion of the Earth and the Luminiferous Ether', reprinted in L. P. Williams (editor), *Relativity Theory: its Origins and Impact on Modern Thought* (Wiley, 1968), p. 34.
7. Quoted by R. S. Shankland, 'The Michelson–Morley Experiment', *American Journal of Physics* 32 (1964), 16–35.
8. G. Holton, 'Einstein, Michelson and the "Crucial" Experiment' [abbreviated below as 'EMCE'], *Isis* 60 (1969), 133–97, p. 160. In much of section 2 of this chapter I have closely followed Holton's analysis of the Einsteinian revolution.

9. 'On the Electrodynamics of Moving Bodies', reprinted in Williams (editor), op. cit., pp. 49–50.
10. Ibid., p. 50.
11. Einstein, 'H. A. Lorentz, Creator and Personality', in *Ideas and Opinions*; quoted by Holton, 'EMCE', op. cit., p. 192.
12. 'Physics and Relativity', in *Physics in My Generation* (Longman, 1970), pp. 103 and 105. For a further discussion of the implications of the Einsteinian space-time framework, see H. Bondi, *Relativity and Common Sense* (Anchor Books, 1964). Note that Bondi's interpretation of the conceptual status of relativity is very different from the interpretation developed here.
13. Quoted in S. Goldberg, 'In Defense of Ether', in R. McCormack (editor), *Historical Studies in the Physical Sciences* (Pennsylvania, 1970), Vol. 2, p. 107.
14. Quoted by Holton, 'Mach, Einstein and the Search for Reality', *Daedalus* 97 (1968) 636–73, p. 651.
15. Ibid., pp. 651–2.
15n. Quoted by S. Goldberg, 'The Abraham Theory of the Electron: The Symbiosis of Experiment and Theory', *Archive for History of Exact Sciences* 7 (1970), pp. 24 and 23.
16. H. A. Lorentz, 1906 lectures, published as *The Theory of Electrons* (1909: 2nd edition 1915; Dover 1952), p. 230.
17. Quoted by Holton, 'EMCE', op. cit., p. 140.
18. Ibid., p. 140n.
19. *The Theory of Electrons*, p. 321.
20. 'The Primary Concepts of Physics', in Williams (editor), op. cit., pp. 118–19.
21. 'The Theory of Relativity', in Williams, pp. 121–2.
22. Ibid., p. 122.
23. Ibid., pp. 122–3.
24. Ibid., p. 123.
25. 'EMCE', p. 185.
26. R. S. Shankland, 'Conversations with Einstein', *American Journal of Physics* 31 (1963), pp. 47–57.
27. Ibid., p. 53. See also R. S. Shankland, S. W. McCuskey, F. C. Leone and G. Kuerti, 'New Analysis of the Interferometer Observations of Dayton C. Miller', *Reviews of Modern Physics*, 27 (1955), 167–78.
28. 'EMCE', p. 195.
29. 'The Origins and Spirit of Logical Positivism', in P. Achinstein and S. F. Barker (editors), *The Legacy of Logical Positivism* (Johns Hopkins, 1969), p. 7.
30. 'The Interpretation of Scientific Theories', in Achinstein and Barker (editors), op. cit., p. 68.
31. O'Rahilly, op. cit., p. 501.
32. Ibid.
33. J. G. Fox, 'Evidence Against Emission Theories', *American Journal of Physics* 33, (1965), p. 16.
34. 'On a Heuristic Point of View about the Creation and Conversion of Light', *Annalen der Physik* 17 (1905), pp. 132–48. See translation by A. B. Arons and M. B. Peppard, *American Journal of Physics* 33 (1965), pp. 367–74.
35. Quoted by W. H. Cropper, *The Quantum Physicists* (Oxford, 1970), pp. 33–4.
36. Quoted by M. Born, *Physics in My Generation*, p. 104.
37. Quoted by A. Ellegard, 'Darwin's Theory and Nineteenth-Century Philosophies of Science', in P. P. Wiener and A. Noland (editors), *Roots of Scientific Thought* (Basic Books, 1957), p. 551.
38. Quoted by Born, op. cit., p. 160.
39. Reprinted in K. Przibram (editor), *Letters on Wave Mechanics* (Philosophical Library, 1967), p. 31.
40. Quoted by M. Born, *Natural Philosophy of Cause and Chance* (Dover, 1964), p. 122. See also *The Born-Einstein Letters* (Macmillan, 1971), p. 149.
41. 'Physics and Reality', *Journal of the Franklin Institute* 221 (1936), p. 349.
42. 'Discussion with Einstein on Epistemological Problems in Atomic Physics', in P. A. Schilpp (editor), p. 235.
43. Ibid., p. 666.
44. Quoted by Born, *Natural Philosophy of Cause and Chance*, p. 123.
45. Reprinted in K. Przibram (editor), p. 36.
46. 'The Philosophy of Experiment', *Nuovo Cimento* 1, (1955), p. 5.
47. Quoted by M. Born, 'The Interpretation of Quantum Mechanics', *British Journal for the Philosophy of Science* 4 (1953–5), p. 95.
48. In P. A. Schilpp (editor), op. cit., p. 672.

NOTES

49. 'The Development of the Space-Time View of Quantum Electrodynamics', *Physics Today* (August, 1966), pp. 31–4.
50. *The Evolution of Physics* (Cambridge, 1961), p. 296.
51. *Physics and Beyond* (Harper and Row, 1971), p. 68.
52. 'The Evolution of the Physicist's Picture of Nature', *Scientific American* (May, 1963).
53. 'Quantum Mechanics and the Aether', *Scientific Monthly* 58, (1954), p. 142.
54. 'Science, Art and Play', reprinted in E. C. Schrödinger, *Science, Theory and Man* (Dover, 1957), p. 29.

Chapter 4 Political Commitments in the Ages of the Scientific and Industrial Revolutions

1n. *The Metamorphoses of Ovid*, trans. M. M. Innes (Penguin, 1955), pp. 31–2.
2. Quoted by T. S. Kuhn, *The Copernican Revolution* (Harvard, 1957), p. 128. *On the Sun* is translated in abridged form in A. B. Fallico and H. Shapiro, *Renaissance Philosophy* (Modern Library, 1967), Vol. 1.
3. Quoted by N. Abbagnano, 'Italian Renaissance Humanism', in G. S. Metraux and F. Crouzet (editors), *The Evolution of Science* (Mentor Books, 1963), p. 242. See also the translation in A. B. Fallico and H. Shapiro, op. cit.
4. See, for example, More's *Utopia* in H. Morley (editor), *Ideal Commonwealths* (London, 1885).
5. Andreae, *Christianopolis*, trans. F. E. Held (New York, 1916); quoted by N. Eurich, *Science in Utopia* (Harvard, 1967), p. 236, and by M. L. Berneri, *Journey through Utopia* (1950: Schocken, 1971), pp. 113–14.
6. Campanella, *City of the Sun*, in H. Morley (editor), op. cit., pp. 229–30.
7. Quoted by C. Hill, *Intellectual Origins of the English Revolution* (Oxford, 1965), pp. 128–9n.
8. Quoted by Eurich, p. 261; see also H. Morley (editor), *The New Atlantis*, pp. 211–12 and M. E. Prior, 'Bacon's Man of Science', in L. M. Marsak (editor), *The Rise of Science in Relation to Society* (Collier Macmillan, 1964).
9. See L. S. Feuer, *The Scientific Intellectual* (Basic Books, 1963), pp. 204, 25 and 58.
10. C. Hill, *Reformation to Industrial Revolution* (Penguin, 1969), p. 253.
11. *The Theory of Moral Sentiments* (6th edition, London, 1790), Vol. 1, p. 393.
12. In A. Skinner (editor), *The Wealth of Nations*, Books I to III (Penguin, 1970), pp. 109–10.
13. *The Wealth of Nations*, Books IV and V (Methuen, 1950), p. 267.
14. Ibid., p. 273.
15. G. E. Mingay, *Enclosure and the Small Farmer in the Age of the Industrial Revolution* (Macmillan, 1968), p. 19; M. Kranzberg, 'Prerequisites for Industrialization', in M. Kranzberg and C. W. Pursell (editors), *Technology and World Civilization* (New York: Oxford, 1967), p. 217.
16. L. Radzinowicz, *A History of English Criminal Law* (Stevens, 1948), Vol. 1, p. 4.
17. Quoted by Marx, *Capital* (Moscow: Foreign Language Publishing House, 1961), Vol. 1, pp. 646–7.
18. For a discussion of the Poor Laws see J. D. Marshall, *The Old Poor Law 1795–1834* (Macmillan, 1968).
19. Quoted by J. L. and B. Hammond, *The Bleak Age* (London, 1934), p. 45.
19n. A. E. Musson and E. Robinson, *Science and Technology in the Industrial Revolution* (Manchester, 1969), p. 7.
20. Quoted by J. D. Bernal, *Science in History* (Penguin, 1969), Vol. 2, p. 540.
21. Quoted by J. Bronowski, *William Blake and the Age of Revolution* (Harper and Row, 1965), p. 149.
22. A. Ure, *The Philosophy of Manufacturers* (London, 1835), p. 368.
23. Quoted by A. Briggs, *Victorian Cities* (Odhams, 1963), p. 112.
24. J. J. Tobias, *Crime and Industrial Society in the Nineteenth Century* (Penguin, 1972), p. 295.
25. *Principles of Political Economy* (Longman, 1909), Book 2, Chapter 1, p. 202.
26. Ibid., p. 211.
27. Ibid., Book 4, Chapter 6, p. 751.
28. Ibid., pp. 750–1.
29. Ibid., Book 4, Chapter 7, p. 754.

30. Ibid., pp. 760–4.
31. Ibid., pp. 789–91.
32. Ibid., pp. 792–4.
33. Ibid., Book 2, Chapter 1, pp. 210–11 and 208.
34. Ibid., Preface to 3rd edition (1852), p. xxix.
35. Ibid., Book 2, Chapter 1, pp. 216–17.
36. Ibid., Bibliographic Appendix, pp. 986–9.
37. *Capital*, Vol. 1, pp. 611 and 15–16.
38. *Manifesto of the Communist Party* (Moscow: Progress Publishers, 1965), p. 47.
39. *Capital*, op. cit., p. 645.
40. D. J. Struick (editor), *The Economic and Philosophic Manuscripts of 1844* (International Publishers, 1964), p. 118.
41. *Capital*, p. 592.
42. *The Economic and Philosophic Manuscripts*, p. 147.
43. Ibid., p. 150.
44. Ibid., p. 169.
45. *The Communist Manifesto*, p. 75.
46. E. Fromm, *Marx's Concept of Man* (Frederick Ungar, 1961), p. 3.
47. *Capital*, p. 8.
48. Ibid., p. 20.
49. *The Communist Manifesto*, pp. 58–9.
50. *Socialism: Utopian and Scientific* (Allen and Unwin, 1950), p. 45.
51. Ibid., pp. 47–8.
52. Ibid., pp. 74–5.
53. Ibid., pp. 86–7.
54. Ibid., p. 69.
55. Ibid., pp. 80–1.
56. Ibid., p. 73.
57. Ibid., p. 82.
58. *The Life of Thomas Cooper, Written by Himself*; quoted by E. J. Hobsbawm, *Industry and Empire* (Weidenfeld and Nicolson, 1968), p. 103.
59. 'Introduction', *Socialism: Utopian and Scientific*, pp. xxxv and xxxviii.
60. Quoted by W. J. Barber, *A History of Economic Thought* (Penguin, 1967), p. 193.

Chapter 5 Capitalism in Crisis and the Keynesian Revolution

1. Quoted by P. d'A. Jones, *The Consumer Society* (Penguin, 1965), p. 283.
2. R. G. Lipsey, *An Introduction to Positive Economics* (Weidenfeld and Nicolson, 2nd edition 1966), p. 59.
3. R. L. Heilbroner, *The Making of Economic Society* (Prentice-Hall, 2nd edition 1968), p. 6.
4. C. Becker, *Progress and Power* (1936; Vintage Books, 1965), p. 102.
5. P. A. Samuelson, *Economics: An Introductory Analysis* (McGraw-Hill, International Student Edition, 6th edition 1964), p. 5.
6. T. F. Dernburg and D. M. McDougall, *Macro-Economics* (McGraw-Hill, International Student Edition, 3rd edition 1968), p. 40.
7. P. d'A. Jones, op. cit., p. 320.
8. A. H. Hansen, 'The Theory of Technical Progress and the Dislocation of Employment', *American Economic Review* (March, 1932).
9. 'A Short View of Russia' (1925), reprinted in *Essays in Persuasion* (Macmillan, 1933), p. 300.
10. *The General Theory of Employment, Income and Money* (Macmillan, 1936), p. 381.
11. Quoted by M. Stewart, *Keynes and After* (Penguin, 1967), p. 70.
12. G. L. S. Shackle, *The Years of High Theory* (Cambridge, 1967), p. 4.
13. Quoted by E. A. G. Robinson in R. Lekachman (editor), *Keynes' General Theory* (Macmillan, 1964), p. 50.
14. *The General Theory*, p. 32.
15. Ibid., preface and p. 20.
16. A. C. Pigou, 'Mr. Keynes's General Theory of Employment, Interest and Money', *Economica* (May, 1936).
17. P. A. Samuelson, 'The General Theory', in R. Lekachman (editor) op. cit., pp. 315–16.
18. Ibid., p. 316.
19. Ibid., pp. 317–18.
20. Ibid., p. 318.
20n. 'Mr Keynes on Underemployment Equilibrium', *Journal of Political Economy* (Oct., 1936), p. 686.
21. *A Guide to Keynes* (McGraw-Hill, 1953), pp. 6 and 11.
22. 'Real and Money Wage Rates in Relation to Unemployment', *Economic Journal* (Sept. 1937), p. 405.
23. *The General Theory*, pp. 18 and 26.

24. See, for example, Stewart, op. cit., pp. 35–47, together with his subsequent discussion of Keynesian economics. Another very useful discussion is by J. Pen, *Modern Economics* (Penguin, revised edition 1969).
25. L. Robbins, *The Great Depression* (Macmillan, 1934), pp. 186–8.
26. D. Winch, *Economics and Policy* (Hodder and Stoughton, 1969), p. 185.
27. *The New Republic* (July 29, 1940).
28. *The General Theory*, p. 34.
29. Ibid., p. 210.
30. A. P. Lerner, 'The General Theory', in R. Lekachman (editor), op. cit., p. 211.
31. This and the preceding discussion are taken from Dernburg and McDougall, pp. 145 and 149, and from G. Ackley, *Macroeconomic Theory* (Collier Macmillan, 1961), pp. 389–90.
32. Quoted by R. Lekachman, *The Age of Keynes* (Penguin, 1969), pp. 46 and 38.
33. See R. L. Heilbroner, op. cit., chapter 7.
34. *The Consumer Society*, p. 292.
35. Ibid., p. 302.
36. H. W. Arndt, *The Economic Lessons of the Nineteen-Thirties* (Frank Cass, 1963), p. 15.
37. W. Arthur Lewis, *Economic Survey 1919–1939* (Allen and Unwin, 1949), p. 52.
38. See R. L. Heilbroner, op. cit., chapter 8.
39. Winch, op. cit., p. 242.
40. Quoted by M. Stewart, op. cit., p. 146.
41. *The General Theory*, p. 129.
42. Arndt, op. cit., p. 175.
43. Ibid., p. 173.

Chapter 6   Objectivity and Commitment in the Social Sciences

1. Barrington Moore Jr, *Political Power and Social Theory* (Harvard, 1958), Chapter IV, reprinted in M. Stern and A. Vidich (editors) *Sociology on Trial* (Prentice-Hall, 1963), pp. 75 and 76.
2. R. K. Merton, *On Theoretical Sociology* (Free Press, 1967), p. 47.
2n. J. Ben-David, *The Scientist's Role in Society* (Prentice-Hall, 1971), p. 1.
3. T. Schroyer, 'The Critical Theory of Late Capitalism', in G. Fisher (editor), *The Revival of American Socialism* (Oxford, 1971), p. 313.
4. Engels' Funeral Speech, in E. Fromm, *Marx's Concept of Man* (Frederick Ungar Publishing Co., 1961), pp. 259–60.
5. 'Am I a Liberal?' (1925), reprinted in *Essays in Persuasion*, p. 324.
6. Ibid., 'A Short View of Russia' (1925), p. 300.
7. *The General Theory*, p. 383.
8. Ibid., p. 381.
9. *The Great Depression*, pp. 199–200.
10. *The New Statesman* (April 1, 1933).
11. Quoted by Winch, *Economics and Policy*, p. 186.
12. *The Great Depression*, p. 193.
13. *An Essay on the Nature and Significance of Economic Science* (Macmillan, 1935), preface.
14. Ibid., p. 1.
15. Quoted by Winch, op. cit., p. 221.
16. *An Essay*, p. 16.
17. *The Economic Problem in Peace and War*; quoted by Winch, op. cit., p. 267.
18. *The Great Depression*, p. 199.
19. 'The Great Slump of 1930', in *Essays in Persuasion*, p. 146.
20. 'John Maynard Keynes' (1946), in R. Lekachman (editor), op. cit., p. 304.
21. *Capital*, Vol. 1, p. 178.
22. *Economics and Policy*, p. 349.
23. 'The ABC of Socialism', in L. Huberman and P. M. Sweezy, *Introduction to Socialism* (Monthly Review, 1968), pp. 53–4. A footnote to 'The ABC of Socialism' states: 'The Material presented here was condensed from Leo Huberman's book, *The Truth About Socialism* and edited by Sybil H. May.'
24. *Capital*, Vol. 1, p. 763.
25. *A Contribution to the Critique of Political Economy*, Preface in E. Fromm, op. cit., pp. 217–18.
26. *The German Ideology*; quoted by E. Kamenka, *Marxism and Ethics* (Macmillan, 1969), p. 5.
27. Theses on Feuerbach, III; see, for example, L. S. Feuer (editor) *Marx and Engels: Basic Writings on Politics and Philosophy* (Doubleday Anchor, 1959), p. 244.

28. 'Reply to Mikhailovsky'; quoted by D. McLellan in *The Thought of Karl Marx* (Macmillan, 1971), p. 136.
29. *Manifesto of the Communist Party* (Moscow: Progress Publishers, 1952), pp. 39–40.
30. Quoted by S. Stoyanović, 'Marx's Theory of Ethics', in N. Lobkowicz (editor), *Marx and the Western World* (Notre Dame, 1967), p. 163.
31. Quoted by M. Rubel in N. Lobkowicz, op. cit., p. 51.
32. D. Horowitz, *Imperialism and Revolution* (Allen Lane, 1969), p. 31.
33. J. F. Cuber, *Sociology: A Synopsis of Principles* (Appleton-Century, 1947), p. 21; quoted by F. W. Matson in *The Broken Image* (Doubleday Anchor, 1966), p. 271.
34. 'The Study of Organizations—Objectivity or Bias?', in J. Gould (editor), *Penguin Social Sciences Survey* (Penguin, 1968), pp. 147 and 165.
35. *The Social Order* (McGraw-Hill, 1957), p. 20; quoted by Matson, op. cit., p. 66.
36. *Sociology* (McGraw-Hill, 1956), pp. 8 and 10; quoted by Matson, op. cit., p. 269.
37. *Economics of Welfare* (Macmillan, 3rd edition 1929), p. 5.
38. *An Introduction to Positive Economics*, p. 7.
39. G. Myrdal, *Asian Drama* (Allen Lane, 1968), p. 31.
40. A. W. Gouldner, 'Anti-Minotaur: The Myth of a Value-Free Sociology', in I. L. Horowitz (editor), *The New Sociology* (Oxford, 1965), p. 196.
41. R. S. Lynd, *Knowledge for What?* (1939: Evergreen Black Cat edition, 1964), p. 183.
42. M. Blaug, *Economic Theory in Retrospect* (Heinemann, 2nd edition 1968), p. 683.
43. *An Introduction to Positive Economics*, p. 59.
44. Ibid., p. 62.
45. 'Value Judgements and Economics', *British Journal for the Philosophy of Science* 15 (1964–5), pp. 97–114.
46. *The Structure of Science*, p. 495.
47. Gouldner, op. cit., p. 216.
48. G. A. Lundberg, *Foundations of Sociology* (David McKay, 1964), pp. 29–30 and 28.
48n. *The Poverty of Historicism* (1957: Routledge paperback edition, 1961), p. 89.
49. R. A. Dahl, *Modern Political Analysis* (Prentice-Hall, 1963), pp. 98–9. These remarks are omitted in the 2nd edition 1970.
50. *The Structure of Scientific Revolutions*, p. 87.
51. *The Logic of Scientific Discovery*, p. 280.
52. 'The Menace of Liberal Scholarship', *New York Review of Books* (Jan. 2, 1969), p. 38.
53. R. S. Lynd, op. cit., p. 249.
54. R. P. Feynman, *The Character of Physical Law* (MIT paperback edition, 1967), p. 173.

Chapter 7 Problems of Post-War Capitalism in the United States

1. *Macro-Economics*, p. 293.
2. L. R. Klein, *The Keynesian Revolution* (Macmillan, 1966), p. 69.
3. *The Accumulation of Capital* (Macmillan, 2nd edition, 1965), p. 54.
4. 'Expansion and Employment', *American Economic Review* 37 (1947), pp. 34–55.
5. W. Allen Wallis, 'Economic Growth; What, Why, How', in R. E. Slesinger, M. Perlman, A. Isaacs (editors), *Contemporary Economics* (Allyn and Bacon, 1967), p. 343.
6. *The General Theory*, p. 324.
7. See J. Henry, *Culture Against Man* (Tavistock, 1966), p. 93.
8. D. A. Schon, *Technology and Change* (Pergamon Press, 1967), p. 198.
8n. W. F. Stolper, *Planning Without Facts* (Harvard, 1966), p. 186. I am indebted to John Adams for bringing this quotation and work to my notice.
9. R. L. Heilbroner, 'The Future of Capitalism', *Commentary* (April, 1966), p. 28.
10. See the discussion by J. K. Galbraith, *The New Industrial State* (Hamish Hamilton, 1967), esp. Chapter 18.
11. *Journal of Retailing* (Spring, 1955), p. 1; (Winter, 1955–6), p. 166; quoted by V. Packard, *The Waste Makers* (Penguin, 1963), p. 33.

12. R. Theobald, *Free Men and Free Markets* (Doubleday Anchor edition, 1965), p. 107.
13. F. M. Fisher, Z. Grilliches and C. Kaysen, 'The Costs of Automobile Model Changes Since 1949', *Journal of Political Economy* (Oct., 1962). See P. A. Baran and P. M. Sweezy, *Monopoly Capital* (Penguin, 1968), pp. 138–40.
14. 'Science, Technology and Human Growth', in N. Kaplan (editor), *Science and Society* (Chicago Rand McNally, 1965), p. 526.
15. K. K. Kurihara, *Introduction to Keynesian Dynamics* (Allen and Unwin, 1956), p. 194.
16. *The General Theory*, p. 131.
17. Ibid., p. 105.
18. Ibid., p. 220.
19. *The Keynesian Revolution*, p. 205.
20. Ibid., p. 189.
21. *The Standards We Raise* (New York, 1953), p. 32; quoted by Baran and Sweezy, op. cit., p. 128.
22. See, for example, P. R. and A. H. Ehrlich, *Population, Resources, Environment: Issues in Human Ecology* (San Francisco: Freeman, 1970). In disagreement with them over the causes of the 'ecological crisis' is B. Commoner, *The Closing Circle* (Cape, 1972). P. R. Ehrlich and Commoner debate the issues in *Bulletin of the Atomic Scientists* (May, 1972). In disagreement with both Ehrlich and Commoner is J. Maddox, *The Doomsday Syndrome* (Macmillan, 1972).
23. *Technics and Civilization* (Routledge 1934), pp. 390 and 399.
24. 'Technology, Life and Leisure', *Nature* CC (1961), p. 513.
25. *Roots of Economic Growth* (Varanasi Ghandian Institute of Studies, 1962) p. 13.
26. *Free Men and Free Markets*, p. 9.
27. *The Costs of Economic Growth* (1967: Penguin, 1969), p. 175.
28. *The New Industrial State*, p. 292.
29. Ibid., p. 272.
30. Ibid., p. 409.
31. 'Putting Marx to Work', *New York Review of Books* (Dec. 5, 1968).
32. Heilbroner, *The Limits of American Capitalism* (Harper Torchbook, 1967), p. 47.
33. *The General Theory*, op. cit., p. 130.
34. A. M. Schlesinger Jr, *The Politics of Upheaval* (Heinemann, 1960), p. 191.
35. Ibid., p. 192.
36. Ibid., p. 637.
37. Ibid., p. 634.
38. Quoted by P. d'A. Jones, *The Consumer Society*, p. 324.
39. D. Perkins, *The New Age of Roosevelt* (Chicago, 1957), p. 69.
40. Ibid., p. 134.
41. Ibid., p. 172.
42. *The Consumer Society*, pp. 326 and 333.
43. F. Cook, *The Warfare State* (New York: Macmillan, 1962), pp. 55–63. In much of section 2 of this chapter I have closely followed Cook's description of American militarism.
44. Ibid., p. 63.
45. Ibid., pp. 64–5.
46. D. Macdonald, 'Comment', *Politics* (March, 1944), pp. 36–7. See also J. M. Swomley, Jr, *The Military Establishment* (Beacon Press, 1964), p. 100.
47. Quoted by Cook, op. cit., p. 72.
48. Ibid., p. 92.
49. Ibid., pp. 106–11.
49n. A. M. Schlesinger Jr, *The Crisis of Confidence* (André Deutsch, 1969), p. 143. See, in particular, the exchange of letters between Schlesinger and Chomsky in *The Listener* (Dec. 18, 1969) and (Jan. 15, 1970).
50. Quoted by W. H. Neblett, *Pentagon Politics* (New York: Pagent Press, 1953), p. 127 and by Cook, op. cit., pp. 111–12.
51. Quoted by Cook, op. cit., p. 97.
52. Reported in the *New York Times* (April 27, 1951); quoted by Cook, op. cit., p. 114.
53. In O. Nathan and H. Norden (editors), *Einstein on Peace* (Schocken, 1968), pp. 538 and 557.
54. G. Owen, *Industry in the U.S.A.* (Penguin Books, 1966), p. 22 and D. A. Schon, op. cit., p. 198.
55. Quoted by L. C. Lewin, *Report from Iron Mountain* (Macdonald, 1968), p. 69.
56. Quoted by Swomley, op. cit., p. 111.
57. See S. Melman, *Pentagon Capitalism: The Management of the New Imperialism* (McGraw-Hill, 1970), Appendix A.

58. See Winch, *Economics and Policy*, pp. 300–1 and 303.
59. R. E. Lapp, *The Weapons Culture* (Norton, 1968), p. 38.
60. McNamara's speech is reprinted in Ibid. The quotation is from p. 211.
61. Ibid., p. 209.
62. Quoted by T. C. Sorenson, *Kennedy* (Bantam, 1966), p. 679.
63. Ibid., p. 686.
64. S. Melman, 'Key Problems of Industrial Conversion to Civilian Economy', in G. L. Mangum (editor), *The Manpower Revolution* (Doubleday Anchor, 1966), pp. 317 and 305.
65. *The Limits of American Capitalism*, p. 105.
66. Melman, 'Key Problems', op. cit., p. 310.
67. Ibid., p. 307.
68. Ibid., p. 306.
69. See M. L. Weidenbaum, 'The Transferability of Defence Industry Resources to Civilian Uses', in *The Manpower Revolution*, p. 328.
70. *The New Industrial State*, p. 392.
71. See Cook, op. cit., pp. 22–3.
72. *The New Industrial State*, p. 229.
73. *The General Theory*, p. 378.
74. R. Lekachman (editor), *Keynes' General Theory* (St. Martin's Press, 1964), p. 9.
74n. See V. Perlo, *Militarism and Industry* (Lawrence and Wishart, 1963), p. 142.
75. Ibid., p. 147.

Chapter 8 Problems of Underdeveloped Countries

1. M. Barratt Brown, *After Imperialism* (Heinemann, 1963), pp. 43, 47 and 48.
2. B. Davidson, *Which Way Africa?* (Penguin, 1964), pp. 37 and 40.
3. Quoted by C. Oglesby, *Containment and Change* (Collier Macmillan, 1967), p. 52.
4. R. W. van Alstyne, *The Rising American Empire* (Oxford, 1960), p. 184.
5. Ibid.
6. Ibid., p. 193.
7. Oglesby, op. cit., p. 53.
8. R. L. Heilbroner, 'Counterrevolutionary America', *Commentary*, April, 1967.
9. R. F. Mikesell, *The Economics of Foreign Aid* (Weidenfeld and Nicolson, 1968), p. 23.
10. 'America Power in the Twentieth Century', *Dissent*, (Sept.–Oct., 1967), pp. 661 and 628.
11. C. Oglesby, 'Liberalism and the Corporate State', in P. Jacobs and S. Landau (editors) *The New Radicals* (Penguin, 1966).
12. D. Wise and T. B. Ross, *The Invisible Government* (Mayflower, 1968).
13. R. J. Barnett, *Intervention, and Revolution: The United States in the Third World* (New American Library, 1968).
14. D. Tobis, 'Foreign Aid: The Case of Guatemala', *Monthly Review* (Jan., 1968).
15. Quoted by H. Magdoff, *The Age of Imperialism* (Monthly Review, 1969), p. 49. In this chapter I have drawn substantially on information and arguments presented by Magdoff in his analysis of Western imperialism.
16. Quoted in ibid., p. 51.
17. Quoted in ibid., p. 54.
18. D. J. Patton, *The United States and World Resources* (Van Nostrand, 1968), pp. 14 and 18.
19. T. S. Lovering, 'Mineral Resources from the Land', in *Resources and Man*; National Research Council (Freeman, 1969), p. 120.
20. J. L. Fisher, in W. R. Ewald Jr, (editor), *Environment and Policy* (Indiana, 1970), p. 333.
21. R. L. Meier, *Science and Economic Development* (MIT, 1956; 2nd edition, 1966), p. 24.
22. R. Bailey, *Problems of the World Economy* (Penguin, 1967), p. 47.
23. B. R. Morris, *Problems of American Economic Growth* (Oxford, 1961), p. 191.
24. H. H. Landsberg, *Natural Resources for United States Growth* (Johns Hopkins, 1964), p. 81.
25. Ibid., p. 250.
26. R. Aron, *The Industrial Society* (Weidenfeld and Nicolson, 1967), pp. 28–9.
27. J. O. Coppock, 'Minerals and Metals', in J. F. Dewhurst (editor), *Europe's Needs and Resources* (Macmillan, 1961), pp. 637 and 625.

28. *Neo-Colonialism, The Last Stage of Imperialism* (Nelson, 1965), p. 65.
29. Quoted by Magdoff, p. 129.
30. W. and P. Paddock, *Famine 1975* (Little, 1968), p. 224.
31. H. H. Landsberg, L. L. Fischman and J. L. Fisher, *Resources in America's Future* (Johns Hopkins Press, 1963), p. 468.
32. J. McHale, *The Ecological Context* (Studio Vista, 1971), p. 149.
33. Quoted by W. A. Williams, *The Tragedy of American Diplomacy* (Cleveland World Publishing Co., 1959), pp. 148 and 169.
34. Quoted by D. W. Eakins, 'Business Planners and America's Postwar Expansion', in D. Horowitz (editor), *Corporations and the Cold War* (Monthly Review, 1969), p. 151.
35. Ibid., p. 156.
36. Ibid., p. 164.
37. *The General Theory*, pp. 381–3.
38. *US Bureau of the Census, Pocket Data Book, USA 1969*, p. 332.
39. Magdoff, p. 178.
40. Ibid., p. 194.
41. Ibid., p. 59.
42. H. L. Jacobson, 'Export Opportunities for Developing Countries', *Progress Unilever Quarterly* 4 (1969).
43. Magdoff, pp. 160 and 162.
44. Ibid., p. 152.
45. G. D. Woods, *Foreign Affairs* (January, 1966).
46. Magdoff, p. 198.
47. Quoted by R. Segal, *The Race War* (Penguin, 1967), p. 176.
48. E. Flores, 'Land Reform and the Alliance for Progress', in L. Randall (editor), *Economic Development: Evolution or Revolution* (Boston: Heath, 1964), p. 52.
49. Quoted by Magdoff, pp. 121–2.
50. L. A. Frank, *The Arms Trade in International Relations* (Praeger, 1969), p. 32.
51. Quoted by Magdoff, p. 143.
52. Ibid., p. 137.
53. Quoted in ibid., pp. 139–40.
54. Oglesby, *Containment and Change*, p. 95.
55. See ibid., pp. 90–4.
56. Quoted by Magdoff, p. 125.
57. Frank, op. cit., pp. 30–1.
58. Quoted by Magdoff, p. 128.
59. Ibid., p. 140.
60. 'Philosophers and Public Policy', *Ethics* 79 (1968), 1–9, p. 1.

Chapter 9 Socialism in Crisis: the Russian Revolution and the Origins of the Cold War

In this chapter I have drawn very substantially on information and arguments presented in the works listed below, and abbreviated in the notes as follows:
CWO   D. F. Fleming, *The Cold War and its Origins* (Doubleday, 1961), Vol. 1.
IR   D. Horowitz, *Imperialism and Revolution* (Allen Lane, 1969).
LLS   M. Lewin, *Lenin's Last Struggle* (Faber, 1969).
SA   J. P. Nettl, *The Soviet Achievement* (Thames and Hudson, 1967).
EH   A. Nove, *An Economic History of the USSR* (1969: Penguin, 1972).

1n. E. Signer and A. W. Galston, 'Education and Science in China', *Science* 175 (1972), p. 15.
2. IR, p. 15.
3. Quoted by G. Barraclough, *An Introduction to Modern History* (Penguin, 1967), p. 217.
4. I. Deutscher, *The Unfinished Revolution* (Oxford, 1967), p. 12.
5. Quoted by Barraclough, p. 218.
6. This is the description given in IR, p. 137.
7. LLS, pp. 4 and 19.
8. IR, p. 26.
9. LLS, pp. 106 and 172–3.
10. Ibid., pp. 111 and 112–13; EH, pp. 84–8.
11. Quoted by R. Dunayevskaya, *Marxism and Freedom* (New York: Twayne, 1964), pp. 241 and 176.
12. LLS, p. 87.
13. SA, p. 94.
14. Quoted by I. Deutscher, *Stalin, A Political Biography* (Vintage Books, 1960), p. 550.

15. EH, p. 221.
15n. See C. K. Wilber, *The Soviet Model and Underdeveloped Countries* (North Carolina, 1969), p. 128.
16. EH, p. 207.
17. Quoted by F. L. Carsten, *The Rise of Fascism* (Methuen, 1967), p. 119.
18. CWO, p. 107. Much of the remainder of this chapter follows very closely Fleming's account of the development of the Cold War.
19. Ibid.
20. IR, p. 82.
21. F. Brockway and F. Mullaly, *Death Pays a Dividend* (Gollancz, 1944), p. 108.
22. CWO, p. 78.
23. Ibid., p. 81.
24. Ibid.
25. Ibid., p. 92.
26. Ibid., p. 102.
27. *New York Times* (July 24, 1941); quoted by D. Horowitz, *From Yalta to Vietnam* (Penguin, 1967), p. 59.
28. EH, p. 273.
29. IR, p. 94.
30. SA, p. 192.
31. Ibid., p. 172 and IR, p. 87.
32. CWO, p. 182.
33. Ibid., p. 282.
34. Quoted by G. Alperovitz, *Atomic Diplomacy* (Secker and Warburg, 1966), pp. 229 and 232.
35. Ibid., p. 229.
36. Quoted by Horowitz, *From Yalta to Vietnam*, p. 57.
37. Ibid., p. 53.
38. Alperovitz, pp. 237–8.
39. Quoted by P. M. S. Blackett, *Military and Political Consequences of Atomic Energy* (Turnstile Press, 1948), p. 122.
40. Alperovitz, p. 242.
41. Ibid., pp. 277–9.
42. A. M. Schlesinger, *The Crisis of Confidence*, p. 123.
43. CWO, p. 348.
44. Ibid., p. 392.
45. Ibid., p. 420.
45n. O. Nathan and H. Norden (editors), *Einstein on Peace* (Schocken Books, 1968), p. 389.
46. CWO, p. 446.
47. Quoted by D. Horowitz (editor), *Containment and Revolution* (Anthony Blond, 1967), p. 12.
48. CWO, p. 391.
49. Blackett, op. cit., pp. 70 and 179.
50. CWO, p. 252.
51. I. Deutscher, 'Myths of the Cold War', in D. Horowitz (editor), op. cit., pp. 13–14.
52. Quoted by Cook, *The Warfare State*, p. 171.
53. Editorial, *The Nation* (Oct. 7, 1961). See also Cook, p. 255.

Chapter 10 On the Ethical Neutrality of Science—The Indictment of the Scientific Mentality

1. S. Drake (editor), *Discoveries and Opinions of Galileo* (Doubleday Anchor, 1957), p. 63.
2. *Science and Society* (Allen Lane, 1969), p. 260.
3. *The Making of a Counter Culture* (Faber, 1970).
4. G. S. Stent, *The Coming of the Golden Age* (New York: Natural History Press, 1969), p. 79.
5. Ibid.
6. M. M. Turmin, 'In Dispraise of Loyalty', in J. D. Douglas (editor), *The Relevance of Sociology* (Appleton Century Crofts, 1970), p. 161.
7. *de Augmentis Scientiarum*, Book III, Chapter V.
8. *The New Organon*, Aphorism III, Book I.
9. *Discourse on Method*, Part 6; quoted by P. Rossi, *Philosophy, Technology and the Arts in the Early Modern Era* (Harper Torchbook, 1970), p. 104.
9n. *The Examinations of the Academies*, in A. G. Debus (editor), *Science and Education in the Seventeenth Century: The Webster-Ward Debate* (Macdonald, 1968), p. 68.
  For a discussion of the 'witch-craze' see, for example, H. R. Trevor-Roper, *The European Witch-Craze of the 16th*

and 17th Centuries (1967: Penguin, 1969) and K. V. Thomas, *Religion and the Decline of Magic* (Weidenfeld and Nicolson, 1971).
10. *Oration on the Dignity of Man*, trans. A. R. Caponigri (Chicago: Regnery, 1956), p. 31.
11. *On the Sense and Feeling of All Things and on Magic*, in A. B. Fallico and A. Shapiro, *Renaissance Philosophy*, Vol. 1 (Random House, 1967), p. 344.
12. *The Assayer*, in S. Drake (editor), op. cit., p. 274.
13. Quoted by A. Vartanian, *Diderot and Descartes* (Princeton, 1953), p. 47.
14. See R. Lenoble, *Mersenne, ou la Naissance du Mechanisme* (Paris, 1943), pp. 532–41; trans. E. Cohen and R. Olson, reprinted in R. Olson, *Science as a Metaphor* (Wadsworth, 1971).
14n. *European Culture and Overseas Expansion* (Penguin, 1970), p. 15.
15. Quoted by Rossi, op. cit., p. 141.
16. Ibid.
17. *Experimental Philosophy*, in M. Boas Hall (editor), *Nature and Nature's Laws* (Harper paperback, 1970), p. 129.
18. Quoted by C. Hill, *The Intellectual Origins of the English Revolution* (Oxford, 1965), p. 90.
19. See Pauli, 'The Influence of Archetypal Ideas on the Scientific Theories of Kepler', in C. G. Jung and W. Pauli, *The Interpretation of Nature and the Psyche* (Routledge, 1955).
20. *The New Organon*, Aphorism 85, Book 1.
21. *Vindiciae Academiarum*, in Debus, op. cit., p. 36.
22. D. Lee, *Freedom and Culture* (Prentice-Hall, 1959), p. 163; quoted by Roszak, op. cit., p. 245.
23. *De Augmentis Scientiarum*, Book 4, Chapter 1.
24. This is the hypothesis D. E. Wooldridge makes in his *The Machinery of the Brain* (McGraw-Hill, 1963) in order to cope with the 'awkward and embarrassing' problem of consciousness (p. 219). He has the opinion that (p. 240): 'No useful purpose has yet been established for the sense of awareness that illumines a small fraction of the mental activities of a few species of higher animals. It is not clear that the behaviour of any individual or the course of world history would have been affected in any way if awareness were nonexistent.'
25. Quoted by F. W. Matson, *The Broken Image*, p. 13.
26. Holbach, *Systeme de la Nature*, Vol. 1, Chapter VI.
26n. R. M. Young, 'Evolutionary biology and ideology: Then and now', in W. Fuller (editor), *The Social Impact of Modern Biology* (Routledge, 1971), p. 201. J. Lederberg, 'The Perfection of Man', in *Nobel Symposium 14, The Place of Value in a World of Facts* (Stockholm: Almquist and Wiksell, 1970), p. 55.
27. *Of Molecules and Men* (Washington, paperback edition, 1967), p. 10.
28. Ibid., p. 16.
29. Ibid., pp. 98–9.
30. Ibid., pp. 93 and 88.
31. D. E. Wooldridge, *Mechanical Man* (McGraw-Hill, 1968), p. 168.
32. Ibid., pp. 202 and 182.
33. Ibid., p. 174.
34. J. B. Watson, *Behaviourism* (Chicago, 1958), p. 11; quoted by F. W. Matson, op. cit., p. 40.
35. B. F. Skinner, *American Scholar* 25 (1955–6), pp. 52–3.
36. *Science and Human Behaviour* (Free Press paperback edition, 1965), p. 447.
37. 'The Problem of Consciousness: A Debate', in J. H. Gill (editor), *Philosophy Today No 2* (Macmillan, 1969), p. 193.
38. *Man-Machine Communication* (Wiley, 1970), preface.
39. Ibid., p. 412.
40. E. D. Weitzman and G. S. Ross, 'A Behavioural Method for the Study of Pain Perception in the Monkey', *Neurology* 12 (1962), p. 264. See also C. Roberts, *The Scientific Conscience* (New York: Braziller, 1967).
40n. *Science, Industry and Society* (Allen and Unwin, 1970), p. 51.
41. *The Coming of the Golden Age*, p. 82.
42. *The Scientific Outlook* (Allen and Unwin, 2nd edition, 1949), p. 51.
43. See, for example, essays by B. Wiley in *The Eighteenth Century Background* (1940: Penguin, 1962) and *Nineteenth*

Century Studies (1949: Penguin, 1964). The lines quoted are taken from Wiley, *Nineteenth Century Studies*, p. 35.

43n. 'On Goethe's Scientific Researches', in *Popular Scientific Lectures 1853* (Dover edition, 1962), p. 20.

44. See, for example, J. F. Lively, *The Enlightenment* (Longmans, 1966), pp. 83–5 and C. C. Gillispie, *The Edge of Objectivity* (Princeton, 1960), pp. 192–201.

45. *Technics and Civilization* (Routledge, 1934), pp. 49 and 51.

46. *Life Against Death* (Sphere, 1968), p. 276.

47. *The Scientific Outlook*, p. 270.

48. Ibid., p. 272.

49. Ibid., p. 267.

50. *The Making of a Counter Culture*, pp. 21 and 8.

51. Ibid., p. 47.

52. Ibid., pp. 208–9.

53. Ibid., pp. 221–2.

54. Ibid., pp. 231–2.

54n. *Technological Society* (Cape, 1965), p. 411.

55. *The Making of a Counter Culture*, p. 219.

56. Ibid., p. 229.

57. Ibid., p. 232.

58. Ibid., p. 244.

59. Ibid., pp. 249 and 268.

60. George Orwell, *1984* (Penguin, 1954), pp. 214–15 and 164.

61. Ibid., p. 164.

62. H. Marcuse, *Eros and Civilization* (1955: reprinted by Vintage Books, 1961), p. 131.

63. E. M. Forster, *Collected Short Stories* (Penguin, 1954), p. 131.

63n. B. Commoner, *Science and Survival* (Ballantine Books, 1971), p. 3.

64. Quoted by M. Greene (editor), *Anatomy of Knowledge* (Routledge, 1969), p. 97.

65. *Problems of Life* (Harper Torchbook, 1960), p. 202.

66. C. R. Rogers and B. F. Skinner, 'Some Issues Concerning the Control of Human Behaviour', *Science* 124 (1956), 1057–66.

67. 'The Problem of Consciousness: A Debate', in J. H. Gill (editor), op. cit., p. 208.

68. See, for example, E. P. Wigner, 'Two Kinds of Reality', *The Monist* 48 (1964), pp. 248–64.

69. 'Life's Irreducible Structure', *Science* 160 (1968), pp. 1308–12.

69n. *A New View of Society* (Penguin, 1970), pp. 94 and 96.

70. In which the Concorde is one of the most outrageous. See esp. J. A. Adams and N. Haigh, 'Booming Discorde', *The Geographical Magazine* (July, 1972).

70n. *The Advancement of Science* 26 (1969), pp. 1–9.

71. For example, in his article 'Quantum Physics and Philosophy', Niels Bohr has emphasized that 'The essentially new feature in the analysis of quantum phenomena is . . . the introduction of a fundamental distinction between the measuring apparatus and the objects under investigation', (reprinted in *Atomic Physics and Human Knowledge: Essays 1958–1962* (Interscience, 1963), p. 3). Bohm and Bub, having presented a hidden-variable theory *which contains quantum mechanics as a special case*, conclude: 'Since our theory explicitly couples the large-scale and the small-scale levels together, all macro-quantities cannot be calculated completely from the micro-laws. In fact, according to the Copenhagen interpretation, this was always implicit in quantum mechanics, but with the unacceptable qualification that such a multi-level theory dealing with the totality, in which quantum and sub-quantum movements depend on the large scale movement and vice-versa, is in principle incapable of being developed' ('A Proposed Solution of the Measurement Problem in Quantum Mechanics by a Hidden Variable Theory', *Reviews of Modern Physics* 38 (1966), 453–69, p. 465).

71n. *The Physicist's Conception of Nature* (Hutchinson, 1958), pp. 15–16.

72. O. Nathan and H. Norden (editors), *Einstein on Peace* (Schocken, 1968), p. ix.

73. Quoted by Shankland, 'Conversations with Einstein', *American Journal of Physics* 31 (1963), pp. 47–57.

74. *Life and Letters* (Murray, 1887), Vol. III, pp. 200 and 205.

75. *New Scientist* (January 28, 1971), pp. 174–6.
76. *The Scientific Outlook*, p. 270.
77. N. Barlow (editor), *The Autobiography of Charles Darwin* (Collins, 1958), pp. 138–9.
78. *Life and Letters*, op. cit., Vol. II, p. 114.
79. 'Further Remarks on Order', in C. H. Waddington (editor), *Towards a Theoretical Biology* (Edinburgh, 1969), Vol. 2, p. 51.
80. *Life Against Death*, p. 276.
81. *An Essay on Liberation* (Allen Lane, 1969), p. 19.
82. *One Dimensional Man* (Routledge, 1964), p. 231.
83. *The Advancement of Science* 26 (1969), pp. 1–9.
84. 'Science: Endless Horizons or Golden Age', *Science* 171 (1971), pp. 23–9.

Chapter 11   The Scientific Community in Crisis—Scientists as Agents of Change?

1. Introduction by H. B. G. Casimir in A. de Shalit, H. Feshbach and L. van Hove, *Preludes in Theoretical Physics in Honour of V. F. Weisskopf* (North Holland, 1966), p. vii.
2. See esp. S. A. Lakoff's discussion of Condorcet in 'The Third Culture', in Lakoff (editor), *Knowledge and Power* (The Free Press, 1966), pp. 14–19.
3. Quoted by F. E. and F. P. Manuel, 'Sketch for a Natural History of Paradise', *Daedalus* (Winter, 1972), p. 87.
4. Saint-Simon, 'The Rule of the Scientists', in F. E. and F. P. Manuel, *French Utopias* (The Free Press, 1966), p. 263. See H. Markham (editor), *Henri Comte de Saint-Simon, Selected Writings* (Blackwell, 1952).
5. Quoted by Lakoff, op. cit., p. 20.
6. 'Karl Marx's Funeral', in E. Fromm *Marx's Concept of Man*, p. 259; and Struik (editor), *Economic and Philosophic Manuscripts*, p. 142.
7. T. Veblen, *The Engineers and the Price System* (New York: B. W. Huebsch, 1921), p. 58.
8. Ibid. See esp. the discussion by Lakoff, op. cit., pp. 33–6, from which this paragraph is extracted.
9. J. K. Galbraith, *The New Industrial State* (Hamilton, 1967), p. 8.
10. Ibid., p. 328.
11. Ibid., pp. 333 and 339–40.
12. Ibid., p. 372.
13. Ibid., pp. 374 and 376.
14. Ibid., p. 335.
15. Ibid., p. 381.
16. Ibid., p. 335.
17. Ibid., p. 339.
18. Ibid., p. 341.
19. Ibid., p. 353.
20. R. L. Heilbroner, 'The Future of Capitalism', *Commentary* (April, 1966), p. 23.
21. Ibid., p. 30.
22. Ibid., p. 32.
23. Ibid., p. 33.
24. Ibid., p. 34.
25. Ibid., p. 28.
26. R. L. Heilbroner, *The Limits of American Capitalism* (Harper Torchbook, 1967), p. 128. The qualification 'and especially from the young' was omitted in the *Commentary* essay.
27. 'The Future of Capitalism', p. 35.
28. 'Physics in the Contemporary World', *Bulletin of the Atomic Scientists*, March, 1948, p. 66.
29. A. K. Smith, *A Peril and a Hope* (Chicago, 1965), p. 9.
30. Ibid., p. 12.
31. H. L. Stimson, 'The Decision to Use the Atomic Bomb', *Harper's Magazine*, CXCIV (February, 1947), pp. 99–100.
32. R. E. Lapp, *The New Priesthood* (Harper and Row, 1965), p. 76; to be abbreviated below as NP.
33. Smith, p. 44.
34. *The Franck Report*, in Smith, p. 562.
35. Ibid., pp. 563 and 566.
36. Ibid., p. 571.
37. Quoted by Lapp, NP, p. 83.
38. R. Jungk, *Brighter than 1000 Suns* (Penguin, 1960), p. 175.
39. D. Horowitz, *From Yalta to Vietnam* (1965: Penguin, revised edition, 1967), p. 57.

40. I. L. Horowitz (ed.), *Power, Politics and People: The Collected Essays of C. Wright Mills* (New York: Oxford UP, 1963), pp. 61–2; quoted by R. E. Lapp, *The Weapons Culture* (Norton, 1968), p. 177.
41. R. Gilpin, *American Scientists and Nuclear Weapons Policy* (Princeton, 1962), p. 28. Much of the remainder of this section 2 follows very closely indeed Gilpin's analysis.
42. Quoted by Lapp, NP, p. 94.
43. Ibid., p. 103.
44. B. Baruch, 'United States Proposals for the Control of Atomic Energy', in *Disarmament and Security*, United States Government Printing Office (Washington, 1956), pp. 192–3.
45. P. M. S. Blackett, *Political and Military Consequences of Atomic Energy*, p. 188.
46. Smith, op. cit., p. 478.
47. Blackett, p. 189.
48. Gilpin, p. 37.
49. Ibid., p. 66.
50. Ibid., pp. 70 and 72.
51. Ibid., p. 89.
52. Ibid., p. 94.
53. Ibid., p. 100.
54. Ibid., p. 121.
55. Lapp, NP, p. 121.
56. 'Atomic Weapons and American Policy', *Bulletin of the Atomic Scientists* (July, 1953).
57. Gilpin, p. 130.
58. Ibid., p. 132.
59. Ibid., pp. 133–4.
60. L. P. Bloomfield, W. C. Clemens Jr and F. Griffiths, *Krushchev and the Arms Race* (MIT, 1966), p. 25.
61. Ibid., p. 26.
62. Ibid., pp. 32 and 72.
63. Gilpin, p. 178.
64. Ibid., p. 194.
65. Ibid., pp. 192 and 193–4.
66. Lapp, NP, p. 138.
67. See F. Cook, *The Warfare State*, pp. 233–4.
68. Quoted by Cook, p. 234.
69. H. K. Jacobson and E. Stein, *Diplomats, Scientists and Politicians* (Michigan, 1966), pp. 160–1.
70. Ibid., pp. 179–80.
71. Lapp, NP, p. 139.
72. Jacobson and Stein, p. 200.
73. Ibid., p. 219.
74. Gilpin, p. 240.
75. Ibid., p. 243.
76. Ibid., pp. 239–40.
77. 'The Case for Ending Nuclear Tests', *The Atlantic Monthly* (August, 1960), pp. 45–6.
78. Ibid., p. 46; and quoted by Jacobson and Stein, p. 223.
79. 'The Test Ban and the Big Hole', *Scientific American* (June, 1960), p. 81.
80. Quoted by Gilpin, p. 271.
81. 'Alternatives for Security', *Foreign Affairs* (January, 1958), p. 204; quoted by Jacobson and Stein, p. 152.
82. *Our Nuclear Future* (Secker and Warburg, 1958), p. 140; quoted by Jacobson and Stein, p. 152.
83. 'The Case for Ending Nuclear Tests', p. 48; quoted by Gilpin, p. 262.
84. 'The Issue of Peace', Unpublished Paper; quoted by Gilpin, p. 262.
85. Press Release, July 16, 1960: quoted by Gilpin, p. 268.
86. Jacobson and Stein, op. cit., p. 355.
87. Bloomfield et al., op. cit., p. 96.
88. Lapp, *The Weapons Culture*, p. 74.
89. I. F. Stone, 'The Test Ban Comedy', *New York Review of Books* (May 7, 1970), p. 18.
90. E. C. Ladd, Jr, 'Professors and Political Petitions', *Science* 163 (1969), pp. 1425–30.
91. See J. Allen (editor), *March 4: Scientists, Students and Society* (MIT, 1970), p. xxii.
92. *Culture Against Man* (Random House, 1963), p. 476.
93. G. Wald, 'A Generation in Search of a Future', in Allen, op. cit., p. 112.
94. See the Editorial, *Scientific Research* (March 31, 1969).
95. F. Wheeler, 'The Anti-Anti Missile Physicists', *New Scientist* (May 15, 1969).
96. J. Shapiro, L. Eron and J. Beckwith, *Nature* (December 27, 1969), p. 1337.
97. A. D. Sakharov, *Progress, Coexistence and Intellectual Freedom* (André Deutsch, 1968), p. 25.

98. Ibid., p. 84.
99. 'What we must do', *Science* (November 28, 1969), p. 115.
100. Ibid., p. 1116.
101. Ibid., p. 1121.
102. Wald, op. cit., p. 115.
103. D. King-Hele, *The End of the Twentieth Century?* (Macmillan, 1970), p. 1.
104. Ibid., pp. 8, 172 and 35.
105. *Futures* (July, 1970), pp. 114–22.
106. 'The Liberty Machine', *Futures* (December, 1971), pp. 338–48.
107. 'When the Growing has to Stop', *The Observer* (March 19, 1972), pp. 11–12.
108. 'A Conference on Workers' Self Management in Science (March 11–12, 1972) Manifesto.
109. 'The Future of Capitalism', p. 35.
110. *Eros and Civilization* (1955: reprinted by Vintage Books, 1961), p. 165.
111. 'Liberation from the Affluent Society', in D. Cooper (editor), *The Dialectics of Liberation* (Penguin, 1968), pp. 188–9.
112. *An Essay on Liberation*, p. 55n.

Chapter 12 Towards a Beautiful World: Some Concluding Thoughts

1. *The World, The Flesh and the Devil* (1929: Cape paperback edition, 1970), pp. 37–8.
2. Ibid., pp. 26, 38, 39 and 42.
3. Ibid., pp. 53–6.
4. Ibid., pp. 64 and 69.
5. Ibid., p. 69.
6. W. Friedmann, 'Interventionism, Liberalism and Power Politics: The Unfinished Revolution in Power Politics', *Political Science Quarterly* (June, 1968); quoted by N. Chomsky, 'On Changing the World', *Cambridge Review* (Feb. 19, 1971), p. 122.
7. 'Is Biology a Molecular Science?', in M. Grene (editor), *Anatomy of Knowledge* (Routledge, 1969), p. 99.
8. 'Reflections on the End of the Republic', *New York Review of Books* (Sept. 24, 1970), pp. 39–40.
9. *The Roots of American Foreign Policy* (Beacon paperback edition, 1969), pp. xi and 87.
10. Quoted by J. A. C. Brown, *Freud and the Post-Freudians* (Penguin, 1961), p. 139.
11. S. Freud, *Civilization and its Discontents* (1931: Hogarth Press, 1962), p. 59.
12. A Storr, *Human Aggression* (Allen Lane, 1968), pp. 114–15.
13. *Civilizations and its Discontents*, pp. 60–1.
14. A. Storr, op. cit., pp. 52–3.
15. H. Kaufmann, *Aggression and Altruism* (Holt, Rinehart and Winston, 1970), p. 142.
16. R. S. Lynd, *Knowledge for What?*, pp. 71 and 196.
17. *The Sane Society* (Routledge paperback edition, 1963), p. 30.
18. *The Art of Loving*, p. 94.
19. *The Problem of Method* (Methuen, 1963), p. 91.
20. R. C. Tucker, *Philosophy and Myth in Karl Marx* (Cambridge, 1961), pp. 137–8.
21. 'Problems of Interpreting Marx', in I. L. Horowitz, *The New Sociology* (Oxford, 1965), pp. 190–1.
22. Quoted by W. Desan, *The Marxism of Jean-Paul Sartre* (Doubleday Anchor, 1966), p. 218.
23. 'Economic Possibilities for Our Grandchildren', in *Essays in Persuasion* (Macmillan, 1933), p. 368.
24. *The German Ideology* (New York: New World paperback edition, 1963), p. 22.
25. *Eros and Civilization* (1955: Sphere, 1969), Political Preface 1966, p. 13.
26. Lessing, quoted by J. Passmore, *The Perfectibility of Man* (Duckworth, 1970) p. 48.
27. Goethe, *Faust, Part One*, translated by P. Wayne (Penguin, 1949), p. 87.
28. *de Magnete* (Dover, 1958), Book V, Chapter 12.
28n. F. A. Yates, *Giordano Bruno and the Hermetic Tradition* (Vintage Books, 1969), p. 38.
29. *Eros and Civilization* (1955; reprinted by Vintage Books, 1961), pp. 146–7 and 151.

30. Yates, p. 24.
31. Ibid., p. 126.
31n. *de Magnete*, Book II, Chapter 3 and Book V, Chapter 12.
32. *The Social Function of Science* (Routledge, 1940), pp. 379, 380 and 382.
33. *The Scientific Outlook*, p. 275.
34. Ibid.

Appendix   Einstein's Political Struggle

With the exception of reference 20 all references are to: O. Nathan and H. Norden (editors), *Einstein on Peace* (Schocken Books, 1968).

1. pp. 341–2. 2. p. 355. 3. p. 380. 4. p. 401. 5. p. 403. 6. p. 416. 7. p. 436. 8. pp. 465–6. 9. pp. 475–6. 10. p. 487. 11. p. 521. 12. pp. 532–3. 13. p. 536. 14. p. 537. 15. pp. 538–9. 16. p. 553. 17. p. 554. 18. p. 570. 19. p. 613.
20. *The Born-Einstein Letters* (Macmillan, 1971), pp. 231–2.
21. p. 616. 22. p. 621. 23. pp. 629–30.

INDEX – NAMES

Abraham, M., 70–1, 71n
Acheson, Dean, 214–15
Alberti, Leon B., 36
Albrow, M., 168
Alfonso, King, 34, 55
Alperovitz, G., 240
Andreae, Valentin, 91–2, 93, 94–5, 249, 257, 320
Aquinas, St. Thomas, 35
Arbenz, President, 208
Aristarchus, 28, 30–3, 47
Aristotle, 29, 32, 253, 254, 319, 367
Arndt, H. W., 146, 148
Arnold, T., 100
Aron, R., 212

Bacon, Francis, xiv, 44, 93, 95, 149, 150, 151, 249, 253, 256, 257, 284
Bailey, R., 210, 211
Baran, P. A., 186
Bechhoefer, B., 304
Becker, C., 117
Beer, S., 314
Ben-David, J., 152n
Bentley, Bishop, 57
Berkeley, Bishop, 7
Bernal, J. D., 325–7, 340
Bertalanffy, L. von, 273
Bethe, H. A., 299, 302–7
Beveridge, Senator, 205
Bierstedt, R., 168
Blackett, P. M. S., 244, 298
Blake, William, 265, 280
Blanshard, B., 275
Blaug, M., 170
Bodin, Jean, 44
Bohm, D., 83, 283, 362 (n. 71)
Bohr, N., 82, 295, 348, 362 (n. 71)
Bondi, H., 352 (n. 12)
Born, M., 22n, 71n, 83, 347
Bosch, President, 208
Box, S., 263n
Boyle, Robert, 255
Brahe, Tycho, 44, 50–1, 52, 161, 320
Branco, General C., 220–1
Brecht, Bertolt, 286
Brockway, F., 234
Bronowski, J., 100–1, 353 (n. 21)
Brown, M. Barratt, 204–5
Brown, N. O., 266, 284
Bub, J., 362 (n. 71)
Bucherer, A. H., 71
Buridan, J., 55
Bush, V., 301

Butterfield, H., 87–8
Byrnes, J., 239, 240, 240n, 243

Cabot, J. M., 219
Campanella, Tommaso, 91–2, 94, 95, 249, 254, 257, 320
Chamberlain, N., 147, 234–6
Chomsky, N., 177, 223
Churchill, Winston, 227, 235, 236–7, 238–9, 241, 243, 295
Cipolla, C. M., 255n
Clavius, Father C., 44
Clayton, W., 215–16
Coleridge, S. T., 264–5
Comenius, J. A., 93
Commoner, B., 272n, 329
Compton, A. H., 295
Conant, J. B., 300
Condorcet, marquis de, 287
Conger, E. H., 205
Cook, F., 194–7 *passim*, 201, 357 (n. 43)
Cooper, Thomas, 112–13
Copernicus, 36–58 *passim*, 67, 72, 74, 160–1, 174, 319, 322
Coppock, J. O., 212–13
Cotes, R., 55
Cotgrove, S., 263n
Crick, F., 259–60

Dahl, R. A., 174–5
Darwin, Charles, 281, 282–3
Davy, Humphry, 100
Democritus, 254, 275
Dernburg, T. F., 117, 180
Descartes, R., 93, 253, 255
Deutscher, I., 245
Digges, Thomas, 44
Dirac, P. A. M., 85–6
Domar, E., 182
Donne, J., 43
Donner, F. C., Chairman of General Motors, 190
Dostoyevsky, F., 278
Dreyer, J. L. E., 4
DuBridge, L., 310
Duhem, P., 7
Dulles, J. F., 301

Eden, Anthony, 238
Einstein, Albert; Aesthetic criteria and, 84–5, 280; attitude to nature of, 280; political struggle of, 197, 243n, 294, 342–8; quantum mechanics and, 4, 22n, 81–5, 320, 348; theory of relativity and, 4, 59, 66–81 *passim*, 120, 319–20, 322

Eisenhower, Dwight D., 198–9, 200, 201, 240, 303
Ellul, J., 269n
Engels, Frederick, 107–8, 110–12, 113, 115, 155, 226, 287–8, 336
Euler, L., 63

Fabricius, D., 54
Fedorov, E. K., 305–6
Feigl, H., 76
Fermi, E., 295
Feuer, L. S., 93–4, 353 (n. 9)
Feyerabend, P. K., 23
Feynman, R. P., 13, 84, 178
Ficino, Marsilio, 88–9, 254n, 338–9
Fisher, J. L., 209, 214
Fleming, D. F., 233–45 *passim*, 360 (n. 18)
Flores, E., 219–20
Fludd, Robert, 254n, 256
Forrestal, J. V., 196, 197
Forster, E. M., 271–3
Foster, W. C., 245–6
Fourier, Charles, 165
Franck, James, 296, 348
Frank, Jerome, 193
Freud, S., 332–3
Fromm, E., 109, 333–4, 334–5

Gabor, D., 187
Galbraith, J. K., 187, 201–2, 288–91, 292, 294
Galileo, 1, 27, 44–51 *passim*, 52, 54, 55, 56, 57, 161, 177n, 249, 254–5
Gassendi, Pierre, 256
Geminus, 30, 34
Gerth, H. H., 297
Giddy, Davies, 100
Gilbert, William, 337 and 339n
Gilpin, R., 298–307 *passim*, 364 (n. 41)
Gissing, G., 248
Glanvill, Joseph, 94
Glass, Bentley, 285, 340
Goethe, 266, 266n, 337
Goldwater, B., 309
Goodwin, B., 281–2
Goulart, President, 220
Gouldner, A. W., 169
Green, A., 168
Groves, General L., 296
Guye, Ch. E., 71

Haar, D. ter, 22n
Halifax, Lord, 233
Hansen, A., 118, 121, 121n

INDEX

Hanson, N. R., 76
Harrington, M., 207, 219n
Heilbroner, R. L., 116, 183, 189, 190, 200, 206, 291–3, 312, 315
Heisenberg, W., 82, 85, 280n
Heitler, W., 273
Helmholtz, H. von, 266n
Henry, J., 310
Higigns, B., 215
Hitler, Adolf, 233–6
Hobbes, Thomas, 258, 284, 285n
Holbach, Baron d', 259, 266
Holloman, H., 184
Holton, G., 66–7, 70–2, 74–5, 351 (n. 8)
Hook, S., 345
Hoover, President H., 116
Horowitz, David, 166, 227
Huberman, L., 162
Humphrey, Hubert H., 309
Huxley, Aldous, 271–3, 278
Huxley, T. H., 81
Huygens, C., 58, 93

Infeld, L., 59, 84
Inkeles, A., 26

Jacobson, H. K., 305, 307
Johnson, President L., xiii, 209, 309
Jones, P. d'A., 145, 194

Kappel, F. R., 224
Kaufmann, Harry, 333
Kaufmann, W., 70–1
Kelvin, Lord, 63, 70
Kennan, G., 243–4
Kennedy, President, 199–200, 237, 246, 307–9
Kepler, Johannes, 45–6, 51–5, 56, 57, 149, 161, 255, 256
Keynes, J. M., 25, 115, 118–22 *passim*, 130–41 *passim*, 144–5, 147–8, 155, 156–61 *passim*, 180, 182, 185, 192–3, 202, 216, 277n, 317, 321–2, 335
Khrushchev, N., 303, 305, 308
Killian, J. R., 303, 304
King-Hele, D., 313–14
Klappholtz, K., 203
Klein, L. R., 181, 185–6
Kolko, G., 330–1
Kranzberg, M., 98
Kubitschek, President, 221
Kuhn, T. S., 10, 11–18 *passim*, 19, 20, 21, 22, 23–5, 70, 79–81, 102, 154, 175, 318–19
Kurihara, K. K., 184–5

Lagrange, J. L., 12, 58
Lakatos, I., 18–24, 70, 77, 79–81, 318–19

Lall, A. S., 305
La Mettrie, J. O. de, 258–9
Landsberg, H. H., 211, 212, 213–14
Laplace, P. S., 81, 149, 259
Lapp, R. E., 297, 303
Latter, Albert L., 304, 305, 306
Laue, M. von, 71n, 72, 347
Lavanchy, Ch., 71
Lawrence, E., 295
Leahy, Admiral W. D., 240
Lebow, V., 184
Lederberg, J., 259n
Leibniz, G. W., 14
Lekachman, R., 202
Lenin, V. I., 227–30
Lerner, A. P., 136
Lessing, G. E., 336
Lewis, A., 146
Libby, W. F., 309
Liebknecht, K., 227
Lilienthal, D., 343, 345
Lipsey, R. G., 116, 168, 170–1, 175
Litvinov, M., 234
Lloyd George, D., 227
Lodge, Senator Henry Cabot, 205
Long, Breckinridge, 238n
Lorentz, H. A., 65–6, 71–2, 75
Lovering, T. S., 209
Luce, H., 198
Lucretius, 254
Lundberg, G. A., 173
Luxemburg, Rosa, 227
Lynd, R. S., 169, 177, 333–4, 348

Macdonald, D., 195
McDougall, D. M., 117, 180
McHale, J., 214
McNamara, R., xiii, 199, 308
Mach, E., 7
Magdoff, Harry, 216–23 *passim*, 358 (n. 15)
Magie, W. F., 72–3
Marcuse, Herbert, xiii, 252, 267, 284, 316, 336, 338
Marlowe, Christopher, 252, 267
Marshall, A., 113–14
Marx, Karl, xiii, 107–10, 112, 114, 115, 116, 131, 154–5, 158, 160, 162–7 *passim*, 226, 228, 232, 287–8, 321, 334–5, 336
Maxwell, J. C., 63, 149
Mazur, P., 186
Meadow, C. T., 262–3
Medawar, P. B., 277n, 284–5, 340
Medvedev, Z. A., 246
Meier, R. L., 210, 211
Melman, S., 200, 201
Mersenne, M., 255, 256

Merton, R. K., 149
Michelson, A. A., 280
Mikesell, R. F., 206–7
Mill, J. S., 101–7, 101, 112, 154, 187, 330
Miller, D. C., 74–5
Mills, C. W., 297
Mirandola, Pico della, 89, 254
Mishan, E. J., 188
Moore, Barrington, 149
More, L. T., 73–4
More, Thomas, 87, 89–91, 94, 249, 320
Morgenthau, H. J., 330
Morley, E. W., 65
Morris, B. R., 210
Morrison, P., 296
Mullaly, F., 234
Mumford, L., 187, 266
Musson, A. E., 100n
Myrdal, G., 168–9

Nagel, E., 6, 172
Needham, Joseph, 56
Nelson, D., 194–5
Newton, Isaac, xiii, 57–8, 59–64, 149, 256n, 339n
Nixon, Richard, 301, 311
Nkrumah, K., 213
Nobel, Alfred, 342
Nove, A., 231, 232

Oppenheimer, R., 294, 295, 299, 300, 301, 345
Oresme, N., 55
Orwell, George, 16, 171, 270–1, 278
Ovid, 87n
Owen, Robert, 165, 276n, 277n

Patton, D. J., 209
Pauli, W., 83
Pauling, L., 348
Peccei, A., 314–15
Pigou, A. C., 120, 122, 140, 168
Planck, M., 73, 149
Plato, 31, 45
Platt, J. R., 312–13
Poincaré, H., 7
Polanyi, M., 18, 276
Popper, K. R., xiii, 2–10 *passim*, 13, 15, 23, 59, 65, 79, 174n, 175, 251, 318, 337
Porta, G. della, 254n
Power, Henry, 255
Ptolemy, 31, 33–5, 160–1, 319
Pythagoras, 28–32, 46

Rabi, I. I., 300
Radzinowicz, L., 99
Rayleigh, Lord, 65
Ricardo, D., 99, 115, 120, 125
Ritz, W., 77–9, 320
Robbins, L., 129–30, 136, 155–9, 321

Robertson, Dennis, 130, 158
Robinson, E., 100n
Robinson, Joan, 181
Rockefeller, N., 202–3
Rogers, C. R., 273–4
Rogers, E. M., 53–4, 351 (n. 39)
Roosevelt, Eleanor, 345
Roosevelt, President F. D., 117, 130, 146–7, 157, 192–3, 197, 294, 295
Rose, Hilary and Steven, 250
Rostow, W. W., 204, 209
Roszak, T., 250, 268–70, 275, 279–80, 284, 293n, 294, 326, 338
Rothermere, Lord, 233
Rusk, D., xiii
Russell, Bertrand, xiii, 264, 267, 282, 340–1, 348

Saint-Simon, H. C. de, 165, 248, 287
Sakharov, A. D., 311
Samuelson, P. A., 120–1, 199
Sartre, J. P., 334, 335
Scheffler, I., 1–3, 23
Schlesinger, A. M., Jr., 179, 193, 196n, 241
Schon, D. A., 183, 198
Schrödinger, E., 82, 83, 86, 280
Schumaker, E. F., 187

Schumpeter, J., 156
Shackle, G. L. S., 119
Shankland, R. S., 75
Shapiro, J., 311, 364 (n. 96)
Shapley, H., 344
Shaw, G. B., 119
Shonfield, A., 314–15
Skinner, B. F., 262, 273, 275
Slichter, S. H., 245
Smith, Adam, 95–8, 99, 105, 108, 119, 321
Socrates, 30
Somervell, General B. B., 194
Sprat, Thomas, 94
Stalin, J. V., 230–239 *passim*, 324
Stein, E., 305, 307
Steinbeck, J., 171
Stent, G., 250, 264
Stimson, H. L., 240–1
Storr, A., 332–3
Sweezy, P. M., 159–60
Swift, Jonathan, 94
Szilard, L., 240, 296, 348

Teller, E., 299, 302, 304–7, 309
Thales, 4
Theobald, R., 184, 187
Thomson, J. J., 70
Tobias, J. J., 24
Townes, C. H., 26
Townsend, Rev. J., 99

Trotsky, L., 227, 229–30
Truman, Harry S., 198, 236, 239–44, 296, 297, 300
Tzu, Chuang, 27

Ure, A., 101
Urey, H. C., 297

Vavilov, N. I., 177n
Veblen, T., 288, 316

Wadsworth, J., 302
Wald, G., 310, 313
Wallace, Henry, 242–3, 243n
Ward, Seth, 256
Watson, J. B., 262
Webster, John, 254n, 256
Weisskopf, V., 328
Whitbread, Samuel, 100
Wien, W., 72
Wiener, N., 343, 343n
Wilson, Charles E., 194, 195–6, 197
Wilson, Woodrow, 205–6
Winch, D., 130, 147, 161, 199
Wittkower, R., 36
Wooldridge, D. E., 261–2, 361 (n. 24)
Wordsworth, W., 265–6

Young, R. M., 259n

Zamiatin, Y., 271–3, 278

INDEX – SUBJECTS

Absolute space and time, see Newtonian paradigm
Advertising, 145–6, 182–3, 183n, 198, 289, 323
Aggression in man, 225, 332–4
Aristarchus' theory, 28, 30–1, 32–3
Aristotelian paradigm, 32, 48–9
Arms expenditure, 117, 131, 147–8, 193–203, 232

Baruch Plan, 241, 263, 297–9
Beautiful world, 2–3, 86, 87, 251–2, 279, 280–5, 329–41 *passim*, esp. 335–41
Beauty, 36, 46, 55, 56, 67, 80, 85–6, 103, 119, 178, 188, 226, 265, 266n, 280, 283, 320, 338
Brains, disembodied, 261–2, 326

Capital punishment, 98–9, 112, 257
Capitalism,
 built-in obsolescence and, 184
 business cycle in, 110, 128–9

competitive ethos of, 103, 105
dehumanization under, 97, 102, 108, 165–6
division of labour, 96–7
product innovation in, 184
China,
 opium war and, 205
 post-revolutionary
  development of, 225n
 Soviet Union and, 231, 247, 305, 308
 United States and, 205–6, 305, 308
Classical economic paradigm, 122–31, 138–41
Consciousness, the 'problem' of, 258, 260, 262, 274, 279, 335–6, 361 (n. 24)
Copernican system, 36–55 *passim*, 74, 160–1, 319
Counter-culture, xiv, 250, 270, 284, 337–9
Czechoslovakia, xiv, 225, 233–5, 243

Disarmament negotiations, 294–308
Domination of nature, 89, 91, 93, 94, 150–1, 252–7, 263, 267–8, 281–3, 328–9

Ecological crisis, xiv, 103, 181n, 184n, 186–7, 323–4
Einsteinian paradigm, 66–79 *passim*, 120, 319–20

Faust legend, 254n, 267–8, 337
Faustian man, 252, 327, 336–7, 340
Franck Report, 296

Golden Age, 87n, 283–5, 287, 326
Great Depression, 116–18, 129–31, 144–8, 155–60, 179–180, 192–4, 321–3

Hermetic philosophy, 88–9, 91–3
Hermetic texts, 89, 89n, 337n
Hermetic utopias, 91–3, 95, 100, 320
Hiroshima and Nagasaki, atomic destruction of, 239–41, 286, 294–5, 342

INDEX

Imperialism in,
  Africa, 205, 213, 217
  China (see under China)
  Greece, 238–9
  India, 204–5, 222, 225
  Latin America (see under Latin America)
  Persia, 208
  Vietnam (see under Vietnam)
Industrial revolution, 98–101

Kepler's Laws, 8, 53–4, 57
Keynesian paradigm, 131–44; Keynes' rejection of Marxism, 118–19, 155
Kuhn's philosophy of science, 11–18, 23–5, 56, 79–80, 102, 175, 318–19

Lakatos' philosophy of science, 18–24, 56, 79–80, 318–19
Laplacian determinism, 81, 259
Latin America,
  American investment in, 217–18
  external debt of, 218
  invasion of Guatemala, Dominican Republic, 208
  military coup in Brazil, 220–1
  poverty in, 206–7, 225
  sale of arms to, 222
Liberated world, 224, 278–9 (see also under 'beautiful world')
Love, 2, 3, 18, 71n, 89, 91, 109, 165, 187, 250, 251, 264, 267, 270, 272–3, 282, 316, 326, 332, 334, 335, 337, 338, 339, 340
Luddites, 101, 257

Magical image of nature, 257
Marshall Plan, 215–16
Marxist paradigm, 107–112, 162–7
Maxwell's equations, 63, 67, 75–6, 77–9
Michelson-Morley experiment, 5, 64–6, 75–6, 77, 80
Mill: John Stuart Mill's paradigm, 101–7
Miller experiment, 74–5
Missile gap, 199–200, 307–8

Nazi Germany,
  appeasement of, 233–6
  Nazi invasion of Soviet Union, 236–7, 245
  Nazi-Soviet pact, 236
New Deal Policy, 130–1, 146–7, 193–4, 321–2
Newtonian paradigm, 12–13, 21, 55, 59–66

absolute space, 60
absolute time, 61
ether, 62–6

Orphic myth, 284, 338–9

Permanent war economy, 191–2, 193–203, 323
Phases of Venus, 41–2, 49, 161
Piecemeal social engineering, 10, 174n
Poor Laws, 99–100
Popper's philosophy of science, 1–10, 79–80, 175, 317–19
Propensity to consume (save), definition of, 133
Ptolemaic paradigm, 33–5, 37, 160–1
Pyramid building, 185, 277
Pythagorean paradigm, 28–32, 46

Quantum mechanics, 4, 25, 73–4, 81–5, 275, 279, 362 (n. 71)

Resources, natural
  American consumption of, 186, 323
  importance to western economy of, 181n, 208–14, 323
Retrograde motion of planets, 29–32, 35, 37–8
Ritzian paradigm, 77–9, 320
Russian Revolution (see also Soviet Union)
  collectivization and industrialization, 231–2
  Lenin and, 228–30
  Marxism and, 226–7
  Western response to, 227–8

Say's law, 124–5, 129, 131–2
Scarcity, problem of, 116–17, 146, 184–5, 187, 188, 191–2, 276–8, 277n, 290, 323–4, 334–5
Science,
  anti-physicalism, 273–6
  capitalism and, 100–1, 102, 107–8, 201, 276–8, 292–3, 311, 327–9
  dystopias and, 270–3
  incommensurability of paradigms, 14–16, 21, 56–7, 80–1
  Industrial revolution and, 98–100, 100n, 107–8
  instrumentalism, Popper's attack on, 7–8
  liberated society and, 279–85
  magical image of nature attacked, 255–6

mechanical philosophy of 17th century, 254–5
normal science, definition of, 11–13
paradigm, definition of, 11–13
physicalism, 258–63
rationality of, Chapter 1 *passim*
research programmes, definition of, 19–21
revolutionary science, definition of, 13–17
Romantic reaction to, 264–6
Roszak's indictment of, 268–70
social aims of, 253, 340–1
teleology rejected, 253–5
Scientific community,
  arms race and, 201, 289–91, 294–308
  as agency of change, 292–3, 309–16, 328–9, 340–1
Slave trade, 205
Socialism (see also Marxist paradigm),
  definition of, 224, 246–7, 324, 329, 335–6
  'inevitability' of, 109–112, 163, 164–7
  Soviet Union and, 228, 232–3, 246–7
  utopian versus scientific, 110, 166
Soviet Union (see also Russian Revolution),
  Balkans policy of, 238–9
  China and, 231, 247, 305, 308
  Cold War and, 238–47
  disarmament negotiations and, 297–9, 301–8
  invasion of, 236–7, 244–5
  Soviet-Nazi pact, 236

Tychonic system, 50–1, 72, 79, 161, 320

Underdevelopment, problem of, xiv, Chapter 8 *passim*, 225–6, 246–7, 323–4, 330–1, 341 (see also under imperialism)
United States (see also under China, Great Depression, Imperialism, resources, Soviet Union)
  corporate power in, 190–1
  'industrial-military complex' in, 194–203

Value-free social science, 167–78, 274, 315–16, 322
Vietnam war, xiii, xiv, 214n, 239, 309–10, 330